Darwinism, Democracy, and Race

Darwinism, Democracy, and Race examines the development and defense of an argument that arose at the boundary between anthropology and evolutionary biology in twentieth century America. In its fully articulated form, this argument simultaneously discredited scientific racism and defended free human agency in Darwinian terms.

The volume is timely because it gives readers a key to assessing contemporary debates about the biology of race. By working across disciplinary lines, the book's focal figures – the anthropologist Franz Boas, the cultural anthropologist Alfred Kroeber, the geneticist Theodosius Dobzhansky, and the physical anthropologist Sherwood Washburn – found increasingly persuasive ways of cutting between genetic determinist and social constructionist views of race by grounding Boas's racially egalitarian, culturally relativistic, and democratically pluralistic ethic in a distinctive version of the genetic theory of natural selection. Collaborators in making and defending this argument included Ashley Montagu, Stephen Jay Gould, and Richard Lewontin.

Darwinism, Democracy, and Race will appeal to advanced undergraduates, graduate students, and academics interested in subjects including Philosophy, Critical Race Theory, Sociology of Race, History of Biology and Anthropology, and Rhetoric of Science.

John P. Jackson Jr. is a Lecturer in Interdisciplinary Studies, Charles Center for Academic Excellence, College of William and Mary, USA.

David J. Depew is Emeritus Professor of Communication Studies and POROI (Project on the Rhetoric of Inquiry) at the University of Iowa, USA.

History and Philosophy of Biology
Series editor: Rasmus Grønfeldt Winther

This series explores significant developments in the life sciences from historical and philosophical perspectives. Historical episodes include Aristotelian biology, Greek and Islamic biology and medicine, Renaissance biology, natural history, Darwinian evolution, Nineteenth-century physiology and cell theory, Twentieth-century genetics, ecology, and systematics, and the biological theories and practices of non-Western perspectives. Philosophical topics include individuality, reductionism and holism, fitness, levels of selection, mechanism and teleology, and the nature-nurture debates, as well as explanation, confirmation, inference, experiment, scientific practice, and models and theories vis-à-vis the biological sciences.

Authors are also invited to inquire into the "and" of this series. How has, does, and will the history of biology impact philosophical understandings of life? How can philosophy help us analyze the historical contingency of, and structural constraints on, scientific knowledge about biological processes and systems? In probing the interweaving of history and philosophy of biology, scholarly investigation could usefully turn to values, power, and potential future uses and abuses of biological knowledge.

The scientific scope of the series includes evolutionary theory, environmental sciences, genomics, molecular biology, systems biology, biotechnology, biomedicine, race and ethnicity, and sex and gender. These areas of the biological sciences are not silos, and tracking their impact on other sciences such as psychology, economics, and sociology, and the behavioral and human sciences more generally, is also within the purview of this series.

Rasmus Grønfeldt Winther is Associate Professor of Philosophy at the University of California, Santa Cruz (UCSC), and Visiting Scholar of Philosophy at Stanford University (2015–2016). He works in the philosophy of science and philosophy of biology and has strong interests in metaphysics, epistemology, and political philosophy, in addition to cartography and GIS, cosmology and particle physics, psychological and cognitive science, and science in general. Recent publications include "The Structure of Scientific Theories," *The Stanford Encyclopaedia of Philosophy* and "Race and Biology," *The Routledge Companion to the Philosophy of Race*. His book with University of Chicago Press, *When Maps Become the World*, is forthcoming.

Published:

Romantic Biology, 1890–1945
Maurizio Esposito

Natural Kinds and Classification in Scientific Practice
Edited by Catherine Kendig

Organisms and Personal Identity
Individuation and the Work of David Wiggins
A.M. Ferner

The Biological Foundations of Action
Derek M. Jones

Darwinism and Pragmatism
William James on Evolution and Self-Transformation
Lucas McGranahan

Darwinism, Democracy, and Race
American Anthropology and Evolutionary Biology in the Twentieth Century
John P. Jackson Jr. and David J. Depew

Around the mid of the last century, evolutionary biology changed to become compatible with and even enable liberal-democratic and antiracist values. In their important book, Jackson and Depew recount the story of this crucial alliance. At a time of profound changes in both the political arena and the biological understanding of gene functioning and heredity, this alliance may look, in retrospect, more fragile and unstable than what we used to believe. Knowing deeply its contingent making and deep entanglement with wider anthropological and sociopolitical debates remains an essential tool to understand our present.

Maurizio Meloni, author of *Political Biology: Science and Social Values in Human Heredity from Eugenics to Epigenetics, Palgrave*

Science historians have long tended to stop at Darwin, and are only now beginning to open up the last century of the science of human evolution to critical historical analysis. In this literate and accessible new book, Jackson and Depew lead us through a marvelously intricate and intertwined intellectual history involving cultural anthropology, biological anthropology, population genetics, evolutionary biology, and racial studies. They scrupulously analyze the work of scholars like Alfred Kroeber, Ashley Montagu, Sherwood Washburn, and Theodosius Dobzhansky, and challenge the facile alt-histories that circulate in contemporary evolutionary psychology. This is an important addition to the library of anyone seriously interested in how we think about human origins and diversity.

Jonathan Marks, *Professor of Anthropology at the University of North Carolina at Charlotte, USA*

Jackson and Depew have produced an important work: a muscular refutation of scientific racism, grounded in science and deploying the tools of the historian. Through rich new readings of the work of five central geneticists and anthropologists, they show that inoculation with the Modern Synthesis of evolutionary biology immunized biological anthropology against racist genetic determinism, leading this group of scientists toward a more egalitarian human biology. Anyone sympathetic to the idea that racial superiority is "in the genes" needs to confront this book. And those of us who find ourselves repeatedly whacking the mole of racist science now have a solid new mallet.

Nathaniel Comfort, *Professor of the History of Medicine, Johns Hopkins University, USA*

Darwinism, Democracy, and Race

American Anthropology and Evolutionary Biology in the Twentieth Century

John P. Jackson Jr.
and David J. Depew

Routledge
Taylor & Francis Group
LONDON AND NEW YORK

First published 2017
by Routledge

2 Park Square, Milton Park, Abingdon, Oxfordshire OX14 4RN
52 Vanderbilt Avenue, New York, NY 10017

Routledge is an imprint of the Taylor & Francis Group, an informa business

First issued in paperback 2019

British Library Cataloguing in Publication Data
A catalogue record for this book is available from the British Library

Library of Congress Cataloging in Publication Data
A catalog record for this book has been requested

ISBN: 978-1-138-62817-5 (hbk)
ISBN: 978-0-367-35858-7 (pbk)

Typeset in Times New Roman
by Wearset Ltd, Boldon, Tyne and Wear

To Michele Jackson and Mary Depew, who lived with this evolving manuscript far too long, but never wavered in their support and encouragement.

Contents

Acknowledgments

This project would not exist if the authors had not met at a two-week workshop on the history of the disciplines at the Obermann Center for Advanced Studies at the University of Iowa in 2002. We begin by thanking the Obermann Center for bringing us together so we could discover that we think about disciplinarity in different, but complementary ways.

David's archival research was supported by a Mellon Emeritus Fellowship. He thanks the Mellon Foundation and the Graduate College of the University of Iowa, Dean John Keller, and the University's Office of Sponsored Research for supporting his application for these funds. John's research was supported by a Sabbatical leave from the University of Colorado, Boulder in 2010, a Kayden research grant in 2013, and throughout the project by the Department of Communication at Colorado.

The Project on the Rhetoric of Inquiry at the University of Iowa and the Department of Communication at the University of Colorado, Boulder, provided office space in which we were able to work together during several summers.

We are grateful to the staffs of the archives we visited during our research for their assistance and efficient service.

This book has been enriched by comments and critiques from colleagues at a number of conferences: The American Forensic Society's Conference on Argumentation (Alta, UT 2005, 2011); Department of Philosophy and Cognitive Science (Lund, Sweden 2012); Great Plains Society for the Study of Argumentation (Ames, IA 2012); History and Philosophy of Psychology Section of the British Psychological Society (Guildford, UK 2014); History of Science Society (Montréal, Quebec 2010, Philadelphia, PA 2012, San Francisco, CA 2015); the International Society for History, Philosophy, and Social Studies of Biology off-year seminar (London, Ontario 2010) and its biennial meetings (Salt Lake City, UT 2011, Montpellier, France 2013, Montréal, Quebec 2015); Institut d'Histoire et de Philosophie des Sciences (Paris, France 2010); James Madison College of Public Policy, Michigan State University (Lansing, MI 2013); Northwestern University in Qatar (Doha, Qatar 2009); Program for Science and Human Culture at Northwestern University (Evanston, IL 2008); Rhetoric Society of America (Minneapolis, MN 2010, San Antonio, TX 2014); and the University of Minnesota Program in the History of Science and Technology and Medicine (Minneapolis, MN 2008).

Our research assistant, Dr. Deirdre Egan, deserves special thanks. Her doctoral research on how college students process information about the biology of race and gender made her a valued collaborator. The Mellon Foundation provided funding for her position.

We are grateful to the following historians and philosophers of biology for discussion of particular texts and issues and for reading drafts of several chapters: Richard Burian, Jean Gayon, Mark A. Largent, Maurizio Meloni, Staffan Muller-Wille, Greg Radick, Michael Ruse, Betty Smocovitis, Denis Walsh, and Rasmus Gronfeldt Winther. The ground-breaking works of John Angus Campbell, John Lyne, Celeste Condit, and Leah Cecarrelli on the rhetoric of evolutionary biology have been an inspiration.

David is especially in debt to present and former members of the University of Iowa's distinguished four-field Department of Anthropology for their tutelage and tolerance: Michael Chibnik, Virginia Dominguez, Laurie Graham, Drew Kitchen, and Glenn Storey. They should not be held even remotely responsible for the result. We also wish to thank Professor Glenn Ehrstein for help on transliterating and translating some of Boas's correspondence.

1 Introduction

In the footsteps of Franz Boas

Evolution, politics, and race: how things stand

Two conflicting facts lie at the heart of the American experience. First, the United States was for a time the only democratic republic since antiquity to have survived more than a few years. Lincoln may have been right to say that if the Union was not preserved government of, by, and for the people might perish from the earth. Second, no less after the Civil War than before it the policy of this country's regime was racist. The Swedish social scientist Gunnar Myrdal identified this persistent tension as *An American Dilemma* (Myrdal 1944). Any claim to fame that the United States might enjoy – and its oft-asserted "exceptionalism" shows that its citizens do indeed crave fame for their democracy in the eyes of history – has rested on efforts to erase the shame of slavery, segregation, and other forms of racism.

It is a truism that World War II, the end of colonialist imperialism abroad, and the Civil Rights Movement at home changed how we talk about race. Still, racism has survived the revelation that "race" is an ideological (by)product of European globalization, even if it is no longer assumed to be a primordial category of our being or an inference from evolutionary biology. Racism lurks in the social world as "institutional racism" and "racism without racists" (Bonilla-Silva 2014). Periodically, it resurfaces in political life, sometimes speaking the language of contemporary evolutionary theory in order to confer legitimacy on itself, but at the same time muttering under its breath the language of the discarded science of biological racism. On the very day of the sixty-first anniversary of *Brown* v. *Board of Education*, which declared racial segregation of schools unconstitutional, the website of Richard Spencer, the man credited with reinvigorating the "alt-Right" that contributed to Donald Trump's electoral victory, posted an article declaring that muddle-headed liberal social scientists were waging a "war on human nature" in denying racial differences in intelligence. According to its author, the notion of racial equality "emerged in the 1960s and had, by the 1970s, become an unchallengeable orthodoxy" (Roth 2015). The article relied on race/IQ researchers such as Arthur Jensen and Richard Herrnstein as well as the sociobiologist E. O. Wilson and the evolutionary psychologist Steven Pinker to claim that, "Darwinian evolution revolutionized the natural sciences. The social sciences have been immune for far too long." We will encounter these figures again.

Especially under the present circumstances, it is important to understand the scientific basis of racial egalitarianism. *Contra* the article just cited, it arose long before the 1960s and when it did it used the language of Darwinism to undo scientific racism, which was an entrenched feature of late nineteenth century non-Darwinian evolutionary biology. Working with the evolutionary geneticists Theodosius Dobzhansky and Leslie C. Dunn, the anthropologists Ruth Benedict and Ashley Montagu were pivotal during and after World War II in taking scientific credibility away from American "racialists," as they called them (Montagu 1942; Benedict and Weltfish 1943; Dunn and Dobzhansky 1946). They succeeded for at least three reasons. First, centered at New York City's Columbia University, they could leverage new means of influence opened up by the ideological and military victory over Nazi racism, most straightforwardly in the 1950 and 1951 UNESCO *Statements on Race* that Montagu and Dunn drafted with help from Dobzhansky (UNESCO 1950, 1951). UNESCO's first Director, Julian Huxley, a biologist, supported their work, not least because he himself was so deeply immersed in the conceptual framework on which Dobzhansky, Montagu, and Dunn based their arguments that he gave it its name: the Modern Evolutionary Synthesis (Huxley 1942).[1]

Second, this initiative was both interdisciplinary and carried on at the permeable boundary between academic and public spheres of discourse. Dobzhansky's claim that natural selection generally favors the evolution of flexible, even anticipatory, ways of meeting environmental contingencies reinforced the signature proposition of Benedict's and Montagu's mentor Franz Boas, the founding father of American academic anthropology, that the equally distributed capacity of humans for acculturation renders racial differences both mutable and trivial (Boas 1911; Dobzhansky 1937, 1951; Chapter 2 of this book). What could be a more "plastic" way of dealing with changing environments, Montagu and Dobzhansky argued in a 1947 paper in *Science*, than our species' naturally selected cultural way of learning and living (Dobzhansky and Montagu 1947)? Admittedly, there was a lot of "black-boxing" of details in this conjecture. There still is. But in the decades since, this sketch of the process of anthropogenesis has continued to facilitate solid discoveries by blocking assumptions that put human races on different rungs of an evolutionary latter. It placed out of bounds the "stadial" thinking, as we call it, that underpins much racist argumentation.

Boas had long been arguing that refusing to rank-order races carries with it a presumption against legal barriers to the considerable amount of interracial mixing that regularly occurs in societies free of caste-like constraints (Boas 1928, 1940). Dunn, Dobzhansky, and their anthropological collaborators anchored his opposition to anti-miscegenation laws in the population-genetic approach to evolutionary theory of the Modern Synthesis. Their arguments informed Supreme Court decisions against racially segregated schools and laws barring racial intermarriage (*Brown* v. *Board of Education* 1954; *Loving* v. *Virginia* 1967; see also Jackson 2001, 17–42; Pascoe 2009, 124–128). But the collaboration did not end there. The mid-century alliance of American biologists and anthropologists also proposed that natural selection is congenial to racially and culturally pluralist democratic institutions (Dobzhansky 1962a; Beatty 1994). How these claims were articulated and defended is the focus of the five studies comprising this book.

In these studies, we highlight a third reason why these arguments gained a foothold in the postwar period. The arguments were persuasive because they were based on science better than the outdated approaches to classification and tendentious appeals to single genes as fixed determinants of traits they challenged. The Modern Synthesis proposed unifying biology's diverse fields by viewing them in the light of evolution and by viewing evolution as a process in which the combined effects of genetic mutation and recombination, natural selection, and several auxiliary factors interact. Evolutionary biologists and biological anthropologists still use these methods, concepts, mechanisms, and inventory of evolutionary scenarios, even if since the 1940s they have added many more tools to their kit.

According to the Modern Synthesis, the interaction of evolution's various factors comes into view only by way of statistical and probabilistic representations. From the perspective of "population thinking," as the makers of the Synthesis called it, evolution is not development or ontogeny writ large, as many biologists previously thought. Rather, it consists in context-dependent shifts over multi-generational time in the relative proportions of genotypes in races and species. Viewed in this way, neither races nor species can possibly be, or embody, types. Races are biogeographically distinctive populations that contain a great deal of genetic diversity, but can interbreed. Species are populations that, having evolved isolating mechanisms, are reproductively closed (Dobzhansky 1937, 11; 1951, 6, 138, 261). The explanatory power of the Modern Synthesis has enabled anti-racist and anti-eugenic theorizing to flourish in both evolutionary biology and biological anthropology since the 1940s because racialism and eugenics are hostage to the typological thinking that the Synthesis rejects. In the 1950s, the physical anthropologist Sherwood Washburn used this insight to transform biological anthropology from its previous fixation on static types to a dynamic understanding of the "functional complexes" that link our morphology to our behavior as encultured beings (Washburn 1951, 1953). Anxious to maintain the unity of Boas's "four fields" (cultural anthropology, physical anthropology, archaeology, and linguistics) as a bulwark against racialist thinking, Washburn used what he dubbed the New Physical Anthropology to insist that anthropologists of all four flavors share a common goal, form a community of inquiry, and together enable the discipline of anthropology to present our species as a unified "family of man" (Haraway 1989). He largely succeeded.

In recent years, however, the new tools of gene sequencing and "genetic cluster" analysis have revived the notion that the complement of genes each of us has correlates fairly well with the received racial categories we use to socially mark off our own and others' communities of descent. "Racial research" writes sociologist Catherine Bliss, "has reemerged and proliferated to occupy scientific concerns to an extent unseen since early twentieth-century eugenics" (Bliss 2012, 2). The mid-century Synthesizers certainly knew of correlations between races and particular traits, especially differential vulnerability to various hereditary diseases, but they rejected the "essentialist" implication that particular traits reveal racially distinctive profiles that integrate a large array of morphological, physiological, psychological, and behavioral characteristics. They also rejected the deterministic implication that our genes circumscribe our life prospects (Dobzhansky 1962a; Washburn 1963). Yet recently some gene-sequencing scientists and pundits who laud their work as "cutting edge" have preached just such essentialist and deterministic implications. The science journalist Nicholas Wade, for example, has suggested that Japanese, Koreans, Taiwanese Chinese, and Icelanders have successfully embraced advanced capitalism in ways that the transplanted African population of Haiti has not because their genomes have been co-adapted by natural selection to environmental challenges not too different from those they still face (Wade 2014, 14). Wade suspects that transported populations and "mixed races" (as if all races weren't mixed) such as African-Americans labor under a particularly "troublesome inheritance," as the title of his book puts it. "This is just what would be expected," he writes, "for populations that had to adapt to different challenges on each continent. The genes specially affected by natural selection control not only traits like skin color and nutritional metabolism, but also some aspects of brain function" that Wade admits are "not yet well understood" but he assures us soon will be (Wade 2014, 4).

For Wade culture does not play the formative role it does for Montagu, Dobzhansky, and Washburn. His implication is that the greater authority of molecular genetics and computer-assisted analyses of clusters of allelic differences over the older and supposedly cruder methods of genetic analysis to which the Modern Synthesis was confined before Crick, Watson, and the Human Genome Project lends new support to views about racial divisions that Boas's mid-century followers were, it is alleged, too quick to dismiss purely for ideological reasons. Speaking more or less on behalf of the American Anthropological Association (AAA), biological anthropologists such as Augustin Fuentes and Alan Goodman challenge this narrative by defending American anthropology's long and productive alliance with the population biology of the Modern Synthesis (Fuentes 2012; Goodman 2013; Callaway 2014). They labor, however, under a disadvantage: the received view of historians of race, eugenics, and evolutionary science was that, after the war, it became "politically correct" to be anti-racist, but the science itself did not change much (Provine 1973; Samelson 1978; Provine and Russell 1986; Barkan 1992). Implied, if not stated, is a suggestion that even Dobzhansky and Dunn remained eugenicists at heart and, in spite of their good intentions and self-deceptions, showed themselves in the practices and prejudices of everyday life to be almost as racist as the next guy (Paul 1984, 1994, 1998). Those who celebrate correlations between gene sequences and conventionally identified continental races perpetuate this historiographical commonplace.

In embracing the conception of race that the Synthesis sought to evade, Wade is confident that racism will recede as knowledge of racially correlated risks leads to therapies that will help individuals no matter what their race (Wade 2014, 37–38). He seems confident that those who analyze these risks will be more empirically scrupulous than their eugenicist and racist forebears. Those less sure that today's scientists have abandoned their predecessors' prejudices have pointed out, however, that enthusiasts for correlating "races" with allelic clusters and distinctive gene sequences, many with no known biological function or adaptive significance, have no way of calling into question the racial categories with which they begin. If you had only genetic markers for height, cranial capacity, skin coloration, or susceptibility to malaria you would not discover the human species dividing itself even approximately into three, five, or even more continental races (Fuentes 2012, 91–93).

Realizing this, one might alternatively respond to proposals like Wade's by advocating returning to the population genetic approach to the term "race," this time really giving it a try (Fullwiley 2014). The difficulty is that the very idea of race may be too socially laden under any description to have biological significance. Montagu urged this point in a friendly decades-long debate with Dobzhansky that we examine in Chapter 4. He did not question the population-genetic turn in evolutionary biology. On the contrary, Montagu sought out Dobzhansky as a tutor precisely because he took that turn (Montagu 1942; see Gannett 2001, 2003; Chapter 4 of this book). By 1950 he was arguing that use of "race" should be confined to experts looking for populations that "differ in the frequency of one or more genes" (Montagu 1950). By the mid-1960s, however, with opposition to the Civil Rights Movement in full fury, Montagu questioned whether the term could ever be sufficiently freed from essentialist and stadial connotations to have even technical uses (Montagu 1963, 1964).[2]

In assessing Wade's view and responses to it we reject the story that the "retreat of racism" was ideological rather than scientific. Molecular-genetic methods for mapping, sequencing, and clustering genetic differences, many developed in the 1980s and 1990s under the aegis of the Human Genome Project, have been a technical boon to evolutionary biologists and anthropologists. Still, the ascendancy of sequencing technology obscures fundamental insights that population biologists evidenced in the heyday of the Modern Synthesis and its uptake by anthropologists. Molecular genetics is rooted in biochemistry, not evolutionary natural history. Its institutional settings, funding streams, and cognitive interests lean toward technoscientific improvement, especially in agriculture and medicine, rather than toward appreciating nature's diversity for its own sake (Dietrich 1994; Duster 2015; Comfort 2012; Roberts 2011). Accordingly, molecular geneticists working on genetically modified crops and medical geneticists trying to remediate what in 1908 the physician Archibald Garrod called "inborn errors of metabolism" tend to focus on defective departures from normal functioning rather than on the context-dependent and hence presumptively valuable differences between individuals, populations, and lineages on which Boas, Benedict, Montagu, Dobzhansky, Dunn, and Washburn based their enthusiasm for culturally and racially pluralistic forms of democratic life. Consequently, genetic determinism and typological essentialism about "human nature" have resurfaced in molecularly dominated approaches to human evolution, even among those who are committed to tolerance.

In this discursive context, racial differences have once again appeared as racial defects, in turn provoking social constructionist reactions that sometimes go beyond criticizing scientism to criticizing science itself as a hopelessly uncritical technoscientific enterprise – a position we neither hold nor advocate. Instead, we reject the binary between scientism and social constructionism within which contemporary public-sphere discussions of human biology tend to labor. In retelling the tale of twentieth century evolutionary theory's relationship to the human sciences, Maurizio Meloni writes:

> It should be, in light of recent scholarship, obvious that eugenics persisted and race reappeared after World War II. But the sociopolitical values to which the sciences of human heredity and human evolution aligned themselves and partly contributed to produce were profoundly mutated. And to think that this was merely superficial cover for an unchanged scientific content would reflect a lack of dialectical understanding of how science and society, political and epistemic values, are genuinely co-produced in each novel historical phase.
>
> (Meloni 2016, 6)

We agree. In this spirit we invite readers to relive with us a succession of controversies in which mid-twentieth century American biologists and anthropologists exhibited good biological reasons for opposing scientific racism and eugenics and for supporting the kind of democratic equality and empowerment we now called liberal. Our book is based on the belief that by following closely how our focal figures – Boas, Alfred Kroeber, Montagu, Dobzhansky, and Washburn – and their allies and intellectual heirs – notably, Dunn, Benedict, Margaret Mead, Gabriel Lasker, Sol Tax, Stephen Jay Gould, and Richard Lewontin – confronted what they saw as anti-democratic biases in the evolutionary biology and anthropology of their day we can put ourselves in a position properly to judge the resonance of their arguments in our time.

A tour of the argument

To appreciate the mid-century inter-animation between Boasian anthropology and egalitarian evolutionary biology we must first correct a number of errors in the historiography of anthropology and the other social sciences that began to accumulate early in the twentieth century. We do so in Chapters 2 and 3. We begin with misunderstandings of the naturalized American German-Jewish geographer and anthropologist Boas. The historian Thomas Gossett has claimed that he "did more to combat race prejudice than any other person in history" (Gossett 1963, 418). This perhaps overstates the case; Boas's racial views were more complex than simple denial of the existence of race and articulation of "culture" as a replacement for it (Teslow 2014). Nonetheless, Boas presented high-grade, statistically unimpeachable evidence that (phenotypic) variation within races is greater than between them; that skin color is not strongly associated with clusters of other traits; that human beings are so prone to race-mixing that the idea of pure types is mythological except where choice of marriage partners has been artificially constrained by force or custom; that in contrast to highly inbred lines there is no evidence that interracial marriage has any negative effect on fitness; that people of different races can and do participate in the same culture; that conversely members of the same race can live full lives in different cultures; and that in general human beings are too deeply embedded in the cultures that all make good context-dependent use of the same evolved capacities to justify the rank ordering of races encoded in stadial evolutionary thinking (Boas 1928, 1940; Stocking 1968, 194). While the term "Boasian" is commonly employed to identify Boas's students and followers, such as Benedict, Kroeber, Mead, and in a more complicated way Montagu, we sometimes extend it to include a wider circle of advocates of these propositions.

An example of the social and political consequences of the diffusion of Boasian ideas occurred in 1906 when at the invitation of its president and W. E. B. Du Bois, who taught there, Boas delivered the Commencement Address at Atlanta University. Du Bois recalled Boas telling the graduating class, "You need not be ashamed of your African past." "He then recounted," Du Bois continued,

> the history of black kingdoms south of the Sahara for a thousand years. I was too astonished to speak. All of this I had never heard and I came then and afterwards to realize how the silence and neglect of science can let truth utterly disappear or even be unconsciously distorted.
>
> (Du Bois 1939, vii; Boas 1906; for background: Liss 1998; Zumwalt and Willis 2008)

Most historians have located Boas's defense of his points at the nurture end of the nature-nurture binary, portraying him as a radical environmentalist who was hostile to Darwinism and indifferent to evolutionary thought generally (Freeman 1983, 26; Gilkeson 2010, 33, for example). True, Boas opposed nineteenth century evolutionism of the insufficiently evidenced (and materialistically dogmatic) sort that he and his German mentors associated with Ernst Haeckel's *Darwinismus*. He was, however, highly complimentary of Darwin himself, whom he viewed as seeing in natural history the same primacy of particular events over premature generalizations that neo-Kantian philosophers found in all historical sciences and that Boas, who was tutored by these philosophers, saw in anthropology:

> I consider it one of the greatest achievements of Darwin to have brought to light the fact that … former events leave their stamp on the present character of a people.… The character and future development of a biological

> phenomenon is … expressed … by … its whole history.… The outward appearance of two phenomena may be identical yet their immanent qualities may be altogether different.
>
> (Boas 1887b, 589)

Just as importantly, Boas favored even by his high standards of empirical proof the embryologist August Weismann's restriction of evolutionarily significant variation to germinal elements and the Danish Mendelian Wilhelm Johannsen's reconceptualization of races as genetically inbred pure lines (Müller-Wille and Rheinberger 2012, 124). Boas based many of the claims we have attributed to him on these grounds, among them the proposition that except in culturally distorted societies and laboratories pure races are rare, evanescent, and frequently dysfunctional. We will review his way of arguing for this and other points in Chapter 2.

Boas, it should be noted, was not alone in turning to hard heredity at the turn of the twentieth century. So did Darwinism's co-founder, Alfred Russel Wallace. So, too, did the evolutionary psychologist William James. Even before Weismann discredited Herbert Spencer's reliance on the heritability of acquired characteristics in a debate that began in 1892, James had adopted strict germ line inheritance in order to help him press his attack on the passivity built into Spencer's idea that external circumstances directly mold the traits of organisms to fit their requirements, with natural selection serving only to weed out failures. He found especially annoying Spencer's notion that instincts are congealed habits accumulated by the heritability of ancestral responses to environmental pressures (James 1890, 686; 1904).[3] Precisely how to integrate germinal elements, environments, selection, habit, instinct, and the evolution of agency was widely debated around the turn of the century. Still, accustomed as we are to identifying biological determinism with genes, it is hard for us to see that then, and for a considerable time thereafter, it was the heritability of acquired characteristics, not genetics, that caused people to worry about biological determinism (Stocking 1968; Meloni 2016). From a certain cosmic height one might appreciate Lamarckian "striving" to pass one's gains to offspring as the source of evolutionary novelty, direction, and progress. But more closely observed the heritability of acquired characteristics implies that the prospects of individuals are negatively constrained by every bad thing that ever happened to their forebears: enslavement, for example. Better to "give an identical starting point to each generation," Weismann urged – and Boas agreed (Weismann 1891a, 168).

If the Boas we portray has been difficult for historians to see, it is because he does not conform to the simplistic dichotomy between nature and nature in which histories of the relation between the social and the natural science tend to be framed, and even less to his reputation as an enthusiast for "nurture" (Degler 1991; Pinker 2002; for a nuanced corrective, Teslow 2014). Boas suspected that the global dispersion of our species and the diversity of our cultural life reflect the plasticity of species-wide germinal elements that evolved prior to or alongside our geographical diversification (Boas 1911, 66). He hoped that when caste-like barriers fail, as they inevitably must in free societies, blacks and whites, Native Americans, Jews newly arrived from Eastern Europe, and Irish and Italian Catholics will resume their natural tendency to intermarry, thereby embodying the promise of democracy. Until someone can conclusively show that miscegenation has bad effects, he advised, we should not expect the bad consequences of race mixing about which nativists and eugenicists fretted (Boas 1911, 274; 1928, 80; 1940; Farber 2011).

In Chapter 3 we will see how Boas's first Ph.D., Kroeber, argued that failure of anthropologists to accept germ-line inheritance in biology inevitably meant misreading cultures in the way nineteenth century anthropologists did: as evolving from savage to civilized in a crypto-biological and stadial way, thereby allowing supposed racial psychologies to contaminate both the biological and the social sciences. Following the neo-Kantian philosopher Heinrich Rickert, Kroeber took cultures to be value-laden sites of interpretation not just by knowledge-oriented observers, but more fundamentally by action-oriented participants themselves, who are able to pass what they learn to following generations. Social transmission, however, is not for Kroeber a form of heredity, a purely biological concept. It has a "superorganic" relation to biology that defines anthropology's discipline-specific subject matter and determines its distinctive interpretive methods (Kroeber 1917). In insisting on this way of demarcating his field, Kroeber put in place a fundamental point of agreement that our focal figures and their allies share. Anthropology's culture concept, not psychology, even social psychology, is the point at which social science intersects with evolutionary biology. Throughout his six-decade career, Kroeber policed his way of marking off anthropology's mode of inquiry against the reduction of cultural meaning to psychological tendencies of individuals, races, or the species.

Kroeber's superorganic was misunderstood, however, not just by his foes, but also by many of his colleagues, including Boas, as reifying cultures into quasi-substantial objects. Chapter 3 demonstrates how well justified was Kroeber's reiterated protestation against this imputation. He was not trying to find an ontological object that would disempower other human sciences, but only to demarcate his field's unique approach. He repeatedly insisted that Boasian anthropologists could integrate their four fields only if they abandoned Lamarckian ideas about heritability (Kroeber 1916a, b; 1917; 1960; Kronfeldner 2010; Jackson 2010). Far from proclaiming indifference to biological theories, Kroeber, like Boas, defended the autonomy of anthropology by adopting Weismannian and Darwinian assumptions about biology. Accordingly, he was delighted to find that during and after World War II geneticists whose experimentalism served their calling as naturalists – Dobzhansky, for one – were reframing genetics to portray natural selection as continuing to adapt populations to particular environmental conditions without endangering the deeper capacity for enculturation that was fixed in our species at its inception (Kroeber 1960; Dobzhansky and Dunn 1946; Dobzhansky and Montagu 1947; Dobzhansky 1962). The point was timely because by the 1920s and 1930s the Lamarckian environmental determinism that had earlier served to construct racial hierarchies had been replaced by genetic determinism, whose advocates embraced the genetic superiority of some typologically defined races over others and hoped to find genetic bases for state-endorsed eugenic policies that would ensure the continued primacy of their own presumably superior race. Fundamental to the eugenic project were efforts to correlate inherited physical differences between individuals with psychological variables that an increasingly elaborate system of mental testing was supposedly revealing.[4]

In Chapter 4 we turn to Dobzhansky. He once told his fellow geneticists that he was "undeservedly honored by being [wrongly] listed among the students of Boas" (Dobzhansky 1968, 103). Among the effects of this influence was that in his 1962 treatise *Mankind Evolving* he backed Kroeber's superorganic as making the capacity for culture the point of intersection between evolutionary biology, the human sciences, and the liberal politics of the emerging postwar world order (Dobzhansky 1962a, 14, 26, 1973, 105). From his "balancing" interpretation of natural selection, which held that natural selection preserves variation for future adaptation as well as using it to solve current environmental problems, Dobzhansky argued, in ways we will explicate, that free choice of marriage partners and occupations will result in a more eugenically optimal society than the tinkering of eugenicists. Dobzhansky thus put the term "eugenics" to ironically reversed uses that, without entirely repudiating it, distanced him from the "eugenic consensus" that nurtured the rise of academic genetics and continued to dog it after World War II (see Paul 1994 on "eugenic consensus").

Appearances to the contrary and his own lapses notwithstanding, Dobzhansky's definitions of populations as Mendelian gene pools, of races as populations marked off by one or more genetic differences from populations with which they interbreed, of species as populations closed to interbreeding, and of evolution as change in gene frequencies in Mendelian populations were not, we contend, expressions of genetic determinism. Instead, they were meant to erect a protective barrier to keep evolutionary theorizing from regressing to the errors about culture, biology, and their relationship that Kroeber identified. Like Kroeber, Dobzhansky was looking for a happy medium between overstressing culture by denying that genes play any role in the etiology of human traits and, at the other extreme, attributing causality exclusively to genetic factors, which undermines the decisive role of the cultural milieu in which our genes are expressed and our traits develop. In *Mankind Evolving*, he used his definitions of evolutionary biology's key terms to place the cytologist C. D. Darlington at the genetic-deterministic end of a spectrum at the other end of which he situated what he saw as the cultural determinism of the anthropologist Leslie White. He located the geneticist Hermann Muller closer to his *juste milieu* than Darlington, but as tilting too far to the deterministic side. He found his anti-racist and anti-eugenicist co-author Montagu more gene-friendly than White, but leaning too heavily on culture as the cause of human differences. Dobzhansky, like Goldilocks, thought he had it "just right." He didn't quite, however, for reasons Lewontin explained by using his former teacher's own definitions and theoretical framework.

Endorsing Kroeber's superorganic made Dobzhansky's *Mankind Evolving* into something of a high-water mark in integrating the biology of the Modern Synthesis with the Boasian ideas we listed at the outset. It is Washburn, however, who deserves much of the credit for aligning anthropology with the Modern Synthesis. We will see in Chapter 5 that even as a graduate student he chafed against the racialist-tinged, morphologically based, and typologically infected physical anthropology of his Ph.D. advisor at Harvard, Earnest O. Hooton. In his first job, as an assistant professor of anatomy in Columbia University's College of Physicians and Surgeons, Washburn, like Montagu, sought tutoring in population genetics from Dobzhansky. But the street ran two ways. *Mankind Evolving* supported Washburn's population-based New Physical Anthropology not only because Dobzhansky's ideas were baked into it, but also because Washburn, Montagu, and other anthropologists helped Dobzhansky fulfill his lifelong ambition to make his work on the humble fruit fly say something about human evolution (Dobzhansky 1962–1963, 634; 1973, ix). Throughout the 1950s Washburn and Dobzhansky facilitated meetings among evolutionary biologists, anthropologists, and social psychologists. Participants in these gatherings and their published proceedings battled about whether natural selection is merely a mechanism for discarding failed organisms or, as Dobzhansky and other pioneers of the Modern Synthesis held, a "creative" process that enhances the ability of populations of organisms to deal with changing environments.[5] Relatedly, different opinions surfaced between those who stressed selection's role in evolving particular psychophysical traits and those taking selection's finest product to be the species-wide capacity for enculturation that binds us into a family of man.

These conferences culminated in a high-profile celebration of the one hundredth anniversary of the publication of Darwin's *Origin of Species* at the University of Chicago in November, 1959. Organized by Washburn's colleague, the anthropologist Tax, its purpose was to assure the public that in the wake of the Modern Evolutionary Synthesis it had nothing to fear from Darwinism or the genial figure of Charles Darwin (Tax and Callendar 1960). That Julian Huxley's Plenary Address, in which he advocated a professedly non-racist but elitist form of positive eugenics, was received as a *faux pas*, as V. B. Smocovitis shows it was, suggests that the event's main message was that in evolving our capacity for culture balancing natural selection made Darwinism safe for democracy and, in the context of the nuclear terrors of the cold war at their most perilous, afforded a measure of hope for the future of the species (Huxley 1960; Beatty 1994; Smocovitis 1999).

The converging anti-racist messages of the New Physical Anthropology, the Darwin celebration at the University of Chicago, and Dobzhansky's *Mankind Evolving* were brought home in no uncertain terms to members of the AAA in an address by Washburn, its newly elected president, in November, 1962 (Washburn 1963). The address is the focal point of Chapter 6. "The Study of Race," as its published version was called, had been commissioned by the AAA leadership as a means of distancing the Association from claims about human evolution that the segregationist propagandist Carleton Putnam had been secretly appropriating from his cousin Carleton Coon, an eminent University of Pennsylvania physical anthropologist who, like Washburn, had been Hooton's student (Putnam 1961; Jackson 2005; Collopy 2015). Putnam's highly effective propaganda machine ensured that southern segregationists and northern conservatives would have Coon's arguments at hand whenever they wanted to put a veneer of science on their crusade to have the Supreme Court's 1954 *Brown* v. *Board of Education* decision declaring segregated schools unconstitutional reversed. The publication of Coon's *magnum opus*, *The Origin of Races* (1962), only a few weeks before the AAA meeting added urgency to Washburn's intervention. In his book, Coon maintained that the Modern Synthesis affords good grounds for thinking that the continentally distributed races of our species evolved in response to (and so were adapted to) different environmental pressures confronting separate *H. erectus* populations before these races independently and at different times evolved into *H. sapiens*, with sub-Saharan Africans – "Congoids," as Coon called them – bringing up the rear a mere 50,000 years ago (Coon 1962). Washburn, who was unwillingly transformed into a minor public intellectual by these events, took up arms against Coon, whose opinions in his view compromised the unity of our species in ways that Wade faintly echoed half a century later as well as the unity of physical and cultural anthropology that Washburn, following Boas and Kroeber, took to be essential in fending off the racialism that Coon's book encouraged. Washburn argued vigorously that *The Origin of Races* was both scientifically retrograde and politically irresponsible.

The Coon controversy is the subject of Chapter 6 because it marked a watershed in the history of American anthropology and evolutionary biology. It was a defining moment after which racism could no longer overtly present itself as scientific. Washburn was able to make a powerful case against Coon's theory of race partly because he knew the arguments of *Mankind Evolving* and in addition had access to an as-yet-unpublished review of *The Origin of Races* in which Dobzhansky demolished Coon's claim to be a "population thinker" working within the conceptual framework of the Modern Synthesis (Dobzhansky 1962c, 1963a, b). In spite of Coon's protestations of scientific objectivity, Dobzhansky was right to suspect him of complicity with Putnam (Jackson 2005; Collopy 2015; Chapter 6 of this book). Interestingly, however, Dobzhansky's closest collaborators, Ernst Mayr and George Gaylord Simpson, accepted Coon's arguments as falling within the Synthesis, even if his claims were provisional, while Dobzhansky flatly rejected the very possibility. This unexpected split between collaborators whose agreements were as habitual as they were crucial to the success of their new paradigm for biology draws our attention to Dobzhansky's long-standing interactions with anthropologists, who led him to see that our shared capacity for culture casts doubt on the old commonplaces about racial rank ordering that Coon's book encoded, repeated, and reflected. Close study of this episode casts doubt on the widespread view of historians that postwar biologists became anti-racists for ideological, not scientific reasons. Mayr's ideas about race were transformed belatedly (Haffer 2007). But the very fact that he eventually acceded to the consensus that formed around Dobzhansky and Washburn testifies at least as much to the power of scientific arguments that arose at the boundary between population-genetic evolutionary biology and four-field anthropology as it does to post-Holocaust ideology.

The upbeat Darwinism of the Chicago Darwin Celebration was widely adopted by anthropologists in the wake of Washburn's address to the AAA. It dominated professional and public opinion until the late-1960s, when, just as a backlash against civil rights legislation began scoring national electoral gains, genetic determinism returned in new defenses of the racial distribution of intelligence (Jensen 1969; Herrnstein 1971, 1973). Lewontin, Dobzhansky's most verbally agile student, and the invertebrate paleontologist Stephen Jay Gould, who was educated in Simpson's shadow, launched a furious attack on Arthur Jensen, Richard Herrnstein, and other advocates of this claim, which they viewed as scientific racism *redivivus* (Gould 1981; Lewontin 1982; Levins and Lewontin 1985, 89–106). They applied the same lines of argument to the nascent field of Behavior Genetics, in whose affairs, much to Lewontin's displeasure, Dobzhansky entangled himself (Lewontin 1976a). They also applied them to the Sociobiology of their Harvard colleague E. O. Wilson (Wilson 1975; Lewontin 1976b; Gould and Lewontin 1979; Lewontin, Rose, and Kamin 1984). The attack surprised and bewildered Wilson (Wilson 1994; Segerstråle 2000). Far from thinking he was flirting with racism or eugenics, he took himself to be expanding the explanatory power of Modern Synthesis by bringing ethology, the study of animal behavior, under its sway, thereby resolving Darwinism's congenital inability to explain cooperative behavior. Wilson has had some success in painting Lewontin and Gould's attacks on Sociobiology as little more than products of the fashionable neo-Marxism of the time (Lumsden and Wilson 1983, 4; Wilson 1994). Accordingly, we point out in Chapter 7 that Washburn's visceral rejection of Sociobiology and Lewontin and Gould's assimilation of its logic to Jensenism and Behavior Genetics are best understood in the long light of Boas, Kroeber, Dobzhansky, Dunn, Montagu, and Washburn's views, which they saw themselves as defending and, in Lewontin's case, correcting.[6] Their arguments are germane to today's debates for this reason.

Our story ends on this note, but not because disputes about genetics, race, and democracy within anthropology or between it and representatives of nearby fields have ceased. Far from it. Since the Sociobiology Controversy, anthropology has had to work through a number of issues that have tended to drive a wedge between its biological and cultural sides. An advocate of Sociobiology, Derek Freeman, attacked the objectivity of the culture concept by calling Margaret Mead's ethnography into question as not just subjectively biased, but fraudulent (Freeman 1983). Napoleon Chagnon interpreted his own ethnographic portrait of the Yanomami Indians of South American as "the fierce people" in terms of the beneficial effect of male violence on reproductive fitness (Chagnon 1968, 1988). Evolutionary Psychology has intensified the putative reduction of culture to psychology and of psychological traits to "genes for" that Chagnon embraced and that Sociobiology's critics disputed. In response, some leaders of the AAA, whose membership has decreasingly overlapped with that of the American Association of Physical Anthropology, have been tempted to defend their commitment to the culture concept and culturally pluralist conceptions of democratic life by pulling away from anthropology's claim to be a science (AAA 2011). This would have bothered Boas, Kroeber, Washburn, and Dobzhansky. So would the imposition of talk about "genes for" this or that psychological trait onto conventionally construed races. We do not enter into these issues in this book. Rather, our aim is to equip readers to respond to contemporary debates by having relived with us the battles of three generations of closely affiliated scientists whose interdisciplinary efforts consciously contributed to the post-racist democratic pluralist politics that stands in need of defense today.[7] Among the lessons our studies suggest is that anthropologists, who in the last century worked with evolutionary biologists to discredit scientific racism and other forms of anti-democratic ideology, can do their country a favor by sticking together in this one.

Some historiographical astigmatisms

Although bits and pieces of our story are familiar to scholars, gaps appear whenever histories of anthropology and evolutionary biology in twentieth century America are too discipline-specific to appreciate the depth of their mid-century interaction. The historiography of the Modern Synthesis is a case in point. Mayr, one of its founders, and the late Will Provine, whose history of population genetics set a high standard for reconstructing technical debates, framed the canonical account (Provine 1971, 1986; Mayr and Provine 1980). They had their differences. Nonetheless, while their edited book on the formation of *The Evolutionary Synthesis* canvassed the influence of (and on) a wide range of relevant fields, it did not mention anthropology or any other social science (Mayr and Provine 1980). Mayr and Provine were defending the scientific credentials of the Modern Synthesis against molecular imperialists like James Watson, who were trying to expel evolutionary natural history from biology departments. They may have thought that even mentioning the social sciences would undercut this aim. It would have been even more counter-productive to acknowledge that social and political issues might have affected the making of the Modern Synthesis. So concerted has been the effort to present the Modern Synthesis as science pure and simple that even framing the relevant questions has been deferred to historians of science writing within the last decade or so. Only recently have they charted the extent of the inter-animation between the Modern Synthesis and anthropology (Smocovitis 2012). Only recently have detailed studies of Dobzhansky's input into the United Nations' *Statements on Race* appeared (Gayon 2003; Müller-Wille 2005; Gormley 2006; Brattain 2007; Farber 2011; Marks 2010a; Selcer 2012). Only recently, too, has the split between Dobzhansky and his closest colleagues, Mayr and Simpson, in the Coon affair been acknowledged, let alone explained (Jackson 2005, Collopy 2015). In Chapter 6, we show that biologists who signed on to the "same" Modern Synthesis were divided on the meaning of so basic a concept as "population thinking" and spite of their show of unanimity harbored different attitudes toward human equality. These differences become invisible when the history of evolutionary biology is sealed off from the history of the social sciences and from the social-political contestations that formed their context (Meloni 2016). As a result, it becomes difficult to write satisfactory accounts of either field.

It is no longer a radical gesture to claim that histories of science are not Olympian surveys presented without viewpoints (Golinski 2010). Historians uncover true and useful facts, many previously hidden in the recesses of archives, but in doing so they produce partisan documents that intervene in controversies both contemporary and historical. When at a later date this becomes clear these histories cease to serve as authoritative secondary sources and become primary documents for other historians. The point is as applicable to histories of the social sciences as it is to histories of biology such as Mayr's *The Growth of Biological Thought* (Mayr 1982). Marvin Harris's *The Rise of American Anthropology*, for example, is an extended defense of the fruits of the anti-Boas and anti-Kroeber tree planted by Leslie White at the University of Michigan (Harris 1968). The biosocial functionalism of that school pits itself against Boas and Kroeber's historical particularism. White and Harris could portray the Boasians' claims as devoid of scientific significance and generalizability only by ignoring the support Boas, Kroeber, Benedict, Mead, Montagu, and Washburn received from cutting-edge biology. Dobzhansky's name does not even appear in Harris's book, signaling indifference to the population genetic Darwinism that has guided evolutionary biologists and biological anthropologists for over half a century. Instead, Harris's "cultural materialism" reflects fidelity to White's law-governed, naturalized alternative to Kroeber's view of how biology and anthropology relate (Chapter 3 of this book).

George Stocking, American anthropology's historian of record, was another sort of historian altogether (Stocking 1968). He knew that anthropology and biology have been joined at the hip from the beginning and that their interactions have always been attuned to public policy debates and ideological struggles. But while Stocking is trustworthy on how nineteenth century Lamarckism gave way to twentieth century genetic Darwinism in biological anthropology – he follows Kroeber's account (Stocking 1968) – he provides no path to explain why, for example, a population geneticist like Dobzhansky allied himself with Kroeber or why Washburn made common cause with him, or why Coon's *The Origin of Races* was panned by Dobzhansky but applauded by Mayr, or why Lewontin was convinced that Dobzhansky's version of population-genetic Darwinism could not prevent the resurgence of new forms of genetic determinism and racial essentialism in the era of molecular biology.

In redressing the deficit it is not enough to talk in general terms about Darwinism, even of genetic Darwinism. Genetics roughly divides historically into Mendelian genetics, transmission genetics, population genetics, quantitative genetics, molecular genetics, and recently developmental genetics. The same polysemy attends the term "Darwinism" both before and after its turn-of-the-twentieth century re-founding as genetic Darwinism. The genetic Darwinism of

1900 was not the genetic Darwinism of 1942 any more than the genetic Darwinism of 1959 was that of 1975. Wilson's *Sociobiology* reflected an effort beginning in the 1960s to incorporate molecular evolution, behavior, and ecology into the Modern Synthesis.[8] A decade earlier, Washburn had united biological with cultural anthropology on terms favorable to the latter by taking Dobzhansky's population genetics as the key to understanding sapient humans as morphologically distinctive hominids because of their culturally inflected way of life (Washburn 1961). Wilson threatened this hard-won unity by slightly but significantly departing from the mid-century version of the Modern Synthesis. Without a clear picture of the unity between mid-century population genetic Darwinism and mid-century anthropology his seemingly minor shift will make Lewontin and Gould's howls of protest seem excessive, petty, and purely political. In considering the history of such episodes and the controversies they provoked we must, then, closely follow the interaction between the social sciences and evolutionary biology and use terms like "Darwinism" and "genetics" with time-sensitive precision. Consequently, we must attend to technical issues that are the spear tip of larger disputes. In this respect Jonathan Marks's work on the postwar history of American biological anthropology hits the right notes and continues the story whose beginnings Stocking told so well (Marks 1995, 2004, 2010a).[9]

Still, preoccupation with staying within disciplinary boundaries does not fully explain why historians have underplayed the trans-generational affinity between population-genetic Darwinians and Boas-inspired anthropologists. Sweeping histories of the interaction between Darwinian evolutionary biology and the social sciences in the American century do exist. The problem is that they have tended on their biological side to treat genes as simple Mendelian factors, if only to make things clearer to their audience, to favor on their social scientific side psychology over anthropology as the point of intersection between the biological and human sciences, and to presume a dichotomy between nature and nurture in the very gesture of insisting on both.

The most influential history of this sort is Carl Degler's *In Search of Human Nature* (Degler 1991). The scandal of Nazism, he argues, provoked social scientists to abandon genetic explanations of human traits and trait-differences. This was a change from the 1920s and 1930s, when biologists were turning their social scientific colleagues in genetically determinist, eugenicist, and racist directions. In supporting the Immigration Act of 1924, for example, which slowed immigration into the United States to a racially skewed trickle, Columbia's proto-behaviorist psychologist Edward Thorndike became almost as genetically deterministic as the eugenicist Charles Davenport. Degler says there is nothing wrong with the general idea of explaining social facts by biological inheritance. The difficulty is that in Thorndike's day the right biology had not yet arrived. Freed now from the racism, eugenics, and anti-democratic animus that provoked the aberrant mid-century reaction in favor of nurture over nature, Degler suggests that hereditarian biology can now resume its natural ascendancy by calling on Sociobiology to attract the social sciences into its orbit. The sordid past turns out to have been prologue to a glorious future.

Tales like Degler's suggest that what we have denied about the genetics-anthropology relationship – that it was driven more by ideology than science – may be true of the genetics-psychology relationship. Degler's case for nurture over nature rested largely on the insubstantial shoulders of behaviorist operant conditioning, whose weaknesses made it easy for readers to opt for the genetic-determinist side of his binary. Psychology has since shifted from behaviorism toward viewing mental acts as mediating between environments and actions. The so-called "cognitive revolution" may have undermined Degler's argument, but it also provided a more powerful rationale for his hereditarianism. Since the early 1990s, evolutionary psychologists, backed by discoveries about the localization of brain functions, have argued that during the long period of human history before the agricultural revolution natural selection gradually evolved specialized mental modules for evaluating and responding to a range of stereotyped situations. Violations of logic such as assuming the consequent are shortcuts favorable to reproductive success and hence are adaptive in ways that, unlike the atavistic behaviors that sociobiologists like Wilson merely hope we can resist, are still useful in most contexts (Wilson 1978, 167, 207; Barkow *et al.* 1992). Evolutionary Psychology has displaced 70s-style Sociobiology for this reason. Still, it advocates retain, and even intensify, its tendency to psychologize culture and naturalize traditional gender roles (Buss 2003; Thornhill and Palmer 2000, for example). Evolutionary Psychology is understandable as updating Sociobiology's resistance to widespread questioning of received views about gender and sexuality in contemporary society (McKinnon 2005).

In making their case, evolutionary psychologists draw on conventional conceptions of the authority of science. They tell a story according to which social scientists of every stripe have had a vested interest for centuries in shielding their disciplines from real scientific progress by assuming that we are born with a "blank slate" on which society imposes its constructions, thereby ruling out the very possibility that our actions might be explained as law-governed behaviors that eventually will be reduced to the physical workings of the brain (Pinker 2002). Steven Pinker agrees with John Tooby and Leda Cosmides that social scientists have used this putative Standard Social Science Model (SSSM) to cling to a "doctrine of intellectual isolationism" (Tooby and Cosmides 1992, 22). Afraid of being "reduced to more basic sciences," they have viewed

> conceptual unification … [as] an enemy and the relevance of other fields as a menace to their freedom to interpret human reality in any way they chose. Thus, despite some important exceptions, the social sciences have largely kept themselves isolated from the crystallizing process of scientific integration.
>
> (Tooby and Cosmides 1992, 21)

Jerome Barkow does not think cultural anthropology is one of the exceptions. He says, "It has not progressed in the cumulative fashion of evolutionary biology" (Barkow 2006b, 348).

Degler's idea of social science's postwar stress on nurture as an anomalous departure from hereditarianism is historically better grounded than the evolutionary psychologists' evocation of an incorrigible SSSM they suppose to have held sway since the time of John Locke. Still, both take at face value Provine's testimony that the sea change toward nurture after World War II was instigated solely by "feelings of revulsion" at the Nazis' murderous racism, not by any new genetic findings (Degler 1991, 205, quoting Provine 1973, 795; Barkan 1992). They might more readily find this fault in themselves, however, than in the postwar Modern Synthesis. So far no gene sequences "for" the cognitive quirks and gender preferences evolutionary psychologists attribute to human nature have been identified, while in the postwar period experimental evidence from Dobzhansky's and other labs combined with discoveries in the field to provide compelling scientific reasons for holding that in changing environments natural selection favors the evolution of phenotypic plasticity.[10] The cultural way of life of *H. sapiens* evolved from *H. erectus* under just such volatile conditions.

None of our focal figures, moreover, ever claimed that natural selection was stopped dead in its tracks when culture took over, as the SSSM supposes. "It is a fallacy to think that specific or ordinal traits do not vary or are not subject to genetic modification; phenotypic plasticity does not preclude genetic variety" (Dobzhansky 1962a, 74). The position evolutionary psychologists attribute to the SSSM is actually more characteristic of eugenicists, whose interpretation of Darwinism led them to believe that modern society cossets the weak by obstructing the eliminative work of natural selection. Stanford University's first President, David Starr Jordan, made this exact point, complaining that, "[i]n the red field of human history the natural process of selection is often reversed. The survival of the unfittest is the primal cause of the downfall of nations" (Jordan 1902, 25; see Gayon 1995). The population genetic revolution exploded this assumption, even if its eugenics-minded pioneers, notably R. A. Fisher, did not anticipate that the erection of the Modern Synthesis on population-genetic foundations meant that what they set afoot would undermine their reason for initiating it. When Lewontin accused Dobzhansky of genetic determinism he was not holding that human differences are exclusively products of enculturation. He meant that Dobzhansky's interactionist view of genes, environment, and developing organisms, not least encultured organisms, was not interactionist – Lewontin would say "dialectical" – enough (Chapter 4 of this book).

For all of these reasons, we can safely conclude that any comprehensive history of anthropology or biology in the twentieth century must include the tangled dealings of evolutionary biologists and anthropologists with each other. A full account of their interaction will take note on the social scientific side of the views of psychologists, sociologists, linguists, and anthropologists. On the side of biology it will include and differentiate between geneticists of every stripe from Mendelian to molecular. This book is not that comprehensive history. Ours is an episodic history of a particular line of argument that arose at the intersection of population-genetic Darwinism and Boas's approach to anthropology. Its limitation to America issues an invitation to scholars to explore how the Modern Synthesis interacted with anthropology in other countries. We expect that it did so differently, not only because anthropology was not quite the same, but also because the Modern Synthesis wasn't either. In charting the social and political issues that spurred the disputes we revisit, especially about race, we acknowledge that we are intervening in present-day debates. True, we break the historical frame only occasionally by flashing forward to see how well the empirical claims our principals advanced have fared, and even then we do so more often in notes than in the main text. We write this way because we hope readers will immerse themselves deeply in the past to equip themselves to make informed judgments about our interpretations of these episodes and even more about contemporary issues in evolution generally and human evolution in particular.

Toward a rhetorical history of biology and anthropology

A novel aspect of the historical stories we tell is the way we tell them. Ours is a history of a connected series of scientific controversies and of the lines of argument deployed by participants in them. This means that we situate the published and archived texts that are our evidence in the space of reason giving in which they were produced. In that space the results of experiments and observation are important. But the well-known under-determination of theories by fact impels scientists to enroll such evidence as they have in support of their own larger visions of science's future and its relevance to human affairs. Anticipatory commitment before all epistemic considerations are known is especially unavoidable when politically charged beliefs and policies are at stake. Under these conditions facts do not speak for themselves. They need someone to speak for them, and for the larger visions they are asked to stand for. In practice, this means speaking against others who deny either one's facts, or their importance, or their interpretation, or the larger significance someone attributes to them. As debates about these issues ramify they become controversies: temporally extended disputes in which, along with a proliferating arsenal of facts, the assumptions and implications of the original issue are brought into the open and full contestation. Scientific knowledge precipitates out of this discursive process.

Argumentation of this sort is by its very nature rhetorical argumentation: argument viewed as a dynamic, social, value-laden process of communication that occurs between speakers and hearers, writers and readers, senders and receivers of messages at times and places when the need to judge is combined with insufficient grounds for doing so. Aristotle, in his *Rhetoric*, was not the first to point out that in looking for points that will be effective in such situations speakers must find arguments that are timely (*kairos*). He was, however, the first to show that because such situations are rife with uncertainty speakers must also construct their personal authority in order to persuade (*ethos*). In addition they must elicit from their audience appropriate emotional responses to their proposed way of resolving the issue at hand (*pathos*). This is not an irrational activity. Artful speakers can be counted on in such fraught contexts to marshal evidence to support their claims (*logos*). That is why, since antiquity, students of rhetorical argumentation have been cataloging lines of argument (*topoi*) that are likely to be effective under conditions in which information is imperfect, the point of what one asserts, and even what it means, is relative to what one's opponents hold, and judges come to the scene of discourse with passions and predispositions that must be acknowledged, appealed to, and even manipulated.[11]

In each of our chapters we will see our focal figures and their opponents, scientists though they are, succeeding and failing to find persuasive lines of arguments, inventing and projecting authoritative *personae*, and eliciting commitment-inducing emotional responses from their audiences. We will see, too, that each of these figures tended to deploy a characteristic strategy of rhetorical argumentation to meet the exigencies of the rhetorical situation at hand.[12] Boas's anti-racist view could be heard in a racist society, for example, because instead of making dogmatic claims he kept his audiences informed about who should bear the burden of proof and where the presumption lies given the current state of evidence about an issue. In this way, we will show in Chapter 2 that Boas's deployment of the argument from presumption satisfied the requirement of timeliness and, in its seeming modesty, inclined his readers to trust his inkling that further inquiry will support his egalitarian hypotheses. We believe that the success of this strategy had much to do with the resilience of the tradition of anthropological inquiry he initiated.

Kroeber felt the weight of trying to institutionalize anthropology in universities that were sometimes reluctant to make room for a new discipline that spanned the natural and social sciences and, in its meaning-laden conception of culture, transgressed into the territory of the humanities. To do so, he employed what scholars of scientific argumentation have identified as "the rhetoric of demarcation" (Gieryn 1999; Taylor 1996). He tried, that is, to identify his discipline's unique way of approaching human affairs without denying that its objects of study could be validly approached from other perspectives. If biology or psychology exceeded their mandates by colonizing anthropology they could be accused of failing to mind their own business as well as of ham-handedly meddling in anthropology's. Kroeber was especially alert to ways anthropologists themselves strayed from their field's path.

In affirming Kroeber's concept of the superorganic, Dobzhansky used definition as a strategy of argumentation. Argumentative impasses in professional as well as public-sphere discussions of race and eugenics, he sensed, could be transcended only if the technical meanings of 'race' and 'species' implied by population genetics were disseminated to and internalized by the citizens of democratic societies (Dobzhansky 1973; Beatty 1994). It is a testament to his effort that today's biological anthropologists, using molecular genetic evidence, find races in his sense in many species, but not (unlike Dobzhansky himself) in our own (Templeton 2013). Then and now the problem is how to communicate the technical meaning of these terms to a democratic audience long buffeted by the effects of crude conceptions of race, natural selection, and other key terms. Lewontin accepted Dobzhansky's definitions, but believed that to confound Jensen, Herrnstein, behavior geneticists, and Wilson he had to wrest authority from his mentor by calling attention to his own expertise in molecular biology and statistical analysis. His arguments exhibit a strong sense of *ethos*.

Washburn's work as a physical, or as he would prefer biological, anthropologist was focused on reconstructing entire ways of life from bones and shards left behind by our ancestors and from analogies in our extant relatives, especially chimpanzees. In our reading of his career Washburn's technical reports are important not just for their empirical content, but also for their role in building a professional community composed of cultural anthropologists, biological anthropologists, and population-genetic biologists. In the years leading to passage of the Civil Rights Act of 1964 and the Voting Rights Act of 1965, he was a tireless organizer of professional conferences and symposia. As a result, he had the gumption to marginalize colleagues who, like Coon, violated the community's understanding of human equality at the moment when the urgency of the call for civil rights for all Americans was making itself felt. Given this aim, Washburn practiced in his many talks, articles, and short pedagogical books what Aristotle called epideictic rhetoric: the rhetoric of praise and blame displayed on important, often ceremonial, public occasions.

The 1959 Darwin centennial celebration at the University of Chicago was such an occasion. Although organized by Tax, Washburn's epideictic hand was at work in the event's effort to free Darwinism from the blame for atheism and imperialism that it had borne in American public memory since the time of William Jennings Bryan. The Modern Synthesis deserved praise because it gave scientific support to the egalitarian underpinnings of democracy. Washburn's epideictic style of address is directly visible in his Presidential Address to the AAA (Washburn 1963). He blamed Coon for ignoring the genetic foundations of population-based Darwinism and therefore for failing to see how much biology and politics had changed since he formed his opinions, or rather inherited them from Hooton. By contrast, Washburn praised Dobzhansky for staying abreast of changes in both science and society. As a result, the value-laden views Washburn and Dobzhansky shared have become so commonsensical that even sociobiologists and evolutionary psychologists have had to present themselves as agreeing with their tolerant spirit, if not their biological letter. In the Epilogue we note that Washburn was as spontaneously hostile to Sociobiology as Gould, Lewontin, and Montagu. But his particular objection reflected what had always been his main concern. He blamed sociobiologists for creating disunity in anthropology in the name of scientific unification. His idea of scientific unity was forming a community, not theoretically reducing one field of inquiry to another (Washburn 1978a, 1978b; Wilson 1998; Cecarrelli 2001a, b). Had he lived to see it he would have been offended by Wilson's (and evolutionary psychologists') defense of scientific unification as theory reduction (Wilson 1998).

The picture we paint of scientific argumentation and the knowledge production it facilitates is not the mid-twentieth century understanding. That understanding had it that scientific institutions are effective only when their inquiries are sufficiently insulated from the rough and tumble of the world to solve problems by subjecting a steadily accumulating body of empirical data to logical rather rhetorical norms of argumentation (Merton 1973). We do not deny that a certain amount of insulation is necessary if controversies like those we are reliving are to be productive. After all, our principals and their allies worked in well-funded and well-governed universities and museums, published in the peer-reviewed journals of record in their disciplines, and held high offices in their professional associations. They were leaders in their fields. We are far from alone, however, in denying that the disputes that arise in these sites are sealed off from the pressures of their time. History and philosophy of science departments and programs that were founded in the 1950s and 1960s with the aim of confirming the received theory-centered, reductionism-oriented view found just the opposite – so much so that in recent decades some sociologists of science have argued that scientific controversies are so deeply informed by social, political, and ideological factors that sociology is entitled to displace history and philosophy of science in judging why this or that theory prevailed in this or that context (Shapin and Schaffer 1985; Latour 1987). The ensuing "science wars" of the 1990s satisfied nobody, including us. Our approach is to stress reasoning as much as the standard view, but to focus on the informal practices of reason giving that flourish at the boundary between technical and public spheres rather than on rationally reconstructing scientific arguments. The latter are useful in retrospectively revealing the validity and soundness of theories. But as we will repeatedly see they reach that happy condition by a rougher process of winnowing.[13]

Notes

1 Huxley's 1942 book was titled *Evolution: The Modern Synthesis*.

2 This is still a live, if often confused, issue for at least two reasons: first, Wade flies in the face of the belief of many physical anthropologists that by the standards biologists apply to other species there are no human biological races at all (Templeton 2013); and, second, many discussants do not quite see that when cultural anthropologists call human races social constructions they are not implying that these constructions don't have real effects, some of them biological (Goodman 2013).

3 Following Darwin, James thought of instincts as naturally selected shortcuts to effective action (James 1890; Darwin 1859, 209–210). "My quarrel with Spencer is not that he makes much of the environment," he wrote, "but that he makes nothing of the glaring and patent fact of subjective interests which cooperate with the environment in molding intelligence" (James to Charles Eliot, 22 November 1878; in Richards 1987, 426–427, n. 61). James was not alone in critiquing Spencer's inability to explain the evolution of agency. So did British philosophers oriented to Hegel's dialectical way of thinking: Samuel Alexander, T. H. Green, Edward Cairns, and through them the American John Dewey (Pearce 2014).

4 It is well established that eugenics nurtured psychological testing. See, for example, Zenderland 1998; Richards 2012.

5 In calling natural selection the creative factor in evolution the founders of the Modern Synthesis were affirming, *contra* mutationists, that natural populations have (and through evolved mechanisms can generate) enough genetic variation to gradually evolve new adaptations, species, and higher taxa (Huxley 1942, 28; Dobzhansky 1962, 430–431; Gould 1981; Mayr 1980, 18; Beatty 2016). Dobzhansky's collaborators agreed on this principle, but not all of them were as insistent on the role of culture

in human evolution, in part because they were not as conversant with anthropology (Chapter 6 of this book).

6 "In fact, you are (and remain) one of my few 'heroes' for your articulate and uncompromising views on social issues over many years and often in the face of predominantly negative opinion" (Gould to Montagu, April 29, 1974, Montagu Papers; see Lewontin 2000, 29, for his indebtedness to Dobzhansky).

7 The Columbia connection linking these scientists is hard to miss. It was the first Ivy League university to open its doors in more than a token way to the striving minorities that by the turn of the century populated America's urban centers. From Columbia, Boas's egalitarianism began to affect ideas about the nature and prospects of American democracy. Horace Kallen's "cultural pluralism" found a congenial home there (Kallen 1924). It was at Columbia that Boas mentored Kroeber, Mead, Benedict, and Montagu. Dobzhansky tutored Montagu, Washburn, and Lewontin there. Although its president, Nicholas Murray Butler, was an authoritarian who tried to limit Jewish students to no more than 20 percent and thought he could hire and fire professors at will (provoking Dewey, Boas, and other faculty members to found the Association of University Professors to protect academic freedom), he did encourage collaboration across disciplinary and collegiate lines, such as putting in place the first cross-departmental "statistical package" of courses (Camic and Xie 1994). In this way Dewey's evolution-based pragmatic epistemology spread across departments and was taken up to one degree or another by many of our focal figures. Columbia's faculty also involved themselves in public affairs. They supported the bargaining rights of all workers, not just themselves; advocated for the Civil Rights of African-Americans; organized rescues of Jewish academics from Hitler's clutches; and in the 1950s nurtured an anti-Stalinist (and anti-Lysenkoist) left.

8 A trace of the institutional tensions attendant on incorporating these fields into the Modern Synthesis is the existence here and there of separate departments in Behavior, Evolution, and Ecology and in some universities Behavioral Ecology.

9 "I also consider that I follow paths laid as well by other influential scholars – Huxley, Boas, [Bronislaw] Malinowski, Dobzhansky, Simpson, Washburn – who built on Darwin's work, augmented, and to some extent superseded it" (Marks 2004, 191).

10 The importance of phenotypic plasticity in evolutionary biology waned in the 1960s and 1970s with the rise of trait-by-trait adaptationism and the focus on structural gene products, but has returned with new knowledge of the intimate interaction of ecological and developmental processes (Pigliucci 2001; Schlichting and Pigliucci 1993; West Eberhard 2003). The stress no longer falls on the supposed inherent superiority and plasticity of heterozygotes (Chapter 4 of this book). Recent suggestions that forms of heritable variation in regulatory gene sectors and epigenetic side-chains attached to DNA are more open to environmental influences than structural genes have raised the possibility that the Modern Synthesis should be revised or even replaced (Laland *et al.* 2014; Jablonka and Lamb 2005; Gilbert and Epel 2009). One implication might be that the very idea of 'gene for a trait' is incoherent. Meloni 2016 explores implications for human evolution and social policy.

11 For some broad overviews of the tradition of rhetorical argumentation see Cox and Willard (1982), Zarefsky (1995, 2014), and Tindale (2004). Some classical texts are: Aristotle, *Rhetoric*; Cicero, *De Inventione*; and Quintillian, *Instituta Oratorica*. In the nineteenth century, Richard Whately's *Elements of Rhetoric* stands out.

> We hold that argumentative composition as a species of rhetoric … The office of the logician is to infer, but the office of the rhetorician is to advocate by adducing proofs.… Even the philosopher who undertakes by writing or speaking to convey his notion to others assumes for the time being the character of advocate of the doctrines he maintains.
>
> (Whately 1963 [1828], 5)

Groundbreaking twentieth-century works on rhetorical argumentation include Kenneth Burke, *A Rhetoric of Motives* (1950); Stephen Toulmin, *The Uses of Argument* (1958); and Chaïm Perelman and Lucie Olbrechts-Tyteca, *The New Rhetoric: A Treatise on Argumentation* (1969).

12 In his studies of rhetorical argumentation from Plato to Edmund Burke, Richard Weaver showed how neatly the rhetorical strategies of famous orators come into view when we identify their characteristic argumentative forms (Weaver 1953).

13 Our stress on rhetoric as argumentation deemphasizes metaphors and other tropes in scientific argumentation found in other work in the rhetoric of science (Gross 1990, 1996; Fahnestock 1999). The centering of figures of speech in rhetoric hearkens to eras in which rhetoric detached from logic, such as the Second Sophistic in late antiquity and the Renaissance in early modernity. The resulting view of metaphor as decoration, learned display, and a pedagogical crutch for audiences who cannot follow scientific reasoning is what many philosophers and scientists still mean by "rhetoric." It was important in the heyday of logical empiricism for philosophers of science to see much more than that in the use of metaphors in scientific explanation. Explanation, they argued, is a matter of "seeing as," and there is nothing like a good metaphor for doing that (Hesse 1963). Understandably, rhetoricians of science have run with this revelation. Now that the point has been made, however, they have turned their attention to other aspects of rhetorical argumentation in science (Cecarrelli 2001a, b; Keränen 2010; Hartelius 2011; Depew and Lyne 2013).

References

AAA. 2011. AAA Long-Range Plan As Amended by the AAA Executive Board. Available online at www.americananthro.org/ConnectWithAAA/Content.aspx?ItemNumber=1985

Barkan, Elazar. 1992. *The Retreat of Scientific Racism: Changing Concepts of Race in Britain and the United States between the World Wars*. Cambridge: Cambridge University Press.

Barkow, Jerome H. 2006. "Introduction: Sometimes the Bus Does Wait." *In Missing the Revolution: Darwinism for Social Scientists*, edited by Jerome H. Barkow, 3–59. New York: Oxford University Press.

Barkow, Jerome H., Leda Cosmides and John Tooby, eds. 1992. *The Adapted Mind: Evolutionary Psychology and the Generation of Culture*. New York: Oxford University Press.

Beatty, John. 1994. "Dobzhansky and the Biology of Democracy: The Moral and Political Significance of Genetic Variation." In *The Evolution of Theodosius Dobzhansky: Essays on His Life and Thought in Russia and America*, edited by Mark B. Adams, 195–218. Princeton: Princeton University Press.

Beatty, John. 2016. "The Creativity of Natural Selection? Part I: Darwin, Darwinism, and the Mutationists." *Journal of the History of Biology* 49 (4): 659–684.

Benedict, Ruth and Gene Weltfish. 1943. *The Races of Mankind*. New York: Public Affairs.

Bliss, Catherine. 2012. *Race Decoded: The Genomic Fight for Social Justice*. Stanford: Stanford University Press.

Boas, Franz. 1887. "Letter to the Editor: Museums of Ethnology and Their Classification." *Science* 9: 587–589.

Boas, Franz. 1906 [1974]. "The Outlook for the American Negro." In *The Shaping of American Anthropology, 1883–1911: A Franz Boas Reader*, edited by George W. Stocking, 310–303. New York: Basic.

Boas, Franz. 1911. *The Mind of Primitive Man*. New York: Macmillan.

Boas, Franz. 1928. *Anthropology and Modern Life*. New York: Norton.
Boas, Franz. 1940. *Race, Language and Culture*. New York: Macmillan.
Bonilla-Silva, Eduardo. 2014. *Racism without Racists: Color-Blind Racism and the Persistence of Racial Inequality in America*. Fourth edition. Lanham: Rowman & Littlefield.
Brattain, Michelle. 2007. "Race, Racism, and Anti-Racism: UNESCO and the Politics of Representing Science to the Postwar Public." *American Historical Review* 112: 1386–1413.
Brown v. *Board of Education*, 347 U.S. 483 (1954).
Burke, Kenneth. 1950. *A Rhetoric of Motives*. New York: Prentice-Hall.
Buss, David M. 2003. *The Evolution of Desire: Strategies of Human Mating*. New York: Basic.
Callaway, Ewen. 2014 "Geneticists Say Popular Book Misrepresents Research on Human Evolution." *Nature* (Newsblog). Available online at http://blogs.nature.com/news/2014/08/geneticists-say-popular-book-misrepresents-research-on-human-evolution.html
Camic, Charles and Yu Xie. 1994. "The statistical turn in American social science: Columbia University, 1890–1915." *American Sociological Review* 59: 773–805.
Ceccarelli, Leah. 2001a. *Shaping Science with Rhetoric: The Cases of Dobzhansky, Schrödinger, and Wilson*. Chicago: University of Chicago Press.
Ceccarelli, Leah. 2001b. "Uniting Biology and the Social Sciences: A Rhetorical Comparison of E. O. Wilson's *Consilience* and Theodosius Dobzhansky's *Mankind Evolving*." *Poroi* 1.1. Available online at http://ir.uiowa.edu/poroi/vol1/iss1/
Chagnon, Napoleon A. 1968. *Yanomamö: The Fierce People*. New York: Holt, Rinehart and Winston.
Chagnon, Napoleon A. 1988. "Life Histories, Blood Revenge, and Warfare in a Tribal Population." *Science*, 239 (4843): 985–992.
Collopy, Peter Sachs. 2015. "Race Relationships: Collegiality and Demarcation in Physical Anthropology." *Journal of the History of the Behavioral Sciences* 51: 237–260.
Comfort, Nathaniel C. 2012. *The Science of Human Perfection: How Genes Became the Heart of American Medicine*. New Haven: Yale University Press.
Coon, Carleton S. 1962. *The Origin of Races*. New York: Knopf.
Cox, J. Robert and Charles Arthur Willard. 1982. "Introduction: The Field of Argumentation." In *Advances in Argumentation Theory and Research*, edited by J. Robert Cox and Charles Arthur Willard, xiii–xlvii.
Darwin, Charles. 1859. *On the Origin of Species by Means of Natural Selection, Or the Preservation of Favoured Races in the Struggle for Life*. London: John Murray.
Degler, Carl N. 1991, *In Search of Human Nature: The Decline and Revival of Darwinism in American Social Thought.* Oxford: Oxford University Press.
Depew, David and John Lyne. 2013. "The Productivity of Scientific Rhetoric." *Poroi* 9 (1). DOI: 10.13008/2151-2957.1153.
Dietrich, Michael. 1994."The Origins of the Neutral Theory of Molecular Evolution." *Journal of the History of Biology* 27: 21–50.
Dobzhansky, Theodosius. 1937. *Genetics and the Origin of Species*. 1st edition. New York: Columbia University Press.
Dobzhansky, Theodosius. 1951. *Genetics and the Origin of the Species.* 3rd edition. New York: Columbia University Press.
Dobzhansky, Theodosius. 1962a. *Mankind Evolving*. New Haven: Yale University Press.
Dobzhansky, Theodosius. 1962b. "*Mankind Evolving* – A Rejoinder." *Perspectives in Biology and Medicine* 6: 274–275.

Dobzhansky, Theodosius. 1962–63. "The Reminiscences of Theodosius Dobzhansky, conducted by B. Land for the Oral History Research Office of Columbia University." Dobzhansky Papers, American Philosophical Society, Philadelphia.

Dobzhansky, Theodosius. 1968. "More Bogus 'Science' of Race Prejudice." *Journal of Heredity* 59: 102–104.

Dobzhansky, Theodosius. 1973. *Genetic Diversity and Human Equality*. New York: Basic.

Dobzhansky, Theodosius and Ashley Montagu. 1947. "Natural Selection and the Mental Capacities of Mankind." *Science* 105: 587–590.

Du Bois, W. E. B. 1939. *Black Folk Then and Now*. New York: Holt.

Dunn, Leslie C. and Theodosius Dobzhansky. 1946. *Heredity, Race, and Society*. New York: New American Library.

Duster, Troy. 2015. "A Post-Genomic Surprise. The Molecular Reinscription of Race in Science, Law and Medicine." *British Journal of Sociology* 66 (1): 1–27.

Fahnestock, Jeanne. 1999. *Rhetorical Figures in Science*. New York: Oxford University Press.

Farber, Paul L. 2011. *Mixing Races: From Scientific Racism to Modern Evolutionary Ideas*. Baltimore: Johns Hopkins University Press.

Freeman, Derek. 1983. *Margaret Mead and Samoa: The Making and Unmaking of an Anthropological Myth*. Cambridge: Harvard University Press.

Fuentes, Agustin. 2012. *Race, Monogamy, and Other Lies They Told You: Busting Myths about Human Nature*. Berkeley: University of California Press.

Fullwiley, Duana. 2014. "'The Contemporary Synthesis': When Politically Inclusive Genomic Science Relies on Biological Notions of Race." *Isis* 105: 803–814.

Gannett, Lisa. 2001. "Racism and Human Genome Diversity Research: The Ethical Limits of Population Thinking." *Philosophy of Science* 68 (3): S479–492.

Gannett, Lisa. 2003. "Making Populations: Bounding Genes in Space and in Time." *Philosophy of Science* 70 (5): 989–1001.

Gayon, Jean. 1995. *Darwinism's Struggle for Survival*. Cambridge: Cambridge University Press.

Gayon, Jean. 2003. "Do Biologists Need the Expression 'Human Races'? UNESCO 1950–1951." In *Bioethical and Ethical Issues Surrounding the Trials and Code of Nuremberg. Nuremberg Revisited*, edited by J. Rozenberg, 23–48. Lewiston: Edwin Mellen.

Gieryn, Thomas F. 1999. *Cultural Boundaries of Science: Credibility on the Line*. Chicago: University of Chicago Press.

Gilbert, Scott and David Epel. 2009. *Ecological Developmental Biology: Integrating Epigenetics, Medicine, and Evolution*. Sunderland: Sinauer.

Gilkeson, John S. 2010. *Anthropologists and the Rediscovery of America, 1886–1965*. New York: Cambridge University Press.

Golinski, Jan. 2010. *Making Natural Knowledge Constructivism and the History of Science*. Chicago: The University of Chicago Press.

Goodman, Alan. 2013. "Bringing Culture into Human Biology and Biology Back into Anthropology." *American Anthropologist* 115: 359–373.

Gormley, Melinda. 2006. "Geneticist L. C. Dunn: Politics, Activism, and Community." Ph.D. diss., Oregon State University.

Gossett, Thomas F. 1963. *Race: The History of an Idea in America*. Dallas: Southern Methodist University Press.

Gould, Stephen Jay. 1981. *The Mismeasure of Man*. New York: Norton.

Gould, Stephen Jay and Richard C. Lewontin. 1979. "The Spandrels of San Marco and the Panglossian Paradigm: A Critique of the Adaptationist Programme." *Proceedings of the Royal Society of London B* 205 (1161): 581–598.

Gross, Alan G. 1990. *The Rhetoric of Science*. Cambridge: Harvard University Press.

Gross, Alan G. 1996. *The Rhetoric of Science*. Revised Edition. Cambridge: Harvard University Press.

Haffer, Jürgen. 2007. *Ornithology, Evolution, and Philosophy: The Life and Science of Ernst Mayr 1904–2005*. New York: Springer.

Haraway, Donna. 1989. *Primate Visions: Gender, Race, and Nature in the World of Modern Science.* New York and London: Routledge.

Harris, Marvin. 1968. *The Rise of Anthropological Theory: A History of Theories of Culture*. New York: Thomas Crowell.

Herrnstein, Richard J. 1971. "I.Q." *The Atlantic* 228 (3): 43–58.

Herrnstein, Richard J. 1973. *I.Q. in the Meritocracy*. Boston: Little, Brown.

Hartelius, E. Johanna. 2011. *The Rhetoric of Expertise*. Lanham: Lexington Books.

Hesse, Mary. 1963. *Models and Analogies in Science*. London: Sheed and Ward.

Huxley, Julian. 1942. *Evolution: The Modern Synthesis.* London: Allen & Unwin.

Huxley, Julian. 1960. "The Evolutionary Vision." In *Issues in Evolution* (vol. 3 of *Evolution after Darwin*, edited by Sol Tax and Charles Callendar), 249–261. Chicago: University of Chicago Press.

Jablonka, Eva and Marion J. Lamb. 2014. *Evolution in Four Dimensions: Genetic, Epigenetic, Behavioral and Symbolic Variation in the History of Life*. Cambridge: MIT Press.

Jackson Jr., John P. 2001. *Social Scientists for Social Justice: Making the Case against Segregation.* New York: New York University Press.

Jackson Jr., John P. 2005. *Science for Segregation: Race, Law, and the Case Against Brown v. Board of Education.* New York: New York University Press.

Jackson Jr., John P. 2010. "Definitional Argument in Evolutionary Psychology and Cultural Anthropology." *Science in Context* 23(1), 121–150.

James, William. 1890. *Principles of Psychology.* New York: Holt.

James, William. 1904. "Herbert Spencer." *Atlantic Monthly* 94 (1): 99–108.

Jensen, Arthur. 1969. "How Much Can We Boost IQ and Scholastic Achievement?" *Harvard Educational Review* 39 (1): 1–123.

Jordan, David Starr. 1902. *Blood of the Nation: A Study of the Decay of Races through the Survival of the Unfit.* Boston: American Unitarian Association.

Kallen, Horace M. 1924. *Culture and Democracy in the United States*. New York: Boni and Liveright.

Keränen, Lisa. 2010. *Scientific Characters: Rhetoric, Politics, and Trust in Breast Cancer Research.* Tuscaloosa: University of Alabama Press.

Kroeber, Alfred L. 1916a "Inheritance by Magic." *American Anthropologist* 18: 19–40.

Kroeber, Alfred L. 1916b. "Heredity Without Magic." *American Anthropologist* 18: 294–296.

Kroeber, Alfred L. 1917. "The Superorganic." *American Anthropologist* 19: 163–213.

Kroeber, Alfred L. 1960. "Evolution, History, and Culture." In *Issues in Evolution* (vol. 3 of *Evolution after Darwin*, edited by Sol Tax and Charles Callendar), 1–16. Chicago: University of Chicago Press.

Kronfeldner, Maria E. 2010. "Won't You Please Unite? Cultural Evolution and Kinds of Synthesis." In *The Hereditary Hourglass. Genetics and Epigenetics, 1868–2000*, edited by Barahona, Ana, Edna Suarez-Diaz and Hans-Jorg Rheinberger, 111–125. Berlin: Max Plank Institute for the History of Science.

Laland, K., T. Uller, M. Feldman, K. Sterelny, G. Müller, A. Moczek, E. Jablonka and J. Odling Smee. 2014. "Does Evolutionary Theory Need a Rethink? Yes, Urgently." *Nature* 514: 161–162.

Latour, Bruno. *Science in Action: How to Follow Scientists and Engineers Through Society*. Cambridge: Harvard University Press, 1987.

Levins, Richard and Richard C. Lewontin. 1985. *The Dialectical Biologist*. Cambridge: Harvard University Press.

Lewontin, Richard C. 1976a. "Sociobiology – A Caricature of Darwinism." *[Proceedings of the] Philosophy of Science Association: Volume Two: Symposia and Invited Papers* 2: 22–31

Lewontin, Richard C. 1976b. "Review of *Race Difference in Intelligence* by J. D. Loehlin, G. Lindzey, and J. N. Spuhler." *American Journal of Human Genetics* 28 (1): 92–97.

Lewontin, Richard C. 1982. "Organism and Environment." In *Learning, Development and Culture: Essays in Evolutionary Epistemology* edited by H. Plotkin, 151–170. New York: Wiley.

Lewontin, Richard C., Steven P. Rose and Leon Kamin. 1984. *Not in Our Genes: Biology, Ideology and Human Nature*. New York: Pantheon.

Liss, Julia E. 1998. "Diasporic Identities: The Science and Politics of Race in the Work of Franz Boas and W. E. B. Du Bois, 1894–1919" *Cultural Anthropology* 13: 127–166.

Loving v. Virginia, 388 U.S. 1 (1967).

Lumsden, Charles J. and Edward O Wilson. 1983. *Promethean Fire: Reflections on the Origin of Mind*. Cambridge: Harvard University Press.

Marks, Jonathan. 1995. *Human Biodiversity: Genes, Race, and History*. New York: Aldine.

Marks, Jonathan. 2004. "What If Anything Is Darwinian Anthropology?" *Social Anthropology* 12: 181–193

Marks, Jonathan. 2010a. "The Two 20th Century Crises of Racial Anthropology." In *Histories of American Physical Anthropology in the Twentieth Century*, edited by M. Little and K. Kennedy, 187–206. Lanham: Rowman & Littlefield.

Marks, Jonathan. 2010b. "Ten Facts about Human Variation." In *Human Evolutionary Biology*, edited by M. P. Muehlenbein, 265–276. Cambridge: Cambridge University Press.

Mayr, Ernst. 1982. *The Growth of Biological Thought: Diversity, Evolution, and Inheritance*. Cambridge: Harvard University Press.

Mayr, Ernst and William B. Provine, eds. 1980. *The Evolutionary Synthesis: Perspectives on the Unification of Biology*. Cambridge: Harvard University Press.

McKinnon, Susan. 2005. *Neo-Liberal Genetics: The Myths and Metaphors of Evolutionary Psychology*. Chicago: Prickly Paradigm Press.

Meloni, Maurizio. 2016. *Political Biology: Science and Social Values in Human Heredity from Eugenics to Epigenetics*. New York: Palgrave Macmillan.

Merton, Robert. 1973. *The Sociology of Science*. Chicago: University of Chicago Press.

Montagu, Ashley. 1942. *Man's Most Dangerous Myth: The Fallacy of Race*. New York: Harper.

Montagu, Ashley. 1963. "What is Remarkable about the Variety of Man is Likeness, not Differences." *Current Anthropology* 4: 361–363.

Montagu, Ashley. 1964. *The Concept of Race*. New York: Free Press.

Müller-Wille, Staffan. 2005. "Race et Appartenance Ethnique: La Diversite Humaine et l'UNESCO Declarations sur la Race (1950 et 1951)." In *60 Ans d'Histoire de l'UNESCO. Actes du Colloque International, Paris, 16–18* Novembre 2005, 211–220. Paris: l'UNESCO.

Müller-Wille, Staffan and Hans-Jorg Rheinberger. 2012. *A Cultural History of Heredity*. Chicago: University of Chicago Press.

Myrdal, Gunnar. 1944. *An American Dilemma: The Negro Problem and Modern Democracy*. New York: Harper.

Pascoe, Peggy. 2009. *What Comes Naturally: Miscegenation Law and the Making of Race in America*. Oxford: Oxford University Press.

Paul, Diane B. 1984. "Eugenics and the Left." *Journal of the History of Ideas* 45: 567–590.

Paul, Diane B. 1994. "Dobzhansky in the Nature-Nurture Debate." In *The Evolution of Theodosius Dobzhansky: Essays on His Life and Thought in Russia and America*, edited by Mark B. Adams, 219–231. Princeton: Princeton University Press.

Paul, Diane B. 1998. *The Politics of Heredity: Essays on Eugenics, Biomedicine, and the Nature/Nurture Debate*. Albany NY: State University of New York Press.

Pearce, Trevor. 2014. "The Origins and Development of the Idea of Organism-Environment Interaction." In *Entangled Life*, edited by Gillian Barker, Eric Desjardins and Trevor Pearce, 13–32. History, Philosophy and Theory of the Life Sciences 4. Springer Netherlands.

Perelman, Chaïm and Lucie Olbrechts-Tyteca. 1969. *The New Rhetoric: A Treatise on Argumentation*. Translated by John Wilkinson and Purcell Weaver. Notre Dame: University of Notre Dame Press.

Pigliucci, Massimo. 2001. *Phenotypic Plasticity: Beyond Nature and Nurture*. Baltimore: The Johns Hopkins University Press.

Pinker, Steven. 2002. *The Blank Slate: The Modern Denial of Human Nature*. New York: Viking.

Provine, William B. 1973. "Geneticists and the Biology of Race Crossing." *Science* 182: 790–796.

Provine, Willam B. and Elizabeth S. Russell. 1986. "Geneticists and Race." *American Zoologist* 26: 857–887.

Putnam, Carlton. 1961. *Race and Reason: A Yankee View*. Washington: Public Affairs Press.

Richards, Graham. 2012. *"Race", Racism and Psychology: Towards a Reflexive History*. London: Routledge.

Richards, Robert. 1987. *Darwin and the Emergence of Evolutionary Theories of Mind and Behavior*. Chicago: University of Chicago Press.

Roberts, Dorothy E. 2011. *Fatal Invention: How Science, Politics, and Big Business Re-Create Race in the Twenty-First Century*. New York: New Press.

Roth, Byron. 2015. "The War on Human Nature." *The National Policy Institute*. May 17. Available online at www.npiamerica.org/research/category/the-war-on-human-nature

Samelson, Franz. 1978. "From 'Race Psychology' to 'Studies in Prejudice': Some Observations on the Thematic Reversal in Social Psychology." *Journal of the History of the Behavioral Sciences* 14: 265–278.

Schlichting, Carl and Massimo Pigliucci. 1993. "Control of Phenotypic Plasticity via Regulatory Genes." *The American Naturalist* 142: 366–370.

Segerstråle, Ullica. 2000. *Defenders of the Truth: The Battle for Science in the Sociobiology Debate and Beyond*. Oxford: Oxford University Press.

Selcer, Perrin. 2012. "Beyond the Cephalic Index: Negotiating Politics to Produce UNESCO's Scientific Statements on Race." *Current Anthropology* 53 (S5): S173–184.

Shapin, Steven and Simon Schaffer. 1985. *Leviathan and the Air-pump: Hobbes, Boyle, and the Experimental Life*. Princeton: Princeton University Press.

Smocovitis, Vassiliki Betty. 1999. "The 1959 Darwin Centennial Celebration in America." *Osiris* 14: 274–323.

Smocovitis, Vassiliki Betty. 2012. "Humanizing Evolution: Anthropology, the Evolutionary Synthesis, and the Prehistory of Biological Anthropology, 1927–1962." *Current Anthropology* 53 (S5): S108–S125.

Stocking, George W. 1968. *Race, Culture, and Evolution: Essays in the History of Anthropology*. Chicago: University of Chicago Press.

Tax, Sol and Charles Callendar, eds. 1960. *Evolution after Darwin*. 3 vols. Chicago: University of Chicago Press.

Taylor, Charles Alan. 1996. *Defining Science: A Rhetoric of Demarcation*. Madison: University of Wisconsin Press.

Templeton, Alan R .2013. "Biological Races in Humans." *Studies in History and Philosophy of Science Part C: Studies in History and Philosophy of Biological and Biomedical Sciences* 44 (3): 262–271.

Teslow, Tracy. 2014. *Constructing Race: The Science of Bodies and Culture in American Anthropology*. Cambridge: Cambridge University Press.

Thornhill, Randy and Craig Palmer. 2000. *A Natural History of Rape*. Cambridge: MIT Press.

Tindale, Christopher W. 2004. *Rhetorical Argumentation: Principles of Theory and Practice*. Thousand Oaks: Sage.

Tooby, John and Leda Cosmides. 1992. "The Psychological Foundations of Culture." In *Adapted Mind: Evolutionary Psychology and the Generation of Culture*, edited by Jerome H. Barkow, Leda Cosmides and John Tooby, 19–136. New York: Oxford University Press.

Toulmin, Stephen. 1958. *Uses of Argument*. Cambridge: Cambridge University Press.

UNESCO 1950. "Statement on Race." In: *Four Statements on the Race Question*, UNESCO 1969. Available online at http://unesdoc.unesco.org/images/0012/001229/122962eo.pdf

UNESCO 1951. "Statement on the Nature of Race and Race Differences." In *Four Statements on the Race Question*, UNESCO 1969. Available online at http://unesdoc.unesco.org/images/0012/001229/122962eo.pdf

Wade, Nicholas. 2014. *A Troublesome Inheritance: Genes, Race and Human History*. New York: Penguin.

Washburn, Sherwood L. 1951. "The New Physical Anthropology." *Transactions of the New York Academy of Sciences* 13: 298–304.

Washburn, Sherwood L. 1953. "The Strategy of Physical Anthropology." In *Anthropology Today: An Encyclopedic Inventory*, edited by Alfred L. Kroeber, 714–727. Chicago: University of Chicago Press.

Washburn, Sherwood L., ed. 1961. *Social Life of Early Man*. Chicago: Aldine Publishing Co.

Washburn, Sherwood L. 1963. "The Study of Race." *American Anthropologist* 65 (3, Part 1): 521–531.

Washburn, Sherwood L. 1978a. "Animal Behavior and Social Anthropology." *Society* 15 (6): 35–41.

Washburn, Sherwood L. 1978b. "Human Behavior and the Behavior of Other Animals." *American Psychologist* 33 (5): 405–418.

Weaver, Richard. 1953. *The Ethics of Rhetoric*. Chicago: Regnery.

Weismann, August. 1891. *Essays upon Heredity and Kindred Biological Problems*, edited by Edward Bagnall Poulton. Oxford: Clarendon Press.

West-Eberhard, Mary Jane. 2003. *Developmental Plasticity and Evolution.* New York: Oxford University Press.

Whately, Richard. 1963. *Elements of Rhetoric: Comprising an Analysis of the Laws of Moral Evidence and of Persuasion with Rules for Argumentative Composition and Elocution.* Edited by Douglas Ehninger. 7th edition. Carbondale: Southern Illinois University Press.

Wilson, Edward O. 1975. *Sociobiology: The New Synthesis.* Cambridge: Belknap Press of Harvard University Press.

Wilson, Edward O. 1978. *On Human Nature.* Cambridge: Harvard University Press.

Wilson Edward O. 1994. *Naturalist.* New York: Island Press

Wilson, Edward O. 1998. *Consilience: The Unity of Science.* New York: Knopf.

Zarefsky, David. 1995. "Argumentation in the Tradition of Speech Communication Studies." In *Perspectives and Approaches: Proceedings of the Third International Conference on Argumentation*, edited by Frans Van Eemeren, Ron Grootendorst, J. Anthony Blair and Charles A. Willard, vol. 1, 32–52. Amsterdam: SICAT, International Society for the Study of Argumentation.

Zarefsky, David. 2014. *Rhetorical Perspectives on Argumentation: Selected Essays.* New York: Springer.

Zenderland, Leila. 1998. *Measuring Minds: Henry Herbert Goddard and the Origins of American Intelligence Testing.* Cambridge: Cambridge University Press.

Zumwalt, Rosemary Lévy and William Shedrick Willis. 2008. "Boas Goes to Atlanta. Updated from an unpublished text by William S. Willis." In *Franz Boas and W. E. B. Du Bois at Atlanta University, 1906*, 41–77. Philadelphia: American Philosophical Society.

2 Franz Boas and the argument from presumption

Boas in Baffinland

In spring, 1882 Franz Boas, aged 24, graduated with a fairly useless Ph.D. in physics from the University of Kiel. In graduate school his interests kept straying from his assigned task of measuring the intensity of refracted light, in which he used statistical techniques to discount subjective perceptions, to psychometrics, which uses statistics to study perceptual variation in its own right, to geography. After defending his thesis and completing his military service in a cushy posting to his hometown in Westphalia, his interest in geography became ascendant. He went to Berlin not to take up a standing invitation to pursue psychometrics with Hermann von Helmholtz, but to propose, find financing for, and prepare himself to conduct a one-man expedition to study the Eskimos of Baffinland in northeast Canada. The relative homogeneity of the Arctic environment, the newly-minted laboratory scientist reasoned, would enable him to determine whether "the dependence of the present-day Eskimo on the configuration and physical conditions of the land" or diffusion of material resources, tools, social customs, language, and beliefs from group to group along migration and trade routes is (more) responsible for their way of life (Boas to A. Jacobi, April 10, 1882, in Stocking 1974, 44; for Boas's life in Germany, see Cole 1999).

Financing and a plan having been put in place, Boas prepared for his expedition by studying with the cream of the Berlin academic elite. In addition to taking crash courses in cartography, astronomy, photography, and languages at the University of Berlin, he learned anatomy and measurement of the human body – the foundational skill of anthropologists in those days – from the great cell biologist, pathologist, and public health reformer Rudolf Virchow, who became a lifelong role model. The curator of the Royal Ethnological Museum, Adolf Bastian, familiarized him with Eskimo artifacts. Bastian championed cultural diffusion. Hence he would have been pleased that Boas and his school held with ever greater confidence that most new techniques and tools are "explainable only by the introduction of outside influences or the immigration of new people with different arts and crafts" (Nelson 1938, 160). Where they are independently or originally developed it is due to the intelligence we share rather than to the environment.

This result was especially pleasing in the light of the Berlin school's opposition to sweeping ideas such as the evolutionism preached by Ernst Haeckel as "*Darwinismus*." Like most forms of what passed for Darwinism in the late nineteenth century, Haeckel's notion of evolution was both biological and social. Like Spencer, he stressed the role of environments in inducing, naturally selecting, and accumulating the fruits of a predictable, or what we will call stadial, sequence of human forms of life that reach their apex in advanced civilizations. Boas opposed this dogma as fully as his mentors. Nonetheless, he returned to Germany something of a heretic. Bastian and Virchow's skepticism about progressive social evolutionism ran deeper than their methodologically scrupulous empiricism and repugnance for premature generalization. For them anthropology was by definition the study of "natural," meaning "not civilized," man, and so ruled out as a conceptual mistake the idea that human groups evolve toward civilization in a law-governed way. Some human groups do become civilized, but that is contingent, rare, and caused by hard-to-pin down confluences of events and circumstances (Zimmerman 2001). Boas's experiences led him to abandon the postulate that anthropological inquiry assumes a discontinuity between natural, pre-historic and historical or cultured man. Rather than regressing to stadial evolutionism, however, he concluded that all human groups are always already *equally historical* and that anthropology is the study of their highly diverse ways of life and of how these ways of life come to be as historical particulars. Early and late, this is the essential Boas.[1]

By his own account, Boas came to this conclusion when a combination of bad weather, missed connections, and the inability to hire dogs to pull his sled – an epidemic had decreased their numbers and native brokers weren't about to sell any to a rookie like Boas – foiled his plan to explore remote western Baffinland. Holed up for a howling winter with the Inuit, who sheltered him, fed him, and taught him to hunt and fish, Boas summed up his experience as follows:

> After a long and intimate intercourse with the Eskimo, it was with feelings of sorrow and regret that I parted from my Arctic friends. I had seen that they enjoyed life, and a hard life, as we do; that nature is also beautiful to them; that feelings of friendship also root in the Eskimo heart; that, although the character of their life is rude as compared to civilized life, the Eskimo is a man as we are; that his feelings, his virtues, and his shortcomings are based in human nature, like ours.
>
> (Boas 1887d, 402)

His admirers have romanticized Boas's insight that primitive man is not so primitive after all, let alone savage (a term Boas avoided), by projecting onto it the full-blown cultural relativism and anti-racist egalitarianism that eventually grew from it. Still, the young Boas thought his Inuit companions, protectors, informants, and instructors showed what he called "*Herzenbildung*," a natural culturedness that compares favorably with the learned *Kultur* he exemplified. Even this seemingly modest claim so mystified his professional colleagues and patrons, however, for whom the whole point of anthropological research was to study people who by definition did *not* have *Kultur*, that it affected Boas's ability to find a suitable position in German academia (Zimmerman 2001, 45; anti-Semitism played a role). He left for the United States in 1887.

Things weren't easy in America either. Anglophone anthropologists cherished a stadial interpretation of human evolution. The leading British anthropologist of the time, Edward Burnett Tylor, put the view as follows: "The institutions of man are as distinctively stratified as the earth on which he lives. They succeed one another in a series substantially uniform over the globe, independent of what seem comparatively superficial differences of race and language" (Tylor 1889, 259). Tylor thought this pattern almost self-evident. The cultural authority of Darwin (albeit flying his flag falsely (Bowler 1988)) bolstered commonplaces about human nature that had circulated among the Enlightened since John Locke, the French *philosophes*, and the writers of the Scottish Enlightenment. These early figures were not biological evolutionists; they took the fixity of species for granted. But they all thought savage societies are earlier versions of sophisticated European societies and so, like children, will eventually develop true civilization (Jacques 1997). With its explicit comparison to the science of geology and its subtle allusion to the influence of Charles Lyell's geology on Darwin's biology, the statement we have quoted from Tylor captures the law-like necessity stadial thinkers saw in cultural evolution, together with the conviction that anthropologists could safely whistle past any contingent and interfering effects of particular environments and routes of diffusion.

Put schematically, Boas eventually rejected the following lynchpins of this consensus:

1. There is a scale of human evolutionary development that runs from savage to civilized.
2. The units of this scale are races.
3. Advance along the scale is, in the absence of constraint, predictable: The same inventions, modified to fit particular environments, can be expected spontaneously to emerge in the same order.
4. The more savage a race the more instinctual and less rational its beliefs and behaviors.
5. If a race fails over time to advance there must be inherent defects in its hereditary material. It must be degenerate.

Anthropologists seldom stated these assumptions so baldly. Their most glaring logical vulnerability, however, if not their moral astigmatism, was already apparent to the statistically adept Francis Galton, Charles Darwin's half cousin and the father of positive eugenics. Like Boas, Galton appreciated Tylor's pioneering application of statistics to the question of how often patriarchal marriage practices are preceded by matriarchy. He listened attentively to Tylor's paper on the subject, but pointed out that his tabulations do not logically compel the stadial interpretation he placed on them, even when they depart from what might be expected by chance. Tylor had no way of ruling out the possibility that any given transition from matriarchy to patriarchy resulted from the influence of another group. “It is extremely desirable for the sake of those who may wish to study the evidence for Dr. Tylor’s conclusions,” Galton argued, “that full information should be given as to the degree in which the customs of the tribes and races which are compared together are independent” (Tylor 1889, 270).

“Galton’s problem,” as it came to be called, would vex methodologists for a long time (Ember *et al.* 2015; Korotayev and Munck 2003; Strauss *et al.* 1975). Boas’s background in physics, psychometrics, and the cautious epistemology of his Berlin mentors prepared him well to use Galton’s statistical innovations and to appreciate early on that similar effects can have different causes. The possibility that historical diffusion and local environmental conditions jointly cause what stadialists ascribed to laws of social development was for him more than a captious point of logic. It was the pole star that, once he found his footing, guided the inquiries that formed his lifework.

Empiricism, presumption, and burden of proof in Boas’s argumentation

In this chapter, we will show explain why Boas maintained that, although statistical analysis of data can shift the burden of proof from one hypothesis to another, even more data must be amassed before a definitive interpretation emerges in most cases. The underlying reason for his empirical fastidiousness is his insight that the causes of human phenomena are more diverse than the phenomena themselves. He shifted the burden of proof onto the stadial view by claiming presumption for the possibility that in the historical sciences similar effects can have different causes. He argued that this principle can be defeated only by adducing compelling evidence against it. If stadialists failed to produce that evidence, the eventual result would be an anthropology whose subject matter consists of diverse evolved historical particulars.

The context in which Boas undermined stadialism and its accompanying racism is important. Boas, a Jew, had emigrated to a country in which racism was endemic and enforced by law and custom. Judiciously, he did not attempt to directly disprove any of the propositions that underpinned the stadial view of civilization on which scientific racism, such as it was, was predicated (Baker 2010, 7). Instead, he called for a change in probative obligations on the part of those who assumed racially tinged stadialism’s truth. He asked his intellectual opponents to shoulder the burden of proof for *their* starting points. His success was based on the acknowledged reliability of his reports about what the current state of evidence entitles us to say or not to say about a given subject at any point in time.

The concept of “burden of proof” is familiar in legal settings, where defendants are presumed innocent until the prosecution proves them guilty, and where specific criteria determine the strength of the burden that the prosecution must overcome. In his seminal treatise, *Elements of Rhetoric*, Richard Whately extended the jurisprudential concepts of presumption and burden of proof to all argumentative occasions. He described a presumption not in terms of probabilities, but as a “preoccupation of the ground, as implies that it must stand good till some sufficient reason is adduced against it; in short, that the burden of proof lies on the side of him who would dispute it" (Whately 1836, 105). Whately told his readers that if they forget to claim presumption for their own position they "may appear to be making a feeble attack instead of a triumphant defense" (ibid., 107).

Presumption and burden of proof are concepts that most obviously apply in contexts in which arguers must reach a yes-or-no decision even if information is imperfect. Under the powerful influence of the empiricist ideal, according to which correctly processed data will speak for themselves if we have enough of them, we often presume that science has no need for differential probative obligations (Dare and Kingsbury 2008). It has, we think, all the time in the world for data to make their own case. Several decades of work in science studies has shown, however, that this idealized picture is seldom true. In many cases, scientific issues are not isolated in a serenely autonomous technical sphere, but are embroiled in political and social controversies that labor under the urgency of the moment (Proctor 1991, 11; Meloni 2016). Even in its own sphere, science is rife with controversies and progresses by working through them. Evidence does not exist in a vacuum anywhere, even the laboratory. It can be understood and assessed only against a backdrop of expectations about how much of it should be required to force peers, in the name of dispassionate inquiry itself, to opt for one or another side of a still-unfolding issue.

These reflections directly relate Boas's argumentation. He entertained very high standards of empirical proof. But he was also aware that anthropology is so embedded in pressing social and political issues that decision under uncertainty is inescapable and so imposes differential probative obligations on participants in its debates. He made this very point in a 1909 address to the American Association for the Advancement of Science on the topic of immigration: "Under the pressure of these events, we seem to be called upon to formulate definite answers to questions that require the most painstaking and unbiased investigation" (Boas 1909, 839). In this instance, Boas was arguing that the weight of evidence, while incomplete, placed the burden of proof on immigration restrictionists. His strategy called on his opponents to prove *their* assumptions and policy recommendations.

Simultaneously, Boas complemented his critical work with a positive research program working from countervailing non-stadial assumptions. In his research reports to governmental agencies, he stressed that he was open to the possibility that his own egalitarian assumptions and policy advice could be overturned by evidence as inquiry proceeded. Rhetorically, this construction of his authority (*ethos*) disarmed his enemies. Built into it was a reasonable hope that a consistent, coherent, and complete account of cultural dynamics will not in the end (whenever that might be) depend on any unproven assumptions at all. For Boas, the scientist's highest obligation is to move toward a definitive account of a particular issue by astutely managing presumptions and burdens of proof in the light of what is known and unknown at a given time. We will trace how he honored these norms and deployed this argumentative strategy in controversies that by

1911 established him as the leading anthropologist in the country – and as the man who more than any other academic, with the exception of W. E. B. Du Bois, who as we saw in Chapter 1 was inspired by him, turned the tide against scientific racism in America.

Boas vs. Mason on natural history museums

Boas's rhetorical strategy was apparent when soon after arriving in America he entered into a debate in the pages of *Science*, the fledgling journal of the American Association for the Advancement of Science that he found work editing, with museum curator Otis T. Mason of the Smithsonian Institution. The issue concerned the display of Native American artifacts in museums of natural history. This dust-up has been well studied by philosophers, historians, and anthropologists (Buettner-Janusch 1957; Stocking 1994; Verdon 2006). Our focus is on how picking this quarrel enabled Boas to institutionalize his working assumptions with a view to freeing American anthropology from unstated, arbitrary, and racist assumptions. Even at this early date, he showed his sensitivity to presumption and burden of proof by asking Mason to justify his starting point rather than simply accusing him of being wrong.

Museums were the first disciplinary homes of American anthropology. The worldwide growth of natural history museums in the last quarter of the nineteenth century was little short of spectacular (Jenkins 1994, 244). In America as elsewhere, natural history museums nurtured academic departments of anthropology rather than the other way around. Harvard's department, for example, was spun off from the Peabody Museum. It was just opening its doors when Boas and Mason tangled. Until the establishment of natural history museums, what had passed for disciplinary anthropology in the United States was a loose affiliation of amateur naturalists and geographers who studied American Indians even as these peoples were being exterminated or greatly reduced in numbers and quality of life. In one way the new natural history museums were continuous with this work. Their focus, too, was on Native Americans. But museums consciously positioned themselves as sites for the production of rigorously certified scientific knowledge.

When Congress established the American Bureau of Ethnology as part of the Smithsonian Museum in 1879, its leader, John Wesley Powell, the colorful one-armed explorer-of-the-American-West-turned-science-administrator, hoped that Native American languages would be the Bureau's primary focus. Powell himself pioneered the classification of Indian languages, work that would soon be taken further by Boas and his students. But those paying the bills steered him instead to the growing collection of artifacts. This decision guaranteed that the Smithsonian museum would become the generative site of the nascent discipline of American archaeological anthropology. Where else could all those pots, arrowheads, totem poles, and ceremonial garments be collected, sorted, and displayed than in museums? Boas himself made a precarious living displaying artifacts in museums such as Chicago's Field Museum and New York's American

Museum of Natural History before founding the Department of Anthropology at Columbia University in 1896 (Teslow 2014).

The obligation to represent accurate, up-to-date scientific knowledge for public consumption demanded that museum workers attend to practical matters of arranging their ever-growing holdings of artifacts (Jenkins 1994, 253). In his formative writings, which set the agenda for American anthropology, Powell asked: How should anthropology explain, in the very act of displaying them, similar inventions in areas widely dispersed?

At first sight, we might think that arranging public displays should follow what counts as scientific knowledge in anthropology, but the reverse is more nearly true. Practices of display embody presumptions that guide knowledge production. Since the matter, though urgent, remained unsettled, Powell spoke the judicious language of presumption: "With regard … to the arts of life the presumption is in favor of independent origin by concausation." By this he meant that internal developmental tendencies combined with, or were modified by, local environmental conditions such as access to specific natural resources. Only when the "origin of such an art cannot be explained by the principle of concausation," he went on, would "the presumption … be in favor of its origin by acculturation [diffusion]" (Powell 1884, 72–73). Less dogmatic than most evolutionists, Powell still embedded stadial thinking in the new institution.

Powell's decision illuminates a worry about scientific knowledge. There were museums before the later nineteenth century, of course, but usually they were little more than souped-up "cabinets of curiosities" of the sort fancied by gentlemen collectors since the 1700s (Hooper-Greenhill 1992; Daston and Park 1998). Modern anthropologists were anxious to show that they could display more than hodge-podges of trinkets of questionable provenance. The modern museum, they felt, should present an ordered array of facts and theories reflecting the principle that scientific knowledge is *ordered* knowledge. But the desire to order material artifacts raised more questions than it answered. What was the proper ordering? To what end? For whom? Public audiences sought wonders and marvels. Researchers sought knowledge. Policymakers, who held the purse strings, sought guidance for public policy (Kohlstedt 2005). Powell's stadial view of civilization's progress served the interests of all three stakeholders. The viewing public affirmed its superiority over savages. Powell and other researchers "added enlightenment" about "the familiar categories of savagery, barbarism, and civilization" (Lowie 1956, 999; Powell 1887). Policymakers put this enlightenment to work by imagining ways to speed up cultural evolution through scientifically based public policy. Given these convergent self-congratulatory cognitive interests, it is not hard to understand how the stadialist overcame the empiricist in Powell.[2]

Otis T. Mason (1838–1908) was a key figure in articulating this agenda. He had been educated at Columbian College (now George Washington University) and taught there at a time when formal instruction in museum work was just entering college curricula (Kohlstedt 1988; Glenn 2000). Mason firmly believed that disciplinary boundaries required anthropologists to completely abandon the gentleman's cabinet of curiosities. "The anthropologist," he declared, "is not a dilettante philosopher, who inquires into old things because they are old, or into curious things while they are curious." Such efforts were flawed because they "omit[ted] all the great movements and needs of society, and overload[ed] the baggage-train of progress with trumpery picked up along the march." Rather than focusing on the "dust upon the mosaic of civilization," Mason wrote, anthropologists "must include in our science all those natural objects, relations, forces, and human progress" (Mason 1883, 359). He thereby linked exorcising the demon of amateurism with received notions of human and scientific progress. He called for "eliminating those local eddies of thought and action which begin and end with the individual, and which constitute his biography" and for instead taking "notice only of those great currents of human phenomena that echo round the world" (Mason 1882, 26).

Mason did not originate the stadialist commonplaces he circulated. It is just because he didn't that he provides such good testimony about what these commonplaces were. His views about museums were influenced in particular by Gustav Klemm of the Museum of Ethnology in Leipzig, who arranged his museum as a story of human progress. Klemm, who influenced Tylor, drew Virchow and Bastian's fire for the same reason Tylor drew Galton's. As a German scholar, Boas knew the background of the dispute and repeated with it with Mason (Mason 1874; Hinsley 1981, 87–89). Working in the American context, however, Boas was singularly alert to how race entered into Mason's story of civilizational progress.

Committing to a stadial theory of civilization required endorsing some sort of universally shared human nature – a common psychological makeup as the foundation for civilizational evolution. Yet it seemed obvious to many that some societies are not as advanced as others. One answer to this anomaly was to conclude, as many early modern social theorists did, that although all societies proceed along the same developmental path the specifics of that path vary in different environments (Jacques 1997, 204–206; this *topos* framed Jean-Baptiste Lamarck's late eighteenth century environmentalist vision of evolution). By the nineteenth century, races had become the presumed sources and carriers of these differences. Mason re-circulated this commonplace. Even while acknowledging that "the classification of mankind is still an open question," he maintained that, "[l]ooking over the earth, we behold men divided into races or consanguineous groups, filled with race prejudices, and restricted by race *capabilities*" (Mason 1882, 32; 1883, 360, our italics). Mason thought the universality of race prejudice is evidence in itself for differential racial capacities. In discussing comparative mythology, he warned his audience that some of the primitive beliefs he reported were pretty outrageous and invited them to appreciate how far *we*, as opposed to other races, have come. "I have frequently thought," he concluded, "while reading of the bloody and cruel fetish worship of the dark-skinned African that a kind providence had effected the whitening of the human skin coordinately with the purifying of religious conceptions" (Mason 1882, 40).

Mason assimilated this picture to the epistemic triumphs of cosmology, geology, and more recently biology. Anthropology's job, he claimed, is to uncover the natural laws that govern cultural progress, just as natural scientists had done for the physical evolution of the universe, the uniformity of geological strata, and the biological history of life on earth. "As Newton and Laplace grasped the unity and organization of the stellar world," and "Darwin first conceived the consanguinity of all living beings and their mutual help or harm, so the anthropologist seeks to unite all that can be known respecting man into a comprehensive science" (Mason 1883, 359–360). Mason read Darwin as supplying a law (common descent with modification) binding biological phenomena together as Newton's laws had united previously unrelated physical phenomena. It was Spencer, however, not Darwin, who according to Mason had discovered the law of evolution for *cultural* phenomena. "To Herbert Spencer we are indebted for the first effort in this direction respecting human phenomena," Mason claimed. The goal of anthropology should be to fill out the sketch that Spencer started by using "better instruments and more reliable material" supplied by professional anthropologists like himself (Mason 1883, 363). This meant embracing Spencer's principle that, like everything else, societies evolve from "the homogenenous to the heterogeneous." In social evolution, this principle preserves (by Lamarckian inheritance) ever finer divisions of labor. Advance toward private property and free markets is ensured by natural selection's elimination of groups and individuals who are unable to adopt to environmental challenges – or, like American Indians, to use the land as efficiently as white settlers.

At a meeting of the Washington, DC Philosophical Society, Mason explained how anthropology mirrors the same pattern of inductive generalization that had already led to triumphs in historical physics, geology, and biology. "The arts of mankind," he argued, "proceed from the same sources as the genera and species of natural objects" because "we may regard the implements and products of human industry in the light of biological specimens. They may be divided into families, genera, and species" that reflect the process of social evolution (Mason 1887, 44). Because specimen objects exemplify types that fall into species and genera, Mason maintained that we may also presume "the same [inductive] rule [as in other sciences] of proceeding from [homogeneously classified] particulars to generals and from [specialization] to a comprehensive view." Accordingly, he informed his audience that we can safely conclude that "Resemblances in Arts Widely Separated" – the title of his paper – are due to a consistent human nature unfolding in the face of particular contingencies that may be operating here and there or now and then (Mason 1886). When confronted with similar inventions – types of throwing sticks, for example, or specific forms of basketry – Mason discounted the claim that these inventions were early products of "the same race" subsequently diffused by travel or migration. Rather, anthropologists, if they are really scientific, must embrace the principle that human beings "will everywhere, under the same stress and resources, make the same inventions" (Mason 1886, 248). So museums could display a line of progress in this or that sphere of craftsmanship no matter where the artifacts come from.

This argument assumes that like effects have like causes. It accommodates differential racial capacities by recognizing that anthropology is subject to *ceteris paribus* clauses as much as physics, geology, and biology. Admittedly, particular environmental stresses and the availability or lack of resources in specific places will result in variant inventions and practices. These differences, Mason conceded, can be magnified by acquired habits, which in turn can be inherited, in some cases blocking the high road to civilization. "In each age and in each grade," he explained in a Lamarckian and Spencerian way, "natural primeval aptitudes are intensified and warped by inheritance and tuition" (Mason 1886, 248). He even acknowledged that the influence of these factors may suggest that "resemblance by independent invention [is] the least probable" explanation for resemblance, thus requiring "positive information" supporting case studies. Nonetheless, invoking orthodox empiricist methods of induction, especially John Stuart Mill's canon of residues, Mason ruled out the possibility that his prize examples of independent invention – the same kind of basket-weaving is found in the Pacific Northwest and in the Congo and the same kind of stick throwing in Australia, Brazil, and "Eskimoland" – might have resulted from diffusion between groups so widely separated. "The case of independent invention," he concludes, has been "clearly made out" (Mason 1886, 250–251).

Boas visited Mason's collections and saw Klemm's influence. When Mason's lecture was published, he objected. "As this plan [for museum organization] is the outcome of [Mason's] philosophical view of the problems of ethnology," he began, "we must scrutinize these in order to judge as to the merits of his system" (Boas 1887b, 588). Mason's scheme meant arranging "the ethnological collections of the national museum according to objects, not according to the tribes to whom they belong, in order to show the different *species* of throwing-sticks, basketry, bows, etc." (Boas 1887a, 485, our italics). But classifying artifacts in this quasi-Linnaean, typological way is only acceptable, Boas pointed out, if one assumes that like effects always spring from like causes. This assumption underpins the very idea that cultures go through similar, even if not exactly identical, stages on their path to civilization. Boas questioned whether this assumption is reliable enough to support Mason's conclusions. It is not just that local exceptions do indeed exist. Rather, it is

> a very rare occurrence that the existence of like causes for similar inventions can be proved, as the elements affecting the human mind are so complicated and their influence so utterly unknown that an attempt to find like causes must fail, or will be a vague hypothesis. On the contrary, the development of similar ethnological phenomena from unlike causes is far more probable, due to the intricacy of the acting causes.
>
> (Boas 1887a, 485)

Mason's proposal, Boas concluded, boiled down to the banal conclusion that, "The disposition of men to act suitably is the only general cause; but this is so general that it cannot be made the foundation of a system of invention" (Boas 1887a, 485).

In supporting the primacy of particular and presumptively different causes of superficially similar effects, Boas appealed to Darwin's authority, which he thought Mason's Spencerism abused. "Since the development of the evolution theory," he wrote, "it has become clear that the object of study is the individual, not abstractions from the individual.... Anthropologists have to study each ethnological specimen individually in its history and in its medium" (Boas 1887a, 485). Turning Mason's appeal to physics as a model of inductive science on its head, Boas wrote:

> I consider it one of the greatest achievements of Darwinism to have brought to light the fact that ... former events leave their stamp on the present character of a people.... The character and future development of a biological phenomenon is ... expressed ... by ... its whole history.... The outward appearance of two phenomena may be identical yet their immanent qualities may be altogether different: therefore arguments from analogies of the outward appearance, such as shown in Professor Mason's collections, are deceptive. These remarks show how the same phenomena may originate from unlike causes. In my opinion ... the axiom, "Like effects have like causes" ... belongs to the class of axioms that cannot be converted. Though like causes have like effects, like effects have not like causes.
>
> (Boas 1887b, 589)

The argument implies that if Mason was the empiricist he claimed to be he would have seen that Darwinism dictates that particulars lead to other particulars unless and until biologists meet the huge burden of proof placed on leaps from particulars to covering generalizations.

This claim put Powell, who had the institution's interests to protect, in something of a pickle. He agreed with Mason that Boas's proposal would result in repetitious and confusing displays, but divided the question by distinguishing between two missions, public education and research. The research mission was "of prime importance" (Powell 1887, 612). Boas and Mason's dispute was on the secondary question of public displays. Since "few selections are made to be shown to the public the great mass of material is kept ready to do service for the investigator." Boas's ideas might be useful for the scientist, but not for education. "In practical affairs, [Boas's tribal arrangement] would be an impossibility by reason of its magnitude" (Powell 1887, 613).

Boas wasn't buying it. The public, he said, needed to see a museum arranged according to his tribal scheme to teach them that, "Civilization is not something absolute, but ... relative, and that our ideas and conceptions are true only so far as our civilization goes. I believe that this object can be accomplished only by the tribal arrangement of collections" (Boas 1887b, 589). Pointing to the best museum practices of Europe, he brushed aside the objection that his plan was impractical: "Experience shows that this can be done with collections from all parts of the world without over-burdening the collection with duplicates, and without making artificial classifications, by grouping the tribes according to ethnic similarities" (Boas 1887b, 614). His argument bore fruit at the anthropology exhibit at the Columbian World's Exposition in Chicago in 1893. Its spirit also lives on in the famous natural history dioramas that still grace the American Museum of Natural History in New York. Yet the anti-hierarchical, relativistic worldview and egalitarian values implied by such display practices affronted Powell, Mason, and others. For decades, Boas and his students struggled for control of the AAA with what they called "the Washington School."

Method in the historical sciences: Boas, Kant, Darwin

In Boas's response to Mason and an essay he wrote on "The Study of Geography" in the same *annus mirabilis* of 1887, Boas began developing a line of argument in which he attacked not just premature generalization, but the even more deeply rooted assumption that empirical science always aims to establish universal laws. Defending the integrity and autonomy of geography, as well as of what at the time he called "anthropo-geography," Boas wrote:

> As soon as we agree that the purpose of every science is accomplished when the laws that govern its phenomena are discovered, we must admit that the subject of geography is distributed among a great number of sciences. If … we would maintain its independence, we must prove that there exists another object for science besides the deduction of laws from phenomena. And it is our opinion that there is another object, the thorough understanding of phenomena.… All agree that the establishment of facts is the foundation and starting-point of science. The physicist compares a series of similar facts, from which he isolates the general phenomenon which is common to all of them. Henceforth the single facts become less important to him, as he lays stress on the general law alone. On the other hand, the [particular] facts *are* the object of importance and interest to the historian.
>
> (Boas 1887c, 137–141, our italics; see Boas 1887b, 588)

Boas was claiming that empirical science has particularizing as well as generalizing aims. Knowledge of historical particulars is made possible, he says, by bringing to bear on them relevant generalizations from other sciences. Geography cannot be reduced to hydrology, meteorology, or any other science that possesses generalizations reliable enough to satisfy our demand for intellectual elegance. But the historical sciences achieve their aim by making use of whatever laws we have to satisfy our desire to understand the experienced worlds we live in (Boas 1887c, 137). Even if we suppose that specifically historical laws might ultimately be found, we should tarry long enough with particular cases and their differences to avoid the hasty generalizations to which stadial thinking about biology and society is prone.

Boas developed this point further in an invited address to the Congress of Arts and Science at the 1904 St. Louis World Fair. Once again positioning the natural-historical Darwin, whom he dissociated from post-Darwinian stadialists, as the catalyst for recognizing that there are particularizing but still thoroughly empirical historical sciences, he declared:

> The new historical view [of Darwin] … came into conflict with the generalizing method of science. It was imposed upon an older view of nature in which the discovery of general laws was considered the ultimate aim of investigation. According to this view, laws may be exemplified by individual events, which, however, lose their specific interest once the laws are discovered. The actual event possesses no scientific value in itself, but only so far as it leads to the discovery of a general law. This view is, of course, fundamentally opposed to the purely historical view. Here the laws of nature are recognized in each individual event, and the chief interest centers in the event as an incident of the picture of the world.
>
> (Boas 1904, 515–516)

There can be little doubt that Boas's reading of Darwin as natural historian was influenced by the rise of neo-Kantianism in German academia when he was a student at Kiel. Neo-Kantians of all stripes followed Kant in taking metaphysics off the table. Instead, they examined what scientists like Virchow and Helmholtz were actually doing with a view to determining the conceptual and methodological presuppositions that allow them to produce genuinely scientific, if revisable, knowledge. Rather than vainly questing for anything beyond what Kant called "the bounds of sense," that is, metaphysics, neo-Kantians, even more than Kant himself, took philosophy to be epistemology, the study of knowledge, and took epistemology to be the philosophy of science (*Wissenschaft*) in a wide sense that embraced both particularizing (ideographic) historical and law-governed (nomothetic) forms of inquiry (Windelband 1894).

Boas was introduced to neo-Kantianism when Benno Erdmann, his friend, teacher, mentor, and editor of one of the first critical editions of Kant's *Critique of Pure Reason*, put Friedrich Albert Lange's *History of Materialism* (1866) into his hands. Lange inspired Erdmann, Boas, and other budding academics to reject as unacceptably metaphysical and non-scientific both Hegelian idealism and belligerent materialism of the sort found in Haeckel's *Darwinismus*. "Materialism is metaphysics and from the point of view of [Kant's] critical philosophy metaphysics is impossible," wrote Wilhelm Windelband, one of the fathers of neo-Kantianism (quoted Köhnke 1991, 162). Boas told his uncle Abraham Jacobi, a physician in New York, "My previous materialistic *Weltanschauung* … was untenable" (Boas to A. Jacobi, April 10, 1882, in Stocking 1974, 43).

Boas took to neo-Kantianism not just because it nourished his youthful desire to become an empirical scientist, but also because by his own account its political outlook resonated with the "living force of the revolution of 1848" that permeated the home in which he grew up (Boas 1938b, in Boas 1974, 41). Neo-Kantians were for a unified *Reich*, but a constitutionally liberal one; for freedom of inquiry and speech in German universities, at the time the best in the world; for expert-guided reforms based on knowledge produced by these

German universities; for religious tolerance, especially between Christians and Jews; and for freer economic exchange and wider political participation.[3] In order to keep the new German state from regressing to the clerical-theological monarchism that they associated with idealism and from degenerating into the materialist worship of power and unscrupulous willingness to use it that Bismarck shared with his radical socialist opponents, neo-Kantians argued that scientific method can reliably guide public policy only when it is free from political interference (Köhnke 1991, 212). Although Boas's Berlin mentors were not technically speaking neo-Kantians – they were working scientists, not philosophers of scientific method – they fit this political and ethical picture (McNeely 2002). Against this background, we can see why Boas packed Kant's *Critique of Pure Reason* with him on his trip to Baffinland and reported that he regularly studied it in his icy wilderness. It easy to understand, too, why in reviewing a book of Kant selections translated into English for *Science*, he remarked, "Undoubtedly the study of Kant is the best introduction to modern philosophy, and a powerful means of guarding students from falling into a shallow materialism or positivism" (Boas 1888a).

Neo-Kantianians came in various flavors. Philosophers in Southwestern German universities focused on demarcating and defending the cognitive rigor of the particularizing historical sciences (Windelband 1894); in Northwest German universities, including Kiel, the emphasis fell on how the laws of the natural sciences are discovered, justified, and applied. But we shouldn't exaggerate these differences (Köhnke1991). Both tendencies opposed the positivists' belief in a single law-oriented inductive method for all sciences, human and natural. They were equally opposed to Wilhelm Dilthey's way of demarcating the *Naturwissenschaften* from the *Geisteswissenschaften* (Bunzl 1996, 53–55; Lyman and O'Brien 2004). Dilthey argued that, unlike the natural sciences, the sciences of the spirit, mind or *psyche* – the human sciences – require entering into the inner life of intending subjects by projecting our own inner states, including our emotions, onto them. "I regard [Dilhey's] dichotomy … between natural sciences [*Naturwissenschaften*] and sciences of the mind [*Geisteswissenschaften*] as unfortunate," Windelband declared (Windelband 1894, 173).[4] If Dilthey was right, psychology could not possibly become the experimental, law-oriented science that Helmholtz, Fechner, Wundt, Erdmann, and at an earlier point Boas himself were cultivating. For the neo-Kantians a nomothetic science need not be a natural science. Nor are historical sciences restricted to human affairs. Boas's "anthropo-geography" is a historical science. So is Darwin's natural history.

As he had asked Mason to do, Boas solicited his audience at the St. Louis World Fair to heed Darwin's example. By positioning Darwin as exemplary of the new evolutionary natural history as Alexander von Humboldt was of the old, he developed a distinctive, if not entirely unique, view of method in the historical sciences (Cole 1999, 123). Perhaps because of his own interdisciplinary passage from physics to geography to ethnography, he envisioned an ongoing interplay between the search for laws and efforts to understand historical particulars (Stocking 1968, 154–155). To this end, Boas did not think of law-governed and historical physics, biology, or anthropology as different sciences, but as two sides of the same sciences. It is "the personal inclination of the investigator," not switching disciplines, that leads one scientist to reconstruct "the actual history of mankind" while another "attempts to establish the laws of its development" (Boas 1904, 514). The former will succeed if they are alert to laws that illuminate cases, the latter if they remain sensitive to differences from which they are abstracting. For this reason, Boas never denied, as Windelband did, that general laws are in principle possible for historical processes. He merely insisted that, "General laws … cannot be clearly formulated or their relative value appreciated without thorough comparison of the manner in which they assert themselves in different cultures.… The application of the [comparative] method is the indispensable condition of sound progress" toward any laws that might turn up (Boas 1886, 907). An implication is that you are more likely to find laws when you are not too ardently looking for them (Chapter 3 of this book).

The mature Darwin was Boas's hero because he viewed him as constantly shifting in this way between generalizing and particularizing impulses rather than projecting his emotions into particulars, as Dilthey recommended and Alexander von Humboldt (and the self-consciously Humboldtian Darwin of the *Voyage of the Beagle*) actually did, or lunging like Spencer at general laws. Boas's reading of Darwin as a natural historian was doubtless facilitated by his familiarity with Anglo-American ways of thought. Still, he was putting put a German, neo-Kantian, historicist spin on Darwin when in 1887 he attributed to him the view that, "The state of an organism [or a species] at any moment is a function of its whole history" (Boas 1887b, 589) and when he asked his audience in St. Louis to follow Darwin's lead in recognizing that everything that has happened to a people leaves a stamp on its present:

> It is a common feature of all forms of evolutionary theory that every living being is considered as the result of an historical development. The development of ethnology is largely due to the adoption of the evolutionary standpoint, because it impressed the conviction upon us that no event in the life of a people passes without leaving its effect upon later generations.
>
> (Boas 1888b [1940], 633)

Boas's sense that historical particulars, whether biological lineages, distinct cultures, individual organisms, or particular persons or events, just are the snowballing effect of what went into their making probably exceeded Darwin's own, but it captured well what positivistic approaches to Darwinism miss. This historicist sense also pervades Boas's view of scientific knowledge as a revisable, but cumulative process. What experts actually know is the state of a question at any given time, not final answers. Still, in order to acquire even revisable knowledge investigators must believe and act as if at any given moment of inquiry they are moving toward the largely hypothetical day when continued searching will have successfully settled a question.[5]

The main theme of this chapter resurfaces here. To assess the state of inquiry at any time and point the way forward, a scientist must know how judiciously to distribute presumptions and burdens of proof. To do so is to avoid premature closure of still open issues, to apply credible standards of evidence when declaring issues settled at least for a while, and to keep the dialectic between the general and the particular moving along. That is what Boas did throughout his career. He stressed managing the conduct of inquiry by properly assigning presumptions and burdens of proof because his idea of scientific method was informed by neo-Kantianism as refracted through Darwin's example.[6]

Burden of proof and the cephalic index

A short-lived academic post at Clark University, Worchester, Massachusetts, in the eighteen nineties allowed Boas to set up a physical anthropology laboratory. Soon he was measuring the weight, height, head shape, and other variables of the ethnically diverse children of Worchester (Baker 2010). In subsequent decades, he pursued biometrical research in sophisticated statistical analyses of both European immigrants and native Americans. His most telling study was of rapid change in head shape among the children of European immigrants who had not (yet) intermarried with "old American stock." Based on an unimpeachably large sample, he showed that the cephalic index of recent immigrants – the ratio between the length, breadth, and width of their head – differed on average from members of the same ethnicity who had been born in America and that the difference is proportional to the length of time their mothers had been in the United States (Boas 1910). This result challenged a well-entrenched dogma according to which the cephalic index is a reliable marker of fixed, inherited racial types and of concomitant character traits and cognitive (dis)abilities.

Boas secured government funding for his biological studies because craniometry carried profound policy implications for the United States, which was experiencing a wave of immigration from southern and eastern Europe. As a prelude to legislation restricting immigration, which it passed in 1924, Congress commissioned a number of studies. Boas's work on head shape was one of forty-two thick volumes produced for the United States Immigration Commission. In 1911, he brought his results and recommendations to a wide audience in *The Mind of Primitive Man* and in 1912 to fellow anthropologists in a methodologically detailed monograph (Boas 1911, 1912). In 1928, under the pressure of a sea change toward genetic determinism crystallized by the perceived immigration crisis, he made his raw data available. In a tract addressed to the public, he attacked claims that races are or can be made pure, that they are inadvisably mixed, and that they distribute human capacities unequally (Boas 1928a, 1928b).

At the time of Boas's 1911 study most anthropologists assumed Lamarckian heritability. On these terms, they believed it would take at a minimum several generations and some upwardly mobile intermarriage for new immigrants to become even informally assimilated to the "white race." Those taking this view, which seemed liberal at the time, succeeded in getting a hearing at all only by stressing the great differences between Europeans and African-Americans, to whom they were at best condescending and at worst hostile (Ignatiev 1995; Jacobson 1998; Taylor 2005). For their part, "native Americans," as old-stock Protestant elites called themselves, opposed assimilation of any sort. They were convinced on what they took to be scientific grounds that intermarriage was dangerous. They were sure that immigrants from southern and eastern Europe were racially, that is to say *biologically*, inferior and that intermarriage harmed white Protestant American civilization. Miscegenation with African Americans was assumed to be even more harmful. It was illegal in much of the country.

Supposed proofs of racial inferiority relied on the science of measuring head shapes. In 1911, Jean Finot noted (with some skepticism) that "our distinction of races" almost completely rests on "brain measurement and its sister art, head measurement" (Finot 1911, 480). The key marker of race was the cephalic index, a measure either invented or discovered, depending on your point of view, by the eighteenth century Swedish anatomist Andres Retzius (1796–1860). "The shape of the human head," proclaimed the MIT-trained engineer William Z. Ripley in 1899, "by which we mean the general proportions of the length, breadth, and height, is one of the best available tests of race known" (Ripley 1889, 37). Through the cephalic index and a few other measures, Ripley mapped three European races: the dolichocephalic or long-headed Teutons, the mesocephalic Alpines, and the round-headed or brachycephalic Mediterraneans. For many scientists, only the Teutons (sometimes called Aryans, but dubbed Nordics by Madison Grant in 1916 during World War I) were assumed to be capable of advanced civilization. In the increasingly fevered run up to World War I, the French anthropologist Georges Vacher de Lapouge declared that, "First place must be given to *Homo Europus* (the dolichocephalic-blond so-called Aryan), while *Homo Alpinus* … and the Mediterranean probably rank in the order named" (Vacher de Lapouge and Closson 1897, 60). Unless something was done to check the widespread interbreeding of the noble Teuton with inferior racial types, he warned, "people in the next century will be slaughtered by the millions for the sake of one or two degrees on the cephalic index" (quoted in Hecht 2003, 168). To a great and tragic degree what he anticipated occurred sooner than he predicted.

Biological anthropologists placed such stress on the cephalic index because of one of the effects of the diffusion of Darwinism. Darwin taught that species are not fixed types but, being subject to shifting selection pressures, are mutable over time. Henceforth, it seemed necessary for racial essentialists, especially those who thought of themselves as Darwinians, to sustain their main claim by finding race-specific traits that are sheltered from natural selection. The cephalic index filled the bill. "So far as we are aware," wrote Ellsworth Huntington,

> the shape of people's heads cannot be influenced by their food, their occupation, or their social and economic conditions. Nor can we see how climatic selection could weed out one type of head as it weeds out one type of complexion.
>
> (Huntington 1919, 172–173)

The cephalic index was viewed as a glimpse into a primordial past when pure racial types were constituted. Accordingly, in the first half of the new century naturalists generally presumed that species-defining traits are non-adaptive results either of geographical isolation or sudden mutation. This assumption was both a product of and a premise for seeing human racial markers as identifying sub-specific types.

Boas had published on the cephalic index even before his immigrant studies, asserting that it is not a fixed characteristic but changes as an individual ages (Boas 1896). He also played off the cephalic index against another supposedly invariant racial marker, cranial capacity. In a study of almost 300 Sioux, Boas showed that there is no statistical correlation between the two measures (Boas 1899). His comprehensive immigration study of 1910 drew on these studies in a full-bore attack on the notion that the cephalic index could do the work of racial classification it had been called on to do scientifically, legally, and politically.

As in his challenge to Mason, Boas went about his work delicately. Rather than declaring a supposedly established fact false he identified the stability of head form as an assumption. "The general tendency of anthropological inquiry," he wrote, "has been to *assume* the permanence of the anatomical characteristics of the present races, beginning with the European races of the early Neolithic times" (Boas 1910, 44, our italics). He wanted to undermine the free pass that this assumption had granted to the stability of head form and by doing so to shift the burden of proof to those who held it. He urged that unless conclusively shown otherwise investigators should presume that *all* organic traits are mutable. Here, too, Boas drew on Darwin's cultural authority to make his point. "The principles of biological science," he wrote,

> forbid us to assume a permanent stability of bodily form. Our whole modern concept of the development of varieties and of species is based on the assumption of [either] cumulative or sudden variation. The variations that have been found in the human body are quite in accordance with this view.
>
> (Boas 1910, 41)

Boas's findings supported the presumption of evolutionary plasticity. In a single generation the head shape of immigrants to America changes significantly:

> The east European Hebrew who has a very round head becomes more long-headed; the south Italian, who in Italy has an exceedingly long head, becomes more short-headed.... We are therefore compelled to draw the conclusion that if these traits change under the influence of environment, presumably none of the characteristics of the human types that come to America remain stable. The adaptability of the immigrant seems to be very much greater than we had a right to suppose before our investigations were instituted.
>
> (Boas 1910, 5, 2)

This result, Boas showed, was statistically too robust to identify pure chance as its cause.[7] Using as touchstones up-to-date Mendelian genetics, with its anti-essentialist assumption that traits segregate independently, Wilhelm Johannsen's identification of genetically pure strains, and August Weismann's claim that natural selection works on inherited germinal factors rather than on characteristics acquired by individuals in their lifetimes, he batted around various causal hypotheses that might explain his discovery. His aim was not to settle the question. Indeed, he denied it could be settled at present (Boas 1912, 555–556). His goal was to assign proper burdens of proof to various causal hypotheses conforming to up-to-date science in order to inculcate his main message: The cephalic index cannot be used to identify invariant racial characteristics.

Boas reviewed three kinds of putative, but respectable explanation for his results: environmental, natural-selectionist, and genetic-mutationist. He did not think explanations based on the direct effect of the natural environments or social practices such as changes in infant care were currently acceptable (Boas 1911, 51, 63). To statistically assess "the influence of environment requires direct comparison of parents living in one environment with children living in another" (Boas 1911, 53). This information was not at hand. In a later edition of *The Mind of Primitive Man*, re-written under the urgency presented by the rise of Hitler, he remarked that explanations supposing a sudden shape-affecting change from rural to urban life could enjoy little or no standing until studies of the "bodily form of individuals of identical genetic makeup [identical twins] who are living under different types of environment" had been conducted (Boas 1939, 96; these studies have since been carried out, but remain inconclusive).

Boas's response to selectionist scenarios varied with the sorts of stories told. It was implausible and question begging to believe that people who left Europe were self-selectively different from people who stayed, with the result showing up in the next generation (Boas 1911, 58).[8] Even more fanciful was the suspicion that, "Among the descendants of immigrants born in America there are an appreciable number who are in reality children of American fathers" (Boas 1911, 62). Boas was also on solid ground in judging that Weismannian natural selection, which works only on strongly heritable factors, rules out thinking that genetic factors can turn transient events, such as the economic difficulties that followed the financial Panics of 1893 or 1907, into selection pressures strong enough to make the observed rapid change selectively adaptive (Boas 1912, 551, 553). Still, he was unable to rule out some sort of selection after arrival in America simply by stating as a matter of principle that the change in question was too fast to count as adaptive natural selection. A new environment might affect immigrants in ways that quickly change the composition of the population. Boas was not anti-selectionist either in principle or in cases where cultural change creates new selection pressures. But he did impose high burdens of proof on adaptationist scenarios because of their question begging, ad hoc, and *post hoc* tendency:

> It goes without saying that application of unproved though possible theories cannot serve as proof of the effectiveness of selection or of environment in

> modifying types. The effectiveness of selection can be proved only by an investigation of the surviving members of a type as compared to those eliminated by death or of a shifting of population connected with the selection of a certain type. On the whole, it seems to my mind that the burden of proof would lie entirely on those who claim … a correlation between head-index, width of face, etc., and death-rate – a correlation which I think is highly improbable and which could be proposed only to sustain the theory of selection, not on account of any available facts.
>
> (Boas 1911, 63)

Boas himself tended toward the genetic-mutationist end of the spectrum of proposed explanations for his facts.[9] He did not dismiss "the possibility that the breaking of the more or less inbred lines of small European villages after arrival in America" was the cause of the changes in head form he documented (Boas 1912, 555). After all, most European immigrants had long been living in small, highly inbred rural communities, giving rise to "local races" that created the illusion of stability on which studies of the cephalic index were largely based. In America "intermarriages of natives of different villages are much more common than in Europe." The mingling might release a large range of natural variation (Boas 1911, 89–91). He also argued that investigators were also probably underestimating the "plasticity (as opposed to permanence) of types" (Boas 1911, 64; 1912, 557). That is, he recognized that in principle the same inherited factors might express themselves differently in different environments.

Boas's studies of the cephalic index nicely display his method of judiciously sorting knowns from unknowns, probabilities from improbabilities, ideas worth pursuing from ideas to be set aside, and assigning presumptions and burdens of proof accordingly. "If we have succeeded in proving changes in the form of the body," he wrote, "the burden of proof will rest on those who, notwithstanding those changes, continue to claim the absolute permanence of other forms and functions of the body" (Boas 1911, 76). "The fact that anthropologists are in the habit of calling heads of a length/breadth index of eighty and more brachycephalic does not constitute brachycephaly as a distinct biological type, but [only] as a mere convenience of description" (Boas 1912, 542).

Boas's assertiveness on this point strengthened between 1911, when *The Mind of Primitive Man* appeared, and 1928, when he addressed *Anthropology and Modern Life* to an American public whose democratic values were being undercut by nativist and eugenicist ideology. He grew even more insistent in the face of Hitler's racism in the 1930s. In the course of his interventions in public debates, Boas formulated the principles of an anti-racist consensus that deepened during and after World War II and found its biological footing in the Modern Evolutionary Synthesis (Chapter 4 of this book). Expressed in his pre-Synthesis way of thinking about them, these principles can be summarized as follows:

1 The term "race" can scientifically be used to describe populations in which reproduction has been constrained in an inbred caste-like way that

intensifies the recurrent expression of particular heritable characters. By this standard, family lines of inbred outcasts, half-castes, and bastards, which are usually viewed as deviations from a typologically construed dominant race, have a better claim to be races than their ancestral stocks (Boas 1928a, 19, 51, 37; Müller-Wille and Rheinberger 2012, 124).[10]

2 Human populations are pervasively hybridized owing to their geographic mobility. There are no pure races. There was more mixing of races in the American south before the Civil War than after it, when segregationist policies began to be enforced by the threat or use of violence. In New York, by contrast, where nothing prevents intermarriage, "Family lines are so diverse that there is no racial unity and no racial heredity" (Boas 1928a, 25, 27). Even in their European homelands, the pervasiveness of intermarriage shows that the so-called three races of Europe are not races at all but social constructions in the most invidious sense of what in reality is a continuum.

3 Racial markers such as skin color are not statistically associated with other traits, as typological essentialist biology holds (Boas 1928a, 29).

4 The amount of phenotypic variation within lineages, including races, is generally greater than between them. "Differences between types of men are, on the whole small compared to the range of variation in each type" (Boas 1911, 94; see also 22–33; African-Americans are not a strongly marked racial lineage, partly in view of the long-standing practice of white males having offspring by black females, 269).

5 There is no evidence that interracial marriage has any negative effect on fitness by any measure of the latter (Boas 1911, 274; 1928a, 80).

6 In the developmental process, inherited factors thoroughly interact with environmental variables, both physical and social, making human traits markedly plastic in ways that adapt them to a wide range of environments (Boas 1911, 64; 1928a, 47–50).

7 Humans are more like domestic animals than wild types. They have been domesticating themselves by restricting mating since the species first evolved (Boas 1928a, 51). This practice is a function of cultural preferences and varying kinship rules. Even though they have real biological effects, such restrictions are both various and in the long run temporary. People mate in spite of racial, tribal, clan, caste, and other social divides. This tendency may prove to be the salvation of the United States.

8 Enculturation patterns are more reliably transmissible across generations than racial markers (Boas 1928a, 143–153, 204). Members of different races can and regularly do acquire and pass on the same culture (Boas 1911; 1928a, 273). People of the same race can equally well internalize different cultural norms, practices, and beliefs.

9 Given these facts, the very idea of a racial hierarchy, and *a fortiori* an unchanging one with whites on top and blacks on the bottom, is nonsensical (Boas 1928a, 270–273). "The greatest care should be taken to develop the cultural germs that are present everywhere rather than to press all primitive peoples into our own cultural mode of life," Boas told the editor of

The Nation (Boas to H. R. Mussey, December 24, 1918, in Boas 1972. The letter was in connection with formulating allied colonial policy at the Paris Peace Conference that ended World War I). Besides, as he had told Du Bois and his students at Atlanta University in 1906, many of those so-called primitive peoples developed civilizations on their own (Boas 1940, vi; Zumwalt and Willis 2008).

We may contrast these theorems with the list we ascribed to stadialists like Mason at the beginning of this chapter. If Boas sowed seeds of doubt about the stadial consensus, it was in part because he did not talk down to the public. This is, we think, in part because he stated plainly what he was unsure of and challenged other experts to avoid dogmatism. His presumption-assigning and burden-shifting way of arguing was perhaps the greatest source of his lasting authority. By arguing this way, he invited his readers, perhaps strategically, to come to even stronger conclusions in favor of racial equality and cultural pluralism than the ones he overtly put forward. He reinforced this inviting approach by stressing the provisional nature of his findings. "I repeat that I have no solution to offer," he told his colleagues. "I have only stated the results of my observations and considered the plausibility of various explanations that suggest themselves.... Let us await further evidence before committing ourselves to theories that cannot be proven" (Boas 1912, 562).

Boas's judiciousness about burden of proof went hand in hand with mastery of statistical reasoning. His analyses went well beyond primitive techniques for finding averages by employing methods of measuring width of variation and assessing statistical significance developed by Galton and Karl Pearson. His support of Weismannian hard inheritance raised the question of whether mutation or natural selection of germinal factors is the source of evolutionary novelties. Like his geneticist colleague at Columbia Thomas Hunt Morgan, he leaned toward mutationism. His immersion in the statistical-probabilistic revolution, however, also led him to recognize that natural selection can do far more than merely eliminate the antecedently unfit, as Spencer, the defeated advocate of soft inheritance, had it. As W. F. R. Weldon and other Darwinian "biometricians" were beginning to show, natural selection can gradually "shift populations connected with the selection of a certain type" by amplifying heritable variation in adaptive directions (Boas 1911, 63). Recognizing as he did that the same inherited factors can be expressed differently in different environments, Boas came close to anticipating the mid-twentieth century view that natural selection selects for genotypes that underwrite phenotypically plastic responses to environmental change, including the flexibility conferred by the cultural way of life that figured in his own work (Chapter 4 of this book).

That possibility, however, lay further along a line of inquiry whose stirrings Boas already appreciated. His sifting of possible explanations for changes in head shapes faithfully reflects the state of discussion in the emerging discipline of evolutionary science when he wrote *The Mind of Primitive Man.* But even this is enough to see that casting Boas as an anti-evolutionist, anti-Darwinian, anti-Mendelian advocate of the putative Standard Social Scientific Model described in Chapter 1, which positions social scientists as enemies of biology, trades on an egregious confusion between modern evolutionary theory and the typological-stadial *Darwinismus* Boas repudiated. The fact that in the later nineteenth century much of what passed for Darwinism illustrates what Bowler calls "The Non-Darwinian Revolution" should not obscure the fact that on wings supplied by genetics and the statistical-probabilistic revolution biometrical Darwinism was on its way to becoming the dominant paradigm in evolutionary biology (Largent 2009, amending Bowler 1988). As a statistical adept Boas was alive to this shift.

Notes

1 Boas thought Bastian's "elementary thought patterns" (*Elementargedanken*) inscribed questionable psychologial theorems into ethnography and short-circuited efforts to investigate historical differences. He passed this view to Kroeber (Chapter 3 of this book). The theme of psychological generalizations blocking "thick" ethnographic description will surface again and again in this book.

2 On the stadialism of another turn-of-the-century Smithsonian scientific administrator, the geologist-paleontologist Charles Doolittle Walcott, see Gould 1989.

3 The neo-Kantian political vision was undermined by elitist, nationalist, and racist tendencies, with the result that Hitler was able to dismantle neo-Kantian academia with help from within: "As the liberal tenets of Virchow and Bastian were abandoned in German itself, they became the cornerstone of the anthropology developed in the United States by Franz Boas" (Bunzl and Penny 2010, 22).

4 Bunzl's interpretation of the influence of Dilthey and Windelband on Boas is compromised by assimilating defenders of the autonomy of the historical sciences to the affect-oriented "counter-Enlightenment" thinking rather fuzzily postulated by Isaiah Berlin (Bunzl 1996, 15, 22–24, 43, 52, 61). One might precariously pin that charge on Herder, Humboldt, and Dilthey, but not on Windelband.

5 Also argued by C. S. Peirce, a devotee of Kant in an age affected by the non-Newtonian physics Kant wrongly presumed to be apodictic. Peirce inspired pragmatists to embrace inquiry as ongoing and inherently revisable. Boas encountered this view at Columbia in his colleague, co-teacher, and comrade-in-arms in liberal causes, John Dewey, but we do not think their overlap was quite enough to make a pragmatist of Boas (Lewis 2001). To its friends and foes alike, pragmatism, including Dewey's "instrumentalism," meant "the belief that knowledge consists of those general propositions … which have in past experience proved biologically serviceable to those who have lived by them" (Lovejoy 1908, 38). It seems unlikely to us that as a neo-Kantian Boas would think of ideas, provisional as they are, merely as tools for negotiating environments, even social environments. Neither, for that matter, did Peirce.

6 Managing the conduct of inquiry involves distinguishing what issues are at stake. Gaskins argues that Kant's reliance on the language of judicial proceedings to distinguish questions of fact from questions of justification and the latter from questions of jurisdiction comes into its own in neo-Kantianism (Gaskins 1992). It's a good point. It is worth noting, however, that these distinctions did not originate in jurisprudence, but were imported into it from late classical rhetorical theory's doctrine of issues, *stases*, or *questiones* (Hoppman 2014).

7 Boas's data were re-sifted nine decades later to see how well his analysis stands up. Sparks and Jantz (2002) are critical of how statistically significant his result is, but Gravlee *et al.* (2003) reaffirm Boas's conclusions by viewing the argument in its

original context: "Given the prevailing faith in the absolute permanence of cranial form, Boas's demonstration of change – *any change* – in the cephalic index within a single generation was nothing short of revolutionary " (Gravlee *et al.* 2003, 136, our italics).

8 Twenty-five years later Boas's student Otto Klineberg refuted a similar argument offered by psychologists who wanted to explain away his finding that northern African Americans outscored southern whites on IQ tests. Klineberg found no evidence that more intelligent African Americans had migrated north (Klineberg 1935).

9 Natural selection and mutation were at loggerheads until R. A. Fisher reconciled them in 1918 (Fisher 1918; Provine 1971; Chapter 4 of this book).

10 Boas followed Johannsen in identifying races by degree of approximation to a "pure" line that has been systematically inbred until it reliably expresses uniform characters. In the real world rather than in laboratories there are few if any pure lines and hence pure races. There are hybrids – endless mixtures of hypothetical pure lines that in view of their pervasiveness in and adaptability to nature cannot be characterized as monsters, deviants, outliers, or bastards the way folk biology and its theoretical doubles presuppose (Müller-Wille and Rheinberger 2012, 138).

References

Baker. Lee D. 2010. *Anthropology and the Racial Politics of Culture*. Durham: Duke University Press.

Boas, Franz. 1887a. "The Occurrence of Similar Inventions in Areas Widely Apart." *Science* 9: 485–486.

Boas, Franz. 1887b. "Letter to the Editor: Museums of Ethnology and Their Classification." *Science* 9: 587–589.

Boas, Franz. 1887c. "The Study of Geography." *Science: Supplement* 9 (210), 137–141.

Boas, Franz. 1887d. "A Year Among the Eskimo." *Bulletin of the American Geographical Society* 19: 383–402.

Boas, Franz. 1888a. "Review of *The Philosophy of Kant* by John Watson." *Science* 12: 81.

Boas, Franz. 1888b [1940]. "Aims of Ethnology." In Franz Boas, *Race, Language, and Culture*. New York: Free Press.

Boas, Franz. 1896. "The Form of the Head as Influenced by Growth." *Science* 4 (July): 50–51.

Boas, Franz. 1899. "The Cephalic Index." *American Anthropologist* 1 (3): 448–461.

Boas, Franz. 1904. "The History of Anthropology." *Science* 20 (512): 513–524.

Boas, Franz. 1909. "Race Problems in America." *Science* 29: 839–849.

Boas, Franz. 1910. *Changes in Bodily Form of Descendants of Immigrants*. Washington DC: Government Printing Office, 112 (The Immigration Commission, Senate Document No. 208).

Boas, Franz. 1911. *The Mind of Primitive Man*. New York: Macmillan.

Boas, Franz. 1912. "Changes in the Bodily Form of Descendants of Immigrants." *American Anthropologist* 14: 530–562.

Boas, Franz. 1928a. *Anthropology and Modern Life*. New York: W. W. Norton.

Boas, Franz. 1928b. *Materials for the Study of Inheritance in Man*. New York: Columbia University Press.

Bowler, Peter. 1983. *The Eclipse of Darwinism*. Baltimore: Johns Hopkins University Press.

Bowler, Peter. 1988. *The Non-Darwinian Revolution*. Baltimore: Johns Hopkins University Press.

Buettner-Janusch, John. 1957. "Boas and Mason: Particularism Versus Generalization." *American Anthropologist* 59 (2): 318–324.

Bunzl, Matti. 1996. "Franz Boas and the Humboltian Tradition: From *Volkgeist* and *Nationalcharakter* to an Anthropological Concept of Culture." In Volkgeist *as Method and Ethic: Essays on Boasian Ethnography and the German Anthropological Tradition*, edited by George W. Stocking, 17–78. Madison: University of Wisconsin Press.

Bunzl, Matti and H. Glenn Penny, 2010. "Introduction: Rethinking German Anthropology, Colonialism, and Race." In *Social History, Popular Culture and Politics in Germany: Worldly Provincialism: German Anthropology in the Age of Empire*, edited by Matti Bunzl and Glenn Penny. Ann Arbor: University of Michigan Press.

Cole, Douglas. 1999. *Franz Boas: The Early Years 1858–1906*. Seattle: University of Washington Press.

Dare, Tim and Justine Kingsbury. 2008. "Putting the Burden of Proof in its Place: When Are Differential Allocations Legitimate?" *Southern Journal of Philosophy* 46 (4): 503–518.

Daston, Lorraine and Katherine Park. 1998. *Wonders and the Order of Nature*. New York: Zone Books.

Ember, Carol R., Melvin Ember and Peter Peregine. 2015. "Cross Cultural Research." In *Handbook of Methods in Cultural Anthropology*, edited by H. Russell Bernard and Clarence C. Gravlee, 561–599. Lanham: Rowman and Littlefield.

Finot, Jean. 1911. "Long Heads, Short Heads." *Contemporary Review* 99: 479–486.

Fisher, Ronald A. (1918). "The Correlation between Relatives on the Supposition of Mendelian Inheritance." *Philosophical Transactions of the Royal Society of Edinburgh* 52: 399–433.

Gaskins, Richard. 1992. *Burdens of Proof in Modern Discourse*. New Haven: Yale University Press.

Glenn, James R. 2000. "Mason, Otis Tufton." *American National Biography Online*. Oxford University Press. Available online at www.anb.org/articles/14/14-00387.html

Gould, Stephen Jay 1989. *Wonderful Life: The Burgess Shale and the Nature of History*. New York: Norton.

Gravlee, Clarence C., H. Russell Bernard and William R. Leonard. 2003. "Heredity, Environment, and Cranial Form: A Reanalysis of Boas's Immigrant Data." *American Anthropologist* 105 (1): 125–138.

Hecht, Jennifer. 2003. *The End of the Soul*. New York: Columbia University Press.

Hinsley, Curtis M. 1981. *Savages and Scientists: The Smithsonian Institution and the Development of American Anthropology, 1846–1910*. Washington: Smithsonian Institution Press.

Hooper-Greenhill, Eilean. 1992. *Museums and the Shaping of Knowledge*. New York: Routledge.

Hoppmann, Michael J. 2014. "A Modern Theory of Stasis." *Philosophy and Rhetoric* 47 (3): 273–296.

Huntington, Ellsworth. 1919. *World-Power and Evolution*. New Haven: Yale University Press.

Ignatiev, Noel. 1995. *How the Irish Became White*. New York: Routledge.

Jacobson, Matthew Frye. 1998. *Whiteness of a Different Color: European Immigrants and the Alchemy of Race*. Cambridge: Harvard University Press.

Jacques, T. Carlos. 1997. "From Savages and Barbarians to Primitives: Africa, Social Typologies, and History in Eighteenth Century French Philosophy." *History and Theory* 36: 190–215.

Jenkins, David. 1994. "Object Lessons and Ethnographic Displays: Museum Exhibitions and the Making of American Anthropology." *Comparative Studies in Society and History* 36 (2): 242–270.

Klineberg, Otto. 1935. *Negro Intelligence and Selective Migration*. New York: Columbia University Press.

Kohlstedt, Sally Gregory. 1988. "Curiosities and Cabinets: Natural History Museums and Education on the Antebellum Campus." *Isis* 79: 405–426.

Kohlstedt, Sally Gregory. 2005. "'Thoughts in Things': Modernity, History, and North American Museums." *Isis* 96 (4): 586–601.

Köhnke, Klaus Christian. 1991. *The Rise of Neo-Kantianism: German Academic Philosophy between Idealism and Positivism*. Cambridge: Cambridge University Press.

Korotayev, Andrey and Victor de Munck. 2003. "'Galton's Asset' and 'Flower's Problem': Cultural Networks and Cultural Units in Cross-Cultural Research: (Or, Male Genital Mutilations and Polygyny in Cross-Cultural Perspective)." *American Anthropologist*, New Series, 105 (2): 353–358.

Lange, Friedrich Albert. 1866. *Geschichte des Materialismus und Kritik seiner Bedeutung in der Gegenwart*. Iserlohn: J. Baedeker.

Largent, Mark A. 2009. "The So-called Eclipse of Darwinism." In *Descended from Darwin: Insights into the History of Evolutionary Studies, 1900–1970*, edited by Joe Cain and Michael Ruse, 3–24. Philadelphia PA: American Philosophical Society.

Lewis, Herbert. 2001. "Boas, Darwin, and Anthropology." *Current Anthropology* 42: 381–406.

Lovejoy, Arthur O. 1908. "The Thirteen Pragmatisms." *The Journal of Philosophy, Psychology and Scientific Methods* 5 (1): 5–8.

Lowie, Robert H. 1956. "Reminiscences of Anthropological Currents in America Half Century Ago." *American Anthropologist* 58: 995–1016.

Lyman, R. Lee and Michael J. O'Brien. 2004. Nomothetic Science and Ideographic History in Twentieth-century Americanist Anthropology. *Journal of the History of the Behavioral Sciences* 40: 77–96.

Mason, Otis T. 1874. "The Leipzig Museum of Ethnology." In *Annual Report of the Board of Regents of the Smithsonian Institution, Showing the Operations, Expenditures, and Condition of the Institution for the Year 1873*, 390–409. Washington: Government Printing Office.

Mason, Otis T. 1882 "What is Anthropology?" In *The Saturday Lectures Delivered in the Lecture Room of the U.S. National Museum Under the Auspices of the Anthropological and Biological Societies of Washington in March and April 1882*, 25–43. Washington: Judd and Detweiler.

Mason, Otis T. 1883. "The Scope and Value of Anthropological Studies." *Science* 2 (32): 358–365.

Mason, Otis T. 1886. "Resemblances in Arts Widely Separated." *American Naturalist* 20 (3): 246–251.

Mason, Otis T. 1887. "Bowyers and Fletchers." *Bulletin of the Philosophical Society of Washington* 9: 44–45.

McNeely, Ian. 2002. *Medicine on a Grand Scale: Rudolf Virchow, Liberalism, and the Public Health*. London: Wellcome Trust Centre for the History of Medicine.

Meloni, Maurizio. 2016. *Political Biology: Science and Social Values in Human Heredity from Eugenics to Epigenetics*. New York: Palgrave Macmillan.

Müller-Wille, Staffan and Hans-Jorg Rheinberger. 2012. *A Cultural History of Heredity*. Chicago: University of Chicago Press.

Nelson, N. C. 1938. "Prehistoric Archeology." In *General Anthropology*, edited by Franz Boas, 146–237. Boston: Heath.
Powell, John W. 1884. *Third Annual Report of the Bureau of Ethnology to the Secretary of the Smithsonian Institution, 1881–82*. Washington: Government Printing Office
Powell, John W. 1887. "Museums of Ethnology and their Classification." *Science* 9: 612–614.
Proctor, Robert N. 1991. *Value Free Science? Purity and Power in Modern Knowledge*. Cambridge: Harvard University Press.
Provine, William B. 1971. *The Origins of Theoretical Population Genetics*. Chicago: University of Chicago Press.
Ripley, William Zebina. 1899. *The Races of Europe: A Sociological Study*. New York: Appleton.
Sparks, C. S. and R. L. Jantz. 2002. "A Reassessment of Human Cranial Plasticity: Boas Revisited." *Proceedings of the National Academy of Sciences* 99 (23): 14636–14639.
Stocking, George W. 1968. *Race, Culture, and Evolution: Essays in the History of Anthropology*. Chicago: University of Chicago Press.
Stocking, George W., ed. 1974. *The Shaping of American Anthropology, 1883–1911: A Franz Boas Reader*. New York: Basic.
Stocking, George W. 1994. "Dogmatism, Pragmatism, Essentialism, Relativism: The Boas/Mason Museum Debate Revisited." *History of Anthropology Newsletter* 21 (1): 3–12.
Strauss, David J. *et al.* 1975. "Mighty Sifts: A Critical Appraisal of Solutions to Galton's Problem and a Partial Solution [and Comments and Replies]." *Current Anthropology* 16(4): 573–594.
Taylor, Gary. 2005. *Buying Whiteness: Race, Culture, and Identity from Columbus to Hip Hop*. New York: Palgrave Macmillan.
Teslow, Tracy. 2014. *Constructing Race: The Science of Bodies and Culture in American Anthropology*. Cambridge: Cambridge University Press.
Tylor, Edward B. 1889. "On a Method of Investigating the Development of Institutions; Applied to Laws of Marriage and Descent." *The Journal of the Anthropological Institute of Great Britain and Ireland* 18: 245–272.
Vacher de Lapouge, G. and Carlos C. Closson. 1897. "The Fundamental Laws of Anthropo-Sociology." *The Journal of Political Economy* 6 (1): 54–92.
Verdon, Michel. 2006. "The World Upside Down: Boas, History, Evolutionism, and Science." *History and Anthropology* 17 (3): 171–187.
Whately, Richard. 1836. *Elements of Rhetoric*. 5th edition. London: B. Fellowes.
Windelband, Wilhelm. 1894. *Geschichte und Naturwissenschaft*. Strassburg: Heitz.
Zimmerman, Andrew. 2001. *Anthropology and Antihumanism in Imperial Germany*. Chicago: University of Chicago Press.
Zumwalt, Rosemary Lévy and William Shedrick Willis. 2008. "Boas Goes to Atlanta. Updated from an unpublished text by William S. Willis." In *Franz Boas and W. E. B. Du Bois at Atlanta University, 1906*, 41–77. Philadelphia: American Philosophical Society.

3 Demarcating anthropology

The boundary work of Alfred Kroeber

Kroeber in focus: the rhetoric of demarcation

Alfred Louis Kroeber was born to well-to-do German-American parents in Hoboken, New Jersey, in 1876. A childhood friend, the physicist Carl Alsberg, remembered him as both shy and adventurous: diffident in his interpersonal relationships, but uninhibited by convention, and seemingly fearless (Alsberg 1936). Attending public schools and benefiting from private tutoring, he entered Columbia College at the age of sixteen, receiving a B.A. in English in 1896 and an M.A. in Romantic drama in 1897. As an undergraduate, he happened to take a class with Franz Boas. As a result, he decided to dedicate his life to anthropology. Alsberg was horrified at Kroeber's choice of such a "vague, inchoate, and intangible subject." He recalled Kroeber replying that chemistry and physics could not supply information on the real problems that faced humanity, but "anthropology is capable of bringing some degree of clarity into the confused thought of men and of freeing them to some degree from hoary tribal taboos" (Alsberg 1936, xvi).

Upon receiving his Ph.D. in 1901 – he was the first in a long line of Boas's distinguished graduate students – Kroeber went to Berkeley to establish a Department of Anthropology at the University of California. At the beginning of the twentieth century California was winding down but not really coming to terms with the genocide committed on its first peoples that began with the rise and subsequent collapse of the Franciscan mission system and reached a violent climax during and after the Gold Rush and the land grabs that followed (Lindsay 2012; Madley 2015). More than any anthropologist of his generation, Kroeber was directly confronted with the task of "salvage anthropology," as it has been called (Harner, in Wolf 2004, 43). He felt time's winged chariot as he analyzed the rapidly vanishing languages of West Coast tribes, working with his more linguistically gifted friend Edward Sapir and others to complete a taxonomy of North American languages as rigorous, complete, and impressive as their European counterparts had produced for Indo-European tongues (Kroeber 1935, 553).

It is in connection with salvage anthropology that we situate the California Indian Ishi, who is forever linked with Kroeber. Kroeber took Ishi under his personal care after he walked out of the Sierra Nevada foothills in 1911. He was

Kroeber and Sapir's informant on the slightly different Native American languages and dialects spoken sometimes as few as fifty miles from each other.[1] Kroeber was deeply affected when Ishi died of tuberculosis in 1916. Still, he was congenitally reluctant to enter into public controversy about the fate of native peoples or to engage in what is now called advocacy anthropology. Boas had already spotted this standoffishness. 'I am not by any means in favor of an absolute aloofness of scientific work from the interests of daily life," he told his former pupil (Boas to Kroeber, October 6, 1908, in Boas 1972).

Now that cultural anthropologists are more engaged, some contemporary historians of the field accuse Kroeber for doing little to halt the destruction of native populations (Buckley 1996; but see Stewart 1961). In our view, this charge underestimates the persistence of Kroeber's youthful ambition to help liberate his fellow humans from "hoary tribal taboos," the depth of his feelings, and how he thought of anthropology's relation to the public sphere.

After the death of his first wife from tuberculosis in 1913, three years before Ishi succumbed to the same disease, Kroeber almost left anthropology to become a psychotherapist. From that time on his teaching and writing betray a strong element of checked *pathos.* In rigidly empirical ethnographical work charting threatened ways of life that once gone would never return, one senses what the Latin poet Virgil called "the tears of things" (*lacrymae rerum*). In contrast to Darwin's and Spencer's upbeat Victorian belief that encounters between less and more advanced peoples would in the long run elevate savages to civilization, Kroeber stressed that these clashes are uniformly and irreversibly disastrous for the weaker party. His ethnographic work combined the sensibility of a former English major – "It is a safe bet," he told Sapir, "that my actual work will always be literature" (Kroeber to Sapir, November 4, 1917, G243 in Golla 1984)[2] – with Boas's strict empiricism to produce a relish for meaning-laden particulars.

The vast storehouse of facts Kroeber could effortlessly remember and organize amazed his students and colleagues. "[He] had a tremendous capacity for recall," said one. "He could remember the comment of a particular informant, and say, 'On the twelfth of August in 1923, Mary So-and-so told me such-and-such" (Harner quoted in Wolf 2004, 43). Fellow Berkeley anthropologist Robert Lowie claimed in a *Festschrift* dedicated to him that his "versatile mind has concerned itself with a greater variety of subjects than probably any of his coevals" (Lowie 1936, xix–xxiii). After his death, Margaret Mead and Ruth Bunzel remarked that Kroeber was "the best representative of the Golden Age" of American anthropology, noting that his wide interests and depth of knowledge were partly a result of the fact that he was "no respecter of disciplinary labels." In the process of collecting and interpreting particulars he borrowed with "equal ease from history, geography, ecology, and psychology," but had "no hesitation in dropping ideas or methods [of these disciplines] if they proved unrewarding" for his own field (Mead and Bunzel 1960, 477). Kroeber's career illustrates Allan Megill's claim that in the historical sciences "explanation is dependent on recountings" (Megill 1989, 648). While much epistemology has been predicated on the notion that explanation should be privileged over "mere" description,

Kroeber dedicated himself throughout his career to describing cultural details in the way Clifford Geertz, who was influenced by him, famously characterized as "thick" (Geertz 1973). In actuality, much explaining goes on in the very act of describing. Small details about particular cultures, times, and places serve as emblems of more general phenomena without undue abstraction – or the absence of disciplined feeling.

Once he had regained his footing, Kroeber remained deeply engaged with his subject and with those whose experience it recounted. True, he did not publically advocate for the oppressed and marginalized or favor applied anthropology. However, he sought to influence public policy by reforming anthropology itself in order give voice to the voiceless in the long run. He explained to Sapir,

> I *am* trying to reach public opinion. I can't print what I write in the *Atlantic Monthly* or the *New Republic* or a philosophical journal or if I did there wouldn't be the least effect. I've got to hit general sentiment, if it hit it at all, *through our profession.*
>
> (Kroeber to Sapir, July 24, 1917, G 234, in Golla 1984 our italics)[3]

At the time he wrote this, Kroeber was incensed by efforts of Boas's professional enemies to seize upon the chauvinistic frenzy of World War I to expel him from the National Research Council because of his allegedly pro-German statements (Stocking 1968, 270–307). The effort, if successful, would have marginalized Boas's students as well. In Kroeber's view, it would have allowed stadialism, whites-only nationalism, racism, Jim Crow, and, perhaps worst, eugenics to infiltrate a field that was far from solidly established in universities. With more than his usual tone of exasperation, he told Sapir, "I'm tired of anthropology being a charity orphan allowed to pick up a profusion of scraps until biologists or geographers or psychologists or Madison Grants take a fancy to having them again" (Kroeber to Sapir, July 24, 1917, G 234, in Golla 1984).

This chapter is a commentary on this sentence. We claim that an accurate view of Kroeber depends on making central his interest in providing anthropology with defensible disciplinary boundaries. His self-imposed mission was to establish, protect, and grow a Boasian four-field anthropology department at Berkeley and use the field's professional associations and journals, especially *American Anthropologist*, to propagate this model more widely in universities. From there he hoped to drill its value-laden implications into public consciousness.

Demarcating the boundaries of this or that science goes back to Aristotle. Many disciplines he marked off still structure our institutions of learning, among them logic, theology, physics, biology, politics, ethics, poetics, and rhetoric. In each case, Aristotle attempted to identify the first principles (*archai*) on which valid and true propositions about each subject area depend. His expectation that one and only one true theory would mark off each sphere of systematic inquiry has been displaced, however, by recognition that a variety of theories inevitably compete within a single discipline as old paradigms give way to new and are themselves eventually dislodged. In this controversy-rich process, scientists, especially in imperfectly established fields, try to capture disciplines for their preferred theories or research programs by narrowing the boundaries of legitimate, field-specific inquiry to include only their own principles, marginalizing or excluding its rivals. Advocates do not exactly declare rival claims *false*. Instead, they attempt to exile them from the discipline altogether so that insiders will not count them as psychology, philosophy, anthropology, evolutionary biology, or whatever *at all*. Such attempts at exclusion generate displays of technical argument and much philosophical posturing. This process of disciplinary demarcation is irreducibly rhetorical. Hence we view Kroeber's lifework through the lens of what Charles Alan Taylor calls the "rhetoric of demarcation" (Taylor 1996).

Kroeber's attempt to demarcate anthropology was helped most by southwest German neo-Kantian philosopher Heinrich Rickert, who gave him the conceptual and methodological tools for doing the job. Venting in 1917 about Grant and his allies' effort to undermine Boasian anthropology, Kroeber told Sapir that he and Lowie had been reading Rickert. He was impressed enough, he said, to contemplate publishing a collection of translated passages from it (Kroeber to Sapir, November [no date] 1917, G 244, in Golla 1984; Rickert 1889).[4]

More categorically than his teacher Wilhelm Windleband, Rickert rejected Dilthey's way of demarcating *Geisteswissenschaften* from *Naturwissenschaften*. Dilthey predicated the ability of practitioners of the human sciences to penetrate the minds of others, past as well as present, on empathetic identification (*Einfühlung*), thereby making psychology the key to human science, but also undermining the possibility that psychology could ever become what it was in fact becoming: an experimentally grounded empirical science. Rickert agreed with Windleband that the exact sciences, including psychology, are oriented to finding general laws and to rigorously subsuming particulars under them while the *Kulturwissenschaften* interpretively burrow into the endless diversity of particular events. What he added, and Kroeber accepted, is that the particulars in question are constituted by acts of valuation by individuals and the groups to which they belong. According to Rickert, the cultural sciences concern "what is produced by man in accord with valued ends" (Rickert 1962, 19 (Rickert 1889)). The cultural scientist uses what Rickert called "the historical method" to study these "goods and the human beings that value them" (Rickert 1962, 89). Historical method "picks out significant events as invested with [value-oriented or value-laden] meaning" (Rickert 1962, 83). As Rickert himself notes, this means that anthropology cannot be reduced to psychology, even social psychology: "The psychical cannot be employed as the definition of culture" (Rickert 1962, 27). Still, Rickert held that interpretation of value-laden and value-oriented facts and their products, including cultural artifacts, can be studied in rigorously empirical ways. All this meant disagreement with Dilthey, who demarcated sciences by the nature of their objects, not different ways of approaching even the same objects, and in consequence thought of his empathy-projecting method was uniquely appropriate to sciences of the spirit (*Geisteswissenschaften*).

The irreducibility of anthropology to psychology, whether collective or individual, is basic to Kroeber's legacy to anthropology. Still, Rickert left Kroeber with the problem of demarcating anthropology from the biological study of human evolution on one hand and from politics-centered history on the other. They, too, he conceded, are in some sense sciences of culture. His solution was to assign to anthropology the study of cultures as relatively integrated meaning and value-laden wholes that arise from but are irreducible to biology and in this sense are "superorganic" (Kroeber 1917). In his work on Native Americans Kroeber identified some eighty-four such cultures. He grouped them into six "culture zones," each radiating influence from a geographical center (Kroeber 1939). Cultures so construed are reconstructed in successive generations by teaching and learning, but in this very process they change. This fact provides the background for historical inquiry in the restricted sense. "Culture," Kroeber said, "is a series of regularities underlying the multitudinous and varying events of human behavior in what is ordinarily called 'history.' Historiographers as such do not deal with culture. They take it for granted" (Kroeber in Tax and Callendar 1960, III, 236).

Identifying disciplinary demarcating – "boundary work," as Thomas Geiryn calls it (Gieryn 1999) – as the core of Kroeber's lifework helps explain why, despite advancing a number of big ideas, including the idea that anthropologists study the "superorganic," he denied that he had any theoretical ambitions or abilities. Early in Kroeber's career we find Boas urging him to work on theoretical as well as empirical issues; it seems that his mentor was worried that his student possessed a bit *too much* of the orientation toward the particular that he himself prized. Kroeber balked at the suggestion. "I do not know what I can do about theoretical articles," he wrote to Boas in 1908. "As I think you know, I try to avoid theoretical questions on principle so far as I can" (Kroeber to Boas, May 5, 1908, Boas 1972). Later, even as he was publishing manifesto after manifesto on the investigative principles governing anthropology, Kroeber denied that he was theorizing. "I'd be willing [to theorize]," he told Sapir, "but I couldn't if I tried. No more than composing a tune. Where you see philosophy there's only the awkwardness of abstract expression" (Kroeber to Sapir, July 24, 1917, G234 in Golla 1984). Pressed, Kroeber claimed only to be provoking his colleagues to disagree with him, thereby advancing the articulation of the discipline. Urging Sapir to publish privately his objections to the superorganic idea, Kroeber wrote, "I do not know if I am right or wrong … I am sure that the only progress is by forcing issues to a head" (Kroeber to Sapir, July 24, 1917, G234 in Golla 1984). When Sapir did just that, Kroeber tried to take the theoretical wind out of his sails by remarking, "Nobody ever is convinced in these discussions. They are merely opportunities for different people to air themselves. Still, they do have the merit of sharpening one's own views" (Kroeber to Sapir, October 29, 1917, G 242 in Golla 1984). Assembling some of his essays together in a volume half a century later, Kroeber warned readers that he never considered himself a "formal theoretician" of the culture concept. He claimed that his thoughts were only "by-products" of his empirical work, "sweated out piecemeal and slowly over fifty years" (Kroeber 1952, 3).

We suggest that these disclaimers are less disingenuous than they appear. Boundary work is by its very nature gathering together a competent community of inquiry. Considered in this light, what look like claims *within* a field are often actually claims *about* the field's relation to other fields under particular conditions of institutional communication, organization, inquiry, and controversy. Gieryn makes extensive use of spatial metaphors in discussing boundary work (Gieryn 1999). It is a sign that Kroeber was in this business that he habitually employed just such metaphors. He spoke often about "making maps" of "territories," "cultivating a patch of ground," and "conquering new lands." Seizing on the image latent in the idea of a "field" of inquiry, he proclaimed that anthropology had to be "surveyed, fenced, and improved" (Kroeber 1915, 283–284).

Gieryn also makes the interesting suggestion that disciplinary formation exhibits three characteristic lines of demarcating argumentation: expulsion of foreigners, assertions of autonomy, and expansion into neighboring territory (Gieryn 1999, 15–18). These *topoi* are often deployed concurrently. Nonetheless, Gieryn's genres of expulsion, autonomy, and expansion are diachronic enough to afford us a way of tracking the developmental arc of twentieth century American anthropology and, in particular, of following Kroeber's way, first, of securing anthropology's autonomy by extruding it from biology and biology from it, then of fending off anthropologists' own tendency to let psychology back in the door, and, finally, of resisting efforts to expand the explanatory power of anthropology, once it had been secured, by extending the reach of the culture concept back into psychology and biology.

In the following section, we will discover signs of the rhetoric of expulsion in Kroeber's lifelong effort to eliminate biological categories from anthropological descriptions and explanations. Our aim is to dispel misconceptions about this injunction. Kroeber's call to study the superorganic did not mean that societies are superorganisms after the fashion of ant colonies or ecological communities.[5] That would spell the very biologizing of culture he opposed. He meant that anthropology stands (just) *above* the discourses of biology (Kroeber 1917). Writing at a time when racial stadialism was perpetuating itself by commandeering Mendelian genetics to support racially coded eugenics and when, partly in response, Lamarckian ideas about biological heritability had not been expelled from the social sciences, Kroeber insisted that anthropology's way of studying human life begins just where biology ends (Stocking 1968, 258–260). This stipulation did not mean that he was indifferent to developments within biology. On the contrary, we will show how and why his conception of culture rested squarely on Weismannian views about biological heritability (Kronfeldner 2009; Jackson 2010).

Gieryn's second kind of demarcation rhetoric is the maintenance of autonomy. For Kroeber this meant for the most part urging anthropologists to resist inviting psychology to invade their discipline. This theme came into play whenever his colleagues misconstrued what he meant by "superorganic." Sapir, for example, suspected that Kroeber's superorganic was not as ontologically neutral or as innocently pragmatic as he claimed. He believed that a collectivist impulse hostile to individuals and their psychologies lurked in it (Sapir 1917). Other Boasians broadly agreed. Paul Radin, Elsie Parsons, and Alexander Goldenweiser found Kroeber's formulation too reified and too anti-psychological to allow individual agents to innovate in cultural practices. In point of fact, Kroeber did not deny that there are ingenious basket weavers. What worried him was whether attempts to locate the source of their ingenuity in individual psychology would inevitably underestimate the role of culture in realizing whatever innate capacities a creative individual might possess, thereby undermining anthropology's distinctive place in the disciplinary scheme of things. Insisting that he was defending the interests of colleagues whose views, if adopted by the community, would weaken anthropology's fragile autonomy, Kroeber suggested that it was they, not he, whose ontological commitments were too strong. He was not affirming or denying anything about the psychology of individuals. He was merely saying that this was not anthropology's business.

The third dimension of demarcation, says Gieryn, aims at expanding a field's territory in order to impress others with its general importance and deter them from meddling. Gieryn spots the rhetoric of expansion whenever "two or more rival epistemic authorities that have secured their own territory square off for jurisdictional control over a contested ontological domain" (Gieryn 1999, 17). Twentieth century American anthropology was first tempted to become expansionistic when sociologists borrowed its culture concept to study communities like Muncie, Indiana (Lynd and Lynd 1929, 1937). Until then, sociology studied modernizing societies and anthropology tribal ways of life. Soon, the social psychological typologizing of sociologists was imprinting on the ethnographic study of cultures. The tendency began during World War II, when Ruth Benedict, Clyde Kluckhohn, Margaret Mead's husband Gregory Bateson contributed to the war effort by drawing up social-psychological profiles of cultures as modern(ized) as those of Germany and Japan with a view to helping the military analyze the motives of its enemies, instruct it in how to deal with captives, and advise it about how to be effective conquerors (Gilkeson 2010). Postwar anthropology built on their way of psychologically identifying (and dangerously atomizing) each culture in terms based in part on pop-Freudian interpretations of styles of child rearing – "diaperology," as its critics called it (Gilkeson 2010, 149). A social-psychological "culture and personality" movement spread through American anthropology in the 1950s under the postwar impulse to unify the social sciences with each other and the natural sciences.

While many post-war anthropologists welcomed the chance to expand into new territory, Kroeber feared that fraternizing with social psychology would lead to anthropology's cooptation and annexation. He was even more hostile to University of Michigan anthropologist Leslie White's way of resisting the turn toward social psychology by treating cultures "from a zoological standpoint as … means of carrying out the life processes of a particular species" (White 1949, 36). White was no less a boundary worker than Kroeber. Although his way of biologizing culture was not race-based, from Kroeber's position his proposal simultaneously made his biology insufficiently Weismannian and his insistence that all human differences are cultural dogmatic. The two sparred until Kroeber's death in 1960.

What, then, about race? In his perennially best-selling textbook, *Anthropology*, Kroeber accepted physical anthropology's received race concept, dutifully enumerated human races, and moved on as quickly as he could to archeological remains and the cultural lifeways of those whose valuing was impressed into these artifacts (Kroeber 1923, 1948a). In a popular condensation of the book, he omitted the topic altogether (Kroeber 1963). It might be thought that by leaving physical anthropology's race concept where it was, Kroeber opened himself to the same sort of objection he leveled at White. Left intact, an inadequate biological concept of race was likely to infiltrate the culture concept. There is something to this. Unreformed ideas about race persisted in physical anthropology into the postwar period, wreaking havoc on the anti-segregationist cause until population-genetic Darwinians and their allies in physical anthropology redefined the concept of race in Mendelian terms that liberated it from racial typologizing and rank-ordering (Chapters 4 and 6 of this book). Presciently, Kroeber embraced the population-genetic reframing of race as soon as he became aware of it, seeing in it help for his career-long battle to appeal to Weismannian biology to protect the autonomy of cultural anthropology and free it from the old idea of race (Kroeber 1960). Kroeber's superorganic remains powerful enough to refute today's evolutionary psychologists, who claim to have found rampant 'biophobia' in the social sciences and who ask anthropologists to psychologize and geneticize human differences (Daly and Wilson 1988, 154–160; Barkow *et al.* 1992; Ellis 1996). It is also a powerful solvent against the contemporary tendency to align molecular genetics with uncritical ideas about biological races (Chapter 1 of this book).

In opposing both biological and cultural imperialism, Kroeber did not intend to insulate anthropology from neighboring disciplines. By legislating its boundaries he was encouraging dialectical interactions among anthropology, history, and evolutionary biology, and among American anthropology's four subdisciplines. Just as the various cultures he studied exchanged goods in the borderlands that separated and at the same time united them, so Kroeber believed that anthropologists can, do, and should exchange ideas in disciplinary "trading zones" without being expansionist in a hegemony-seeking way. We agree with Maria Kronfeldner that Kroeber's efforts to establish anthropology's autonomy was actually a step toward interdisciplinity (Kronfeldner 2010, 111–125). Julie Thompson Klein puts the point this way:

> Although specialization is often vilified in the discourse as a negative force promoting fragmentation, specialization has in fact fostered a number of interactions as disciplinarians approach each other's borders. The depth of disciplinary study may open up relationships at the intersection of parts of two disciplines, especially when contiguous problems are involved.
>
> (Klein 1990, 43)

The superorganic – a tool for weeding anthropology's garden: 1901–1923

Kroeber's early effort to demarcate anthropology had two overlapping targets: racialist anthropologists like Osborn, Grant, and Davenport, who believed they could decant the old wine of racialist-infected eugenics into the new bottles of genetic discourse, and social scientists who even in the 1910s were clinging to discredited Lamarckian notions about biological and social heritability in order to save the idea of social progress. Kroeber was not alone in adopting hard inheritance in biology and denying that social transmission is a form of heredity at all. Like other opponents of Spencer, not least Weismann himself, he was aware that, talk about social progress aside, thinking of cultural transmission as soft heredity implies that, "[t]he social is always on the brink of becoming biological, [as] habits are turned into instincts, and the life experiences of a previous generation are embedded in the biology of a successive one" (Meloni 2016, 5). Kroeber was aware that Lamarckian inheritance is particularly egregious when is linked to the supposedly disqualifying effects on particular races of multigenerational experiences like enslavement. Still, Stocking has rightly observed that among "left Weismannians," as Meloni calls them, Kroeber "seems to have been virtually alone … in realizing the [desirable] implications of the expulsion of Lamarckism [from biology] for the maturation of the social sciences" themselves (Stocking 1968, 259). He insightfully saw that the Weismannian sea change in evolutionary biology could simultaneously rid anthropology of stadial and race-based notions of human evolution and create a space for a descriptive science of anthropology that replaces biological heredity with learning, tradition, innovation, and diffusion (Kronfeldner 2010; Jackson 2010). The most common way of going astray on this point is to presume that Kroeber thought of cultural transmission as a second kind of heredity. He did not.

Kroeber was already attacking anthropologists' biologizing and psychologizing in his doctoral dissertation (Kroeber 1901). The dissertation examined decorative artwork among the Arapaho. Beyond providing a rich description of their artwork, Kroeber was interested in examining the cultural function of art in general. Is primitive art meant to represent reality, he asked, or is it purely decorative? His answer refuses the binary. Primitive art is bound by aesthetic conventions, but within these conventions it is meant to be realistic. So "the main characteristic of Arapaho art [is] its fusion (which is more truly an undifferentiation) of the realistic and decorative tendencies" (Kroeber 1901, 324).

Kroeber made another general point about so-called "primitive art" (Kroeber 1901, 324). After cataloging examples of the "undifferentiation" of the decorative and realistic functions of art from all over the world and issuing a boilerplate warning about the danger of generalizing from "selected examples such as these," he nonetheless concluded:

> This fusion of two differing tendencies is not merely a frequent or widely distributed occurrence.… It is universal because it is necessary. Both the

> representative tendency and the decorative tendency are deep rooted in the human mind, so that it must be virtually impossible to suppress them for any length of time or among any considerable number of men.
>
> (Kroeber 1901, 326)

He continued in a vein that would have pleased Boas's German mentor Bastian (Chapter 1 of this book). "Every culture," he wrote, "must contain among its motive forces more or less of every tendency, because the tendencies are in the human mind and hence ineradicable" (Kroeber 1901, 327; see Chapter 2 of this book). Bastian would have been less happy, however, to hear that, while it is true that the origin of art can be found in universal psychological mechanisms of the human mind, these can play no explanatory role in anthropological science. Anthropologists searching for the causal origins of art, Kroeber argued, face intractable problems:

> If [art] it is comparatively recent in origin there must until a certain time have been no art among the Arapaho, while at that moment it sprang up full-blown, not as a crude undifferentiated thing, but a highly-specialized pictorial art. Such an event would be extremely remarkable, not to say marvelous, and more in need of an explanation than the phenomenon it explained.
>
> (Kroeber 1901, 329)

Suppose, as an alternative, that pictorial representation is very old and emerged gradually over long periods of time. In that case the origin of art will have been lost to the investigator in the mists of time. The deeper you recede into the past the less likely you are to find *a* reason to stop the causal chain and declare some particular event to be *the* origin of the work of art. To be sure, "[n]o myth, no artistic convention, nor any other thing human, ever sprang up from nothing" (Kroeber 1901, 333). But causal chains, even if they are in part known, are tantamount to infinite when it comes to problems like the origin of art. Hence Kroeber rejected in principle the idea that one could ever hope to explain culture by pointing to a past event and declaring it to be *the* cause. To describe cultural facts accurately you must repudiate causal-explanatory chains altogether and rely on description to do your explaining (Kroeber 1901, 333).

Kroeber used this insight to examine existing literature on the origins of mythology in his dissertation. Reviewing rival accounts of the origins of mythology, he noted that each appeals to one or another supposed psychological capacity or tendency of human beings. Although he conceded that each theory captures *a* tendency of human behavior, he argued that none of them can stand as *the* explanation of the origin of mythology precisely because the others are involved as well. "This multiplicity of tendencies or causative forces," he wrote, "necessarily refutes any explanation that uses and allows only one of them" (Kroeber 1901, 332).

In this elegant inference Kroeber cleared the way for asserting the autonomy of anthropology by defining away the psychological as one of its provinces. The argument was doubtless congenial to Boas, who as Kroeber's advisor was also its primary audience, but it would not have pleased Bastian and others in Germany who were attempting to erect anthropology on the basis of a set of elementary thought patterns and hence on psychological foundations (Chapter 2 of this book). Kroeber argues that the fact that certain psychological tendencies are so "inherent in mind" that they can be said to lie "at the root of all anthropological phenomena does not by that fact alone make them objects of study by anthropologists" (Kroeber 1901, 332). Psychological mechanisms are fixed at the level of the species, but the cultural patterns they produce are infinitely variable and varied. In anthropology, accordingly, it is the mind's products, not the mind itself, that are the proper object of study: "The products of mind (the phenomena studied by anthropologists) are, like mind itself, beginningless (for us)" (Kroeber 1901, 333). The last two words – 'for us' – are significant. It is not that culture is a phenomenon without a beginning. It is only *for us* anthropologists that this is so. The field's objects of inquiry come into view by a pragmatic restriction on its conditions of inquiry. Only by screening off questions about the evolution and differential distribution and temporal ordering of psychological traits – the holy grail of stadial theories of social evolution – can one hope to find the information the anthropologist wants. Look for origins and you will misdescribe the relevant facts.

When Kroeber read Rickert in 1917 he was already prepared to hear his message owing to his early attempt to demarcate his own field. To his earlier contention that anthropologists describe the products of mind, not mind itself, Kroeber added Rickert's view of the value-orientation of the cultural phenomena to derive his signature idea that culture is superorganic. What he meant by "history" and "historical," terms he uses constantly, also follows Rickert: Historical facts are value-oriented and value-laden in ways requiring interpretation of particulars rather than subsumption as instances under laws. Linking Kroeber's work in 1901 to his work in 1917 guides our interpretation of his terms "superorganic" and "historical" and explains his stubborn refusal throughout his life to reintroduce even a little psychology or biology into anthropology.

Kroeber's concern with boundary questions during this period reflected his personal problems. The mid-teens in particular were a time of great difficulty. His second wife called it his "hegira" (T. Kroeber 1970, 292). His first wife had died. Ishi had died. He had undergone psychoanalysis in part because he was losing the youthful hopes for anthropology he had expressed to Alsberg. It is not unreasonable to expect that he would have questioned his field at a time when he was questioning his own role in it. Nonetheless, even as he was wresting with his personal demons, there was also an uncharacteristically public dimension to Kroeber's worries. Suspicion of Germans as subversives when America was moving toward entry into World War I afforded an opportunity, indeed an excuse, to try to expel Boas and his influence from the National Research Council. Kroeber had no illusions about what would happen if the efforts of Davenport, Grant, and others succeeded. The specious Weismannism and Mendelism of the eugenics movement of the day would confer scientific authority on biological stadialism by recasting it in genetic language. With it would come a great wind of pseudo-scientific racism in which it was categorically asserted that biological race is genetically transmitted and genetically threatened by miscegenation with lower races. The subtitle of Grant's widely read *The Passing of the Great Race* is *The Racial Basis of European History* (Grant 1916). As early as 1914, Kroeber was telling audiences that, "So far as civilization is concerned there is no such thing as an Anglo-Saxon breed or a white man's burden" ("Eugenics Called Snare and Joke" 1914, 1). This rhetorical situation, more than loyalty to his mentor, illuminates Kroeber's deeply felt, but academic mode of social and political engagement. When he told Sapir that he had "to hit general sentiment … through our profession" he was implying that his way of demarcating anthropology would frame the arguments needed to foil the expulsion effort, which in the event was blocked by a few votes. But both the stakes and Kroeber's confidence in his demarcation arguments also explain why he was markedly hostile to eugenics.

In taking on eugenics and its alignment with racialism, Kroeber was confronting a large target. In the early twentieth century, every industrialized country embraced one or another form of it (Kevles 1985; Dikötter 1998; Barrett and Kurzman 2004). Eugenics in this sense was "not so much as a clear set of scientific principles as a 'modern way' of talking about social problems in biologizing terms" (Dikötter 1998, 467). Because these terms were overly generous, it was promoted or at least tolerated by practicing biologists, who seldom recognized their own prejudices in it. They thought they were being socially useful and responsible. For Kroeber, however, the eugenics movement was worse than a political mistake. It was a scientific blunder because it confused biological and social phenomena. It was not well demarcated and so was not well confined to its own business, if it had any proper business at all beyond animal and plant breeding:

> Chemists do not feel impelled to expound the rise of genius in chemical terms or explain the variety of moral codes by valences and atomic weights. They therefore leave civilization alone or, if they pronounce judgments in its field, do so avowedly as laymen. But biologists view the province of the social from their very doorsteps.
>
> (Kroeber 1916, 38)

In spite of his inward preoccupations, accordingly, Kroeber became one of the most outspoken scientific critics of eugenics in the United States in the 1910s, chafing at its muddy way of blurring the cultural, the political, and the biological. At the time, few American scientists were willing to say things like this:

> [Eugenics] is more refined but no less vain than the short cut which the savage follows, when, to avoid the trouble and danger of killing his foe in the body, he pierces, in the safety and amid abjurations uttered in the

> convenience of his own home, a miniature image addressed by the name of the enemy. Past ages have had their dragons of superstition to fight. Our battles against this ever re-arising brood dawn no smaller and as unceasing; and it would be shallow to try to defer or soften the inevitable conflict by withholding from this movement its true designation. Eugenics … is a fallacy. It is a mirage like the philosopher's stone, the elixir of life, the ring of Solomon, or the material efficacy of prayer; and to those who are led by its learned modernity to receive it earnestly, it is a destructive snare.
>
> (Kroeber 1917, 188–189)

Unlike some of its early critics – William Jennings Bryan, for example – Kroeber did not oppose eugenics because he rejected evolutionary theory, which eugenicists claimed as their own. He opposed it because he endorsed the *scientific* study of evolution. Evolution as such, he noted, is an ancient idea; he offered the evolutionary myths of many cultures to prove his point (Kroeber 1916). Darwin was not the first evolutionist. His genius was to combine three well-evidenced ideas – variability, heredity, and competition – and insert them into the process of evolution, thereby discovering evolution's primary mechanism: natural selection. Kroeber was sure that Darwin's key idea would undergo further modification as new developments shed further light on the evolutionary process, but he was no less sure that henceforth "the world must probably forever believe that natural selection is of some influence in the shaping of life" (Kroeber 1916, 25). It was left to Weismann's doctrine of hard heredity, he continued, to complete what Darwin began but, because he did not quite break with the "older pseudo-process of Lamarck" of use inheritance, did not finish (Kroeber 1916, 25). Kroeber maintained that Weismann "was as clear a thinker as Darwin; and his accomplishment will in the end be rated in proportion" (Kroeber 1916, 26). Under Weismann's "onslaught," he declared, "the Lamarckian structure" proved

> absolutely hollow. Experiment failed to produce even a scrap of positive evidence in its favor. Renewed examination demonstrated that there was not a single alleged instance which was more than logically possible. Practically every case of use inheritance was explicable by selection.
>
> (Kroeber 1916, 26)

Still, the power of Darwinian natural selection, even when combined with overwhelming evidence for hard heredity, left Kroeber with a puzzle. Although biologists had stopped talking about use inheritance, he thought it odd that few of them and even fewer social scientists were trying to stamp it out as a "pernicious heresy" (Kroeber 1916, 28). "Scarcely anywhere since Weismann," Kroeber noted,

> is there any zeal against the doctrine of acquired heredity as something radically and vitally and destructively wrong. Biology … scarcely professes

> a cardinal article of faith on acquired heredity. What brings it about that there exists so much weak condemnation, half tolerance, and hankering?
> (Kroeber 1916, 28–29)

Kroeber's answer was as subtle as his imputation that anthropologists should take the lead in calling the eugenicists' bluff was bold. There are two evolutionary processes, a biological one in the Darwinian/Weismannian/Mendelian mode and a social-cultural process in which "use modification is permanent and transmittal of the acquired exists" (Kroeber 1916, 31):

> Darwinism is often spoken of as allied to anthropological thought. But there is no specific connection. The one deals with biological phenomena and processes; the other begins where these leave off. The common element is the wholly generic concept of evolution, equally applicable in astronomy and geology. Organic evolution is essentially modificatory, cultural evolution is cumulative. *The one is bound up with heredity, the other in principle is free from it.* The similarity is merely a loose analogy, and the Darwinian point of view has retarded and confused the understanding of culture.
> (Kroeber 1928a, 495, our italics)

Kroeber contended that biologists were as guilty as social scientists of conflating these mutually exclusive sorts of evolution. The cumulative character of cultural evolution tempts them to see progress in Darwinian-Weismannian evolution just as it tempts social scientists to see social inheritance as a form of, or at least a close analogue to, biological heredity. There were several kinds of evolution for Kroeber, but only one kind of heredity.

The slide from social learning to social inheritance to biological heredity, Kroeber argued, perpetuated ill-formed ideas about social evolution even in scientific minds that made them prey to embracing or tolerating eugenics. Natural selection's focus on particular time- and space-bound instances of organism, gene, and environment interaction does not allow new characteristics to accumulate nearly as automatically as evolutionary progressivists and stadialists assumed. It is context-dependent. Inappropriate projections of the accumulative processes of culture onto nature made it difficult for biologists to see the incompatibility between their professed Weismannism and their embrace of eugenics. They saw eugenics as a way in which knowledgeable humans could keep evolution progressive by compensating for cultural practices that protect the unfit from natural selection. "The entire doctrine of eugenics is an endeavor to attain moral ends by biological means," he argued. "Moral of course is social; and yet the open protests have come – strange partnership! – from the orthodoxly religious and the professedly skeptical but rarely from the enlightened camp of science" (Kroeber 1916, 34).[6] As Kroeber saw it, the actual situation is that

> Speech, knowledge, arts, learning, and all our activities except the bare substratum of physiological abilities are not inborn. Heredity gives us the slate

> and the pencil in good working order. Our individual kinds of slates and the sharpness of our pencils are also wholly from heredity. But with the writing on the slate, which is the part we play in civilization, heredity has nothing to do. That comes from [our] social situation, in other words from the existing civilization into which we are born.
>
> (Kroeber 1916, 31)

In spite of its riff on the old Lockian trope of the blank slate, this is *not* an expression of Tooby, Cosmides and Pinker's Standard Social Science Model (SSSM) (Chapters 1 and 7 of this book). Kroeber complained that behaviorists teach that human beings come into the world with practically no equipment at all. If true, this would make acquiring culture entirely a matter of individual life experience, reducing the very idea of culture to "a series of accidental events" and pulling the biological base out from under cultural life (Kroeber 1928b, 326). At the same time, however, our inherited equipment afforded no opening to genetic determinism and hence to eugenics. It marked the divide between biological heredity and social transmission.

While his published papers and private letters raged against popularizers of racism and eugenics and criticized biologists who underestimated the effect of the Weismannian revolution, Kroeber also found targets in the ranks of social scientist who tended both to biologize *and* psychologize social concepts like race. An example is Gustave Le Bon, who in his *Psychology of Peoples* took it as his task "to describe the psychological characteristics which constitute the soul of races, and to show how the history of a people and its civilization are determined by these characteristics" (Le Bon 1912, xvii). Kroeber (showing influence from pragmatic epistemology) was disgusted because "as a scientific concept or tool a 'race soul' is as intangible and useless as any phrase of medieval philosophy" (Kroeber 1917, 185). "If," he argued, Le Bon

> had said "spirit of civilization," or "tendency or character of culture," his pronouncements would have commanded less appeal, because seeming vaguer; but he would not have had to rest his entire thought upon a supernatural idea antagonistic to the body of science to which he was trying to attach his work; and if non-mechanistic, his efforts at explanation would at least have earned the respect of historians.
>
> (Kroeber 1917, 185)

Worse, seeming to misunderstand the science on which his work was supposedly based, Le Bon argued that the progress of civilization depends on the accumulation of heritable traits by favored races. Kroeber refuted him by calling on Weismann's authority to distinguish organic *heredity* from civilizational *accumulation*. "If there is anything that heredity does not do," he declared, "it is accumulate. If, on the other hand, there is any one method by which civilization may be defined as operating, it is precisely that of accumulation" (Kroeber 1917, 186). By refusing to understand the difference between *heredity* in biology and cultural *inheritance*, Le Bon had produced a work that was neither scientific nor historical. Only by keeping these forms of inquiry confined to their own spheres, Kroeber warned, could each produce work worthy of itself.

Because he believed that social scientists had a special responsibility to repair the damage done by the eugenics movement, Kroeber berated Le Bon for failing to heed the lessons of the new biology:

> Biology has been born in the last century or two. It has forged its weapons, taught itself their use, conquered a territory, and stands forth a young giant of prowess. What wonder that it has proceeded by the divine right of power to annex the antiquated realm of history that lay adjacent, and to impose its rule and laws without inquiring whether they were fit. The greater fault is not with the biologists who have explained historical phenomena by organic processes, but with the sociologists who have accepted and welcomed these alien explanations.
>
> (Kroeber 1916, 34)

Kroeber's candidate for a sociologist who almost got it right, but fell short, was Lester Frank Ward, whose views served as a whetstone for honing his own. Like Emile Durkheim, Ward ascribed social evolution, in which he ardently believed, to the spread of religious and national solidarity rather than to racial differences or economic competition. In this respect he was a heretical Spencerian (Ward 1893). As the first president of the American Sociological Association, Ward accepted Weismann's hard heredity in the case of other animals, but maintained that, "When the human species is to be treated the tables are, in a manner, turned" (Ward 1891b, 315; see also Ward 1891a). His reason was that, "Professor Weismann and most of his followers, constituting what is now generally known as the school of neo-Darwinian," imply that "education has no value for the future of mankind, and its benefits are confined exclusively to the generation receiving it." The accumulation of acquired characteristics both socially and biologically seemed to Ward the only way to ensure "permanent progress for humanity" (Ward 1891b, 319). His worry was shared by many social evolutionists in the decade after Weismann bested Spencer. Lamarckism was retained by social scientists well past its sell-by date for just this reason (Stocking 1968; Meloni 2016).

Kroeber sympathized with Ward, but rejected his solution. Ward's view, he said, was "if not a deep view, a common one; however worthless intrinsically, [it is] representative and significant" (Kroeber 1917, 187). It was common because it seemed that "to abandon Lamarck and accept Weismann would be to yield up the social sciences to an unrestricted biological determinism" (Stocking 1968, 256). It was worthless, however, because as an anthropologist Kroeber was in possession of something Ward was not: "a concept of culture severed from all biological connections" (Stocking 1968, 256), including the mistaken notion that cultural transmission is a form of heredity. Kroeber took the term "superorganic" from Spencer. "In spite of his happy coinage," he wrote, Spencer "did not adequately conceive of human society as holding a specific content that is non-organic" (Kroeber 1917, 188). Neither did reform Spencerians like Ward. "All these writers," Kroeber concludes, "failed to adequately recognize that culture could be treated as a completely separate entity.... Mental activity as biologists have dealt with it being organic, any demonstration concerning it consequently proves nothing whatever as to social events" (Kroeber 1917, 1).

Defending anthropology's patch of ground by keeping psychology out: 1923–1944

Kroeber's colleagues did not read his work as a mandate to insulate the social from the biological sciences in the way Evolutionary Psychologists and other myth-makers of the SSSM presume. They read his demarcation of anthropology as it was intended to be read: as an effort to interrelate biology and anthropology in a way that left the autonomy of each intact. Alexander Goldenweiser, for example, characterized Kroeber's writings as exploring "the theoretical relation of the historic to the biological sciences" (Goldenweiser 1920, 26). Still, even if he had got the biology-anthropology boundary right, many of his fellow anthropologists, including Goldenweiser, doubted whether he had mapped the psychology-anthropology boundary correctly. Kroeber aligned the evolution of our psychological capacities with biology, but kept psychology out of anthropology. Having proclaimed himself open to criticism by his colleagues on this point, he certainly got what he asked for.

The most sophisticated as well as one of the earliest objections to his demarcation proposal came from a young student of Boas, Herman K. Haeberlin (1890–1918). His name is almost forgotten today because of his premature death from diabetes at the age of twenty-eight. However, Haeberlin's critique of Kroeber's position was a serious challenge to the superorganic idea and is worth examining in some detail, as is Kroeber's reply. Haeberlin was born in Akron, Ohio to German parents. Boas met him in Berlin in 1913, where the young man was working with the psychologist Wilhelm Wundt (1832–1920), a former student of Helmholtz (Miller 2007). Best remembered as a father of experimental psychology, Wundt actually saw the laboratory work for which he is famous as part of a larger project aimed at mapping all human psychological phenomena around the guiding idea of *Völkerpsychologie* ("the psychology of peoples"). In it he attempted to bridge the gap between the experimental-materialist approach of which he was master and the meaning-laden idealism to which he aspired (Smith 1991, 120–121). Haeberlin inherited Wundt's double ambition, in part because he worked under him while Wundt was producing a massive, ten-volume (still untranslated) *Völkerpsychologie* (Wundt 1900–1920).

Four main ideas recur in this sprawling treatise: (1) Experimental psychology of the mental states of individuals can never reach all psychological questions, even if, *contra* Dilthey, it is valid for the study of individual psychology; (2) It needs supplementing by investigation of "those mental products which are created by a community of human life and are, therefore, inexplicable in terms merely of individual consciousness, since they presuppose the reciprocal action of many" (Wundt 1916, 3); (3) The study of *Volksseelen* (collective identities) through analysis of the language, myths, and customs of a people can be just as objective as the study of individual processes through experimental methods (Danziger 1983, 307); (4) The "goal of *Völkerpsychologie* [as a whole] is to complement individual psychology in the investigation of higher psychological processes" (Wong 2009, 248).

From this it followed that psychology as Wundt demarcated it fused psychology as a natural science and what people like Rickert called the cultural or historical sciences. This ambition entails a conclusion that Haeberlin endorsed: The mind that generates cultural objects is too intimately connected to processes down to the level of basic physiology for the cultural or historical sciences to constitute "a self-standing discipline" separated from psychology (Diriwächter 2004, 99). Haeberlin realized that Wundt's proposal contradicted Kroeber's superorganic demarcation of anthropology. Siding with Wundt, he argued that psychology in his sense, "which deals with the mind and *all* of its expressions, is per se the link between the natural … and the mental sciences (*Geisteswissenschaften*)" (Haeberlin 1915, 759, our italics):

> Wundt states categorically that folk-psychology deals with the psychology of language, religion (*Mythus und Religion*), and custom. These three types of cultural phenomena are the achievements *par excellence* of the folk-mind (*Volksgeist*). Not the individual, but the group (*die Gemeinschaft*, community) is the creator of language, religion, custom. Of course, the group consists physically of a number of individuals; but those folk-psychological phenomena, so argues Wundt, represent a higher synthesis that transcends the scope of individual consciousness.
>
> (Haeberlin 1916b, 287–288)

Given these principles, Haeberlin thought Kroeber's severing anthropology from psychology amounted to "bureaucratic police regulation" (Haeberlin 1915, 756). For him, the fundamental problem was to bring psychology in Wundt's expansive sense and cultural history "into a harmonious relation with each other" so that at last history could become scientific in ways that he did not believe Kroeber's value-oriented interpretive methods allowed (Haeberlin 1916b, 301). He appealed to Wundt's "principle of creative synthesis as the one cardinal principle of psychic life" in order to decry Kroeber's demarcation of anthropology as unscientific (Haeberlin 1916b, 280). Kroeber, he claimed, was committing "the cardinal sin of arbitrary elimination" by his "aprioristic attempt to delimit the scope of history from that of science" (Haeberlin 1915, 756).

Haeberlin was almost alone among Kroeber's anthropological colleagues and students in recognizing that he was not making an ontological claim about the quasi-substantial existence of a superorganic object. On the contrary, Haeberlin harped on the anti-scientific implications of Kroeber's notion of values as constitutive of cultural-historical objects of study by characterizing them as merely subjective. His argument was that investigators could not treat value-laden objects of inquiry without imposing their own values on the data. Kroeber, he said, "clouds our vision by means of a normative formula" that is corrupted empirically by "subjective ideas about what the aim of history *ought* to be" (Haeberlin 1915, 756–757, emphasis in original). Following Wundt, Haeberlin maintained that the study of culture is scientifically accessible only because "culture ... is just as objectively real as ... drums, clubs" and other artifacts of traditional ethnological study (Haeberlin 1916a, 10). In this he read Kroeber right. But he thought that cultural objects do not have an ontological status sufficiently independent of their underlying *psychological causes* to sustain Kroeber's attempt to part anthropology from psychology. For Haeberlin, "The phenomena with which *Völkerpsychologie* deals are objects to be explained (*explananda*) by *explanantia* that lie below the cultural level itself" (Diriwächter 2004, 96). These *explanantia* alone, Haeberlin argued, can make the study of historical processes scientific. He implied that objectivism of this kind makes Kroeber's main point without entangling itself in his errors. If, like Boas, Kroeber wanted to argued that there are no significant differences in intelligence between advanced and less advanced but equally encultured peoples he should have taken an objective, scientific path to prove it (Brock 1992, 210).

In replying, Kroeber deployed his favorite line of argument: Causal-psychological explanations fail to hit their target because they inevitably misdescribe cultural *explananda.* Perhaps because the young man died before the publication of his response, Kroeber refrained from polemics by eulogizing his "high qualities of mind and lovable personality" before launching into a barrage of demarcating:

> Psychology may share with the biologist and chemists a conviction that consciousness rests absolutely on an organic basis and through this on an inorganic basis. But as a psychologist his business is the determination of the manifestations and processes of consciousness as consciousness.
>
> (Kroeber 1918, 635)

Kroeber conceded there may be, as Wundt and Haeberlin alleged, "a social as distinct from an individual psychology that explains in mechanistic terms the [causal] basis of cultural history." But the "depiction of its phenomena," Kroeber says, "is left to cultural history," that is, anthropology (Kroeber 1918, 636). When this fact is ignored or violated mechanistic explanations that belong to the "inferior [=lower] science of psychology" in either its individual or its social dimension distort what is being "depicted." Kroeber pointed to his *bête noirs*, eugenicists and scientific racists, as making this very mistake:

> In spite of [its] solid foundation in the facts and laboratory methods and [its] successful interpretations of genetic and statistical investigation [eugenics] is a program that short circuits itself in proposing to attain social ends by

> organic means.... [So too is] the opinion that races of men differ as potential factors of social effects.
>
> (Kroeber 1918, 647)

Why, Kroeber asked, is so an egregious mistake so common? His reason was that the "depictive" side of physics is so weak and its epistemic reputation so high that scientists overlook the very distinction between depiction and explanation, especially in recent efforts to find methodological and conceptual bases for the human sciences. Spencer's, Comte's, and Ward's conceptions of sociology showed the fallacy at work when they used the term "force" to describe phenomena constituted by intersubjectively interpretable intentional and aspirational meanings. "Social forces" was a conceptual non-starter.

Most of Kroeber's colleagues argued that he was too categorical in excluding psychology from anthropology. They regretted his bracketing of psychology not, however, because they were as enamored of social psychology as Haeberlin – they, too, were cultural descriptivists – but because they thought individual psychology played an important role in ethnographic inquiry itself. Individual psychology explained the origins of innovations within what would otherwise be or be portrayed as sclerotic folkways. Psychology, Kroeber's critics implied, did not disrupt cultural facts from below, as it did in Wundt and Haeberlin. It entered into cultural "depictions" themselves by giving intentional agency to a culture's movers and shakers.

Sapir, for example, maintained that one could not possibly grasp the legal culture of New Orleans without understanding the extraordinary mind of the individual who was Napoleon (Sapir 1917, 44). Bamboozled by his own quasi-nominalized term "*the* superorganic," Sapir argued, Kroeber reified culture into a collective object. "If I understand him rightly," he wrote, "he predicates a certain social 'force' whose gradual unfolding is manifested in the sequence of socially significant phenomena we call history" (Sapir 1917, 443, showing in its use of "force" the very fallacy to which Kroeber objected as well as a gratuitous imputation to him of Hegel-like historical development). Many of Kroeber's favorite examples of cultural facts, Sapir said, minimize the individual in history because they are cases of simultaneous discovery by separate individuals in the sciences, technology, and arts. Such discoveries, as well as the gradual, transformational change in social practices that pervaded cultural life, did not belie the fact that they were products of individual consciousness. Individuals may be products of their culture, but that doesn't mean they and their minds weren't individuals. Goldenweiser agreed. "The civilizational stream," he wrote, "is not merely carried but is also unrelentingly fed by its component individuals (Goldenweiser 1917, 449). But he went even further than Sapir, seeing in Kroeber's supposed treatment of culture an ontological object that exerts "force" on individuals a form of "cultural or civilizational determinism" in which "events in history occur when they must occur" (Goldenweiser 1917, 448).

All this was far off the mark. Notwithstanding, the imputation that Kroeber made the superorganic into a metaphysical reality dogged him throughout, and after, his life. Even Boas, who initially took no part in the piling on, helped circulate what became a stock objection. "It seems hardly necessary," he wrote, "to consider culture a mystic entity that exists outside the society of its individual carriers" (Boas 1932, 245). By the 1950s, David Bidney was claiming that Kroeber had "converted an epistemic or methodological abstraction into a distinct ontological entity, which he understood as an independent emergent level of reality, no longer subject to natural selection and the laws of organic evolution" (Bidney 1953, 51; also see White 1969, 95). This line of argument had by then turned into the opposite of what Kroeber claimed, showing how commonplace the accusation had become. Had not Kroeber himself charged Comte, Spencer, and Ward with thinking of collective forces as impinging on individuals? Was he not defending the contingency of the historical against determinism? How could he possibly be making cultures quasi-substances? His rejoinders, however, had little effect. We can already feel his frustration in his complaint to Sapir that all he was doing was sticking up for the discipline:

> I don't give a red cent whether cultural phenomena have a reality of their own, as long as we treat them as if they had. You do, most of us do largely, but most of [us also] hang back and fear to avow it and let geographers and biologists … walk over us. If we're doing anything right, it deserves a place in the world. Let's take it, instead of being put in a corner. That's not metaphysics: it's blowing your own horn.
>
> (Kroeber and Sapir, no date, probably November, 1917, G 244 in Golla 1984)

To some degree this impasse was rooted in the tin ear of many of Kroeber's colleagues to his effort to identify each field's distinctive conceptual and methodological conditions of inquiry. His colleagues could not imagine that the same objects could be treated differently by different disciplines, even though he could not have been clearer on the point:

> Mechanistic science has accomplished wonders in a brief space by adhering ever more rigidly to its own peculiar methods and allowing no limits to be set to its application of these methods. Yet that a tool has proved its service for a purpose does not affect the value of other purposes or the utility of other tools for other purposes.… The applicability of science to any and all domains of human cognizance must be expressly affirmed. But the same phenomenon can after all be viewed with different ends.
>
> (Kroeber 1917, 207–208)

Kroeber's critics missed the point because they did not adopt a critical perspective in Kant's sense. They treated as first-order ontological commitments what were intended to be discipline-demarcating, second-order epistemological statements about "the conditions of the possibility" of pursuing this or that form of inquiry as (a) science. Understandably, Kroeber stuck to his guns. Not to do so, he believed, would endanger the autonomy of his increasingly flourishing field.

These meta-issues raised their head again in the 1950s when the culture-and-personality movement swept through postwar American anthropology, giving rise to well-funded initiatives aimed at producing a unified social science that would integrate psychology, sociology, and anthropology's culture concept, in part by jettisoning physical anthropology. Harvard's Department of Social Relations is an example of the era's quest for a unified social science that would legitimate the important role of the social, and not just the natural, sciences in guiding postwar society (Gilkeson 2010; see Chapter 5 of this book). Clifford Geertz, a graduate student in the department's early days, characterized Social Relations as an "experiment" in which "cultural anthropology was conjoined not with archeology and physical anthropology, as was and unfortunately still is normally the case, but with psychology and sociology" (Geertz 2000, 7).

It might seem that Kroeber himself was complicit in this enterprise. After retiring from Berkeley in 1946, he spent a year at Harvard working with Clyde Kluckholn, who had joined the Social Relations faculty (Gilkeson 2010, 177).[7] With Geertz's help, they reviewed, categorized, and critiqued every definition of culture they could find in order to formulate the most adequate (Kroeber and Kluckhohn 1952, v). Their aim was to protect anthropology's claim to autonomy from being submerged in and subverted by a new super-field. Accordingly, they excluded social-psychological definitions and refloated Kroeber's, into whose sub-clauses they clumsily inserted implied rejections of a raft of views they examined in the course of their dialectical survey:

> Culture consists of patterns, explicit and implicit, of and for behavior acquired and transmitted by symbols, constituting the distinctive achievement of human groups, including their embodiment in artifacts; the essential core of culture consists of traditional (i.e. historically derived and selected) ideas and especially their attached values; culture systems may, on the one hand, be considered as products of action, on the other as conditioning elements of further action.
>
> (Kroeber and Kluckhohn 1952, 357)

Rather than integrating anthropology with the other social sciences, Kroeber was shoring up anthropology's boundaries as he had conceived them at least since 1917 by using this definition to deconstruct the ill-formed questions preoccupying the culture-and-personality movement and the Social Relations project. An example is asking at what point in its development a child becomes a Zuni or a Samoan or an American and how, once acquired, that identity affects their adult lives. The question, he said, is wrongly posed:

> Human beings *are always culturalized.* That is, they are culturally determined – and heavily determined – by the time they reach the age at which they become potential causes of culture. What is therefore operative is a

> powerful system of circular causality. The human beings who influence culture and make new culture are themselves molded; and they are molded through the intervention of other men who are culturalized and thus products of previous culture. So it is clear that, while human beings are always the *immediate* causes of cultural events, these human causes are themselves the result of antecedent culture situations, having been fitted to the existing cultural forms which they encounter.
>
> (Kroeber 1949b, 193, our italics)

This answer was aimed at sociologists and psychologists, but it also contains a good reply to the old accusations of his friends. If they were anthropological enough to recognize that cultures do not press down like forces on already individuated humans, as they generally were, they would appreciate that culture guides development by a process of "circular causation" in which we become individuated. They would also realize that Kroeber's superorganic demarcation never called into question the existence of geniuses. Admittedly, there is no "Samoan Mozart" in the sense of a little Mozart lurking within a Samoan musician, as an oft-reiterated thought experiment seems to pose the question. Genius, a psychological term, means "potential, not realized genius" when it is treated anthropologically. From that disciplinary perspective it is a "cultural flowering" that allows potential genius to express itself. Therefore, "something in the wave-like character of culture growths [not actualized individual psychology or Sapir's simultaneous discovery] is at the bottom of otherwise unexplainable clusterings of genius" (Kroeber 1946, 15).

Initially, Kroeber was unsuccessful in derailing the Social Relations train. The integration of social scientific fields and expulsion of physical anthropology he feared was institutionalized at Harvard almost as soon as he decamped to Columbia in 1947. Under the dispensation of the sociologist Talcott Parsons, anthropology's role in a putative unified social science was confined to dealing with "ideas, values, and symbols," leaving the rest to sociology and social psychology (Gilkeson 2010, 175). In the end, however, it was four-field anthropology that survived and Social Relations that fell apart. It survived in part because Geertz nicely solved the "What is it?" question about culture by using Ludwig Wittgenstein's notion of shared language (the impossibility of a private language) to define the "forms of life" that constitute cultures (Geertz 1973). The irony is that Geertz himself was sufficiently imprinted with the Social Relations mystique to regard it as "unfortunate" that ethnography continued to be locked up with physical or biological anthropology, with which he believed it had little in common (Geertz 2000, 7). Notwithstanding, reaffirmation of the four-field arrangement was in part due to Geertz's success in reframing Kroeber's definition of culture in ways that immunized anthropology against psychologism.

The limits of expansion – anthropology is not a natural science: 1944–1960

Kroeber objected not only to allowing psychology to disrupt anthropology's unique mission, but just as strenuously to it providing itself with biological foundations. This, too, undermined its autonomy. His expulsion of biological heredity in the 1910s was the centerpiece of this insistence. Kroeber spotted the threat in the functionalist trend that supplied the wings for anthropology's postwar territorial expansion, according to which each and every item in a culture must have a function, as in organisms. But he spotted it even more in Leslie White's grounding of the culture concept in energetic-ecological processes, which are not just biological but biophysical, since they are applications of the laws of energetics and thermodynamics. White used this conceptual apparatus to explain cultural evolution. "I like to regard revolution in culture," White said, "from the standpoint of the thermodynamic nature of sociocultural systems," such as the agricultural revolution of about 10,000 years ago and the extraction of fossil fuels that began only recently (White, in Tax and Callendar 1960, III, 229). Another bridge too far, Kroeber judged. He accepted a sober notion of cultural evolution, but remained wary of formulations of this idea that flirted with reducing the social to the natural sciences.

Kroeber's worry went back to a rather unseemly argument with Boas about natural scientific methods in anthropology. Kroeber's reliance on qualitative-interpretive and Boas's on quantitative-statistical methods reflected their different ways of appropriating neo-Kantian ideas about the laws that both of them presumed anthropology was working toward, if only in eschatological time. The argument arose when Kroeber took exception to his mentor's way of dealing with the subject of his own dissertation, artistic styles. The announced goal of Boas's 1927 book *Primitive Art* was "to determine the dynamic conditions under which art styles grow up," diffuse, and change (Boas 1927, 7). The problem was complicated, Boas argued, because anthropologists could not usually apprehend gapless sequential changes in a style and so were forced to use discontinuous distributions in geographical space as proxies for the continuous process presumed to underlie their diffusion. Boas believed this procedure is reliable only to the extent that "the more limited the distribution of a cultural trait in space … the more we can presume that it is a comparatively recent development in time" (Boas 1927, 5). This principle is no stronger than a presumption, however, and so was hedged about by Boas's customary methodological caveats (Chapter 2 of this book). For this reason, Boas found "quite untenable" Kroeber's elevation of this presumption to a "general principle" according to which artistic styles spread out from culture centers that Kroeber confidently used the principle to identify (Boas 1927, 6). In an echo of the argument he had leveled at Mason half a century earlier, Boas claimed that diffusion could not be a basis for claims about chronological sequence because "the converse is often true" (Boas 1927, 6; Chapter 2 of this book). A recent trait could be diffused widely and *contra* Kroeber the geographical center of an artistic style might not be its actual point of origin.

Kroeber's review called Boas's book the "soundest, most penetrating, and probably the most comprehensive work existing on primitive art" (Kroeber 1929, 139). However, he cautioned that it was not actually an example of "the historical method of which Dr. Boas has so often been considered the avowed and leading exponent" (Kroeber 1929, 139). He conceded that Boas recognized that phenomena as complex as cultural processes undergo historical development. But, he charged, Boas did more than merely caution against premature diachronic reconstructions of their origin and fate. He rejected the very idea. Kroeber thought he did so because his mentor remained under the spell of the methodological presuppositions of the physical sciences in which he was trained (Chapter 2 of this book). The ideal of airtight inferences from a flood of empirical data that are supposed to speak for and aggregate themselves, Kroeber says, "is most easily understood as the deep-seated distrust of a mind schooled in the approach of the inorganic exact sciences" (Kroeber 1935, 545). One formed by the norms of laboratory science cannot readily see how value-laden objects of inquiry reach out, as it were, to sense-making by value-oriented interpreters to produce diachronic "depictions" that are empirically trustworthy if they are drawn from as much empirical data as Kroeber generally provided. According to Kroeber, Boas's "trained incapacity," as Kenneth Burke would have labeled it, reflected the natural scientist's habit of breaking data down more than linking them up:

> When the web of the space and time relations of phenomena as they occur given in nature are torn apart for the examination of dynamic elements as such, the approach, whether apparatus and experiment are used or not, is that of the laboratory instead of that of nature. This procedure may be "scientific" to a higher degree, [but] it is not history or natural history.
>
> (Kroeber 1929, 140)

Kroeber wrote to Boas directly, telling him that "to judge historical efforts in anthropology by [natural] scientific standards" was an inappropriate, overly scrupulous way of treating its subject matter (Kroeber to Boas, August 28, 1935 in Boas 1972). Perhaps remembering his exchanges with Haeberlin, he charged his mentor not just with physics-envy, but also with fraternizing with the psychologizing enemy. "It would be misleading to consider a non-historical method essentially allied to that of Wundt's [psychologism] historical merely because it recognizes the historical complexity of cultural phenomena" (Kroeber 1929, 140). Once one sees experience and inquiry as value-laden in Rickert's sense one can work at finding "a nexus among phenomena" by building context- and data-rich comparative portraits of cultural processes of meaning making that sometimes yield reliable diachronic histories (Kroeber 1935, 546):

> Every anthropologist or historian concerned with culture realizes that cultural situations make more sense, reveal more meaning, in proportion as we know more of their cultural antecedents, or, generically, more total cultural

> context. In other words, cultural forms or patterns gain in intelligibility when they are set in relation to other cultural patterns.
>
> (Kroeber 1948b, 411)

Since he thought that is just what Boas's scientism blocked, we find Kroeber, even long after his teacher died in 1942, psychoanalyzing him "as having some strange parsimony, austerity, or inner compulsion" that "made him … chary of adding interpretations. My best guess is that it was a perverse dread of unscientific subjectivity" (Kroeber 1956, 152).

Predictably, Boas was upset, sensing that Kroeber's lax standards were endangering anthropology's scientific aspirations. It was only after Kroeber took exception to some points he had made in his 1935 AAA Presidential Address, however, that Boas defended himself publicly. Writing in *American Anthropologist*, he confessed that he did not recognize himself in Kroeber's portrait, but characteristically added that he might be wrong about himself (Boas 1936, 137). He agreed with Kroeber, he said, that the distinctive feature of the historical approach is its "endeavor at descriptive integration." He saw himself doing just this in virtually all of his studies, and not only in his early work on *The Central Eskimo*, which Kroeber had taken to be the sole, genuinely historical exception in Boas's work. The fact that he used statistics in constructing descriptions was a red herring, Boas said. It had nothing to do with scientism and everything to do with what Kroeber called "descriptive integration." Nor did Boas deny that full-fledged histories are possible. He merely insisted that a well-evidenced diachronic history is a by-product of descriptive saturation, not something at which an anthropologist can aim or achieve without lowering standards of what counts as success:

> We have descriptions of culture more or less adequately understood. These are valuable material. They yield, if done well, most illuminating material in regard to the working of the culture, by which I mean the life of the individual as controlled by culture and the effect of the individual upon culture. But they are not history. For historical interpretation the descriptive material has to be handled in other ways. For this work archaeological, biological, linguistic, and ethnographic comparisons furnish more or less adequate leads.
>
> (Boas 1936, 137)

Boas took this view because he was a less interpretive neo-Kantian than Kroeber and Rickert (Chapter 2 of this book). He had a balanced idea of how his four-fields inform one another. By contrast Kroeber thought of physical anthropology as a backdrop for interpretive archeology and ethnography. Still, his accusation that this difference meant that Boas was under the spell of scientism seems excessive.

The issue popped up in another form in Boas's 1935 article rebutting Kroeber. In it Boas devotes considerable time to rebuffing a suggestion by the South

African anthropologist A. W. Hoernle, who suggested that, feeling as he did about Kroeber, Boas should actively embrace the natural-scientific ideals to which Kroeber had accused him of clinging. He could do so by becoming an anthropological functionalist (Hoernle 1933). For functionalists, anthropology can be a science without being as epistemologically as subjective, perspectival, or interpretive as history. It treats cultures not as superorganisms, but as sufficiently like organisms to make it true that they come to be by acquiring an integrated set of parts, processes, and behaviors that enable them to maintain their identity over time and space, and sometimes to exert a concerted sway over other cultures. Functionalism's paradigm is Bronislaw Malinowski's *Argonauts of the Western Pacific* (Malinowski 1922). Hoernle read Boas's disparaging remarks on functionalism in his response to Kroeber as a "complete reversal" of what she believed had earlier been his more favorable attitude. She recommended that Boas return to the fold if he wanted to sustain his point against Kroeber (Hoernle 1933, 82). This is because functionalism's advantage over historicism is that it enables anthropologists to identify the role this or that characteristic plays in sustaining a culture without referring to the usually unrecoverable, lost-in-time story of how this or that characteristic got there – the point at issue between Kroeber and Boas – and without placing an undue burden on the psychology of individuals to internalize, modify, and pass on a set of ideas. According to functionalists, if you want to find a cultural function pull a part out and see what happens to the whole. This is what scientific experimentalism requires. Boas, however, politely rejected Hoernle's offer and instead expressed solidarity with Kroeber's historicism. Somewhat ungraciously, Kroeber then appropriated Hoernle's argument to suggest that in fact Boas "was a functionalist, in that his prime interest lay in structural interrelations, change, and process even before Radcliffe-Brown or Malinowski had written a line," and so was a supporter of strict science over historical contingency (Kroeber 1935, 541).

As in functional biology, the history of which stretches back in the medical tradition to antiquity, functionalism in the social sciences introduces the notion of homeostatic self-maintenance and with it a degree of goal-oriented teleology (sometimes called "teleonomy") that retains the idea of scientific lawfulness while at the same time freeing anthropology from its nineteenth century flirtation with impinging mechanistic forces of the sort that Kroeber disdained. The notion that, merely by existing, cultures can be assumed to get and have the functional and goal-oriented characteristics they need carries with it a suspiciously Panglossian presumption that everything works for the best in all actual and perhaps even possible worlds. Its attractions to anthropologists – the postwar "culture and personality" movement, for example – would no doubt have been diminished were it not for the fact that its chief advocate, Malinowski, happened to be a very great ethnographer.

Kroeber believed that functions do evolve in cultures as well as organisms. But he was no friend of functional*ism*, describing it as "verbal wish-fulfillment, bred perhaps in the hope of promulgating a system that was all-embracing and all-explaining" (Kroeber 1949a, 318). He took even greater exception to Leslie White's effort to bring law-like evolutionism back into anthropology shorn of stadialist racism, eugenics, and Panglossian circular reasoning. White called on physical principles as deep as the Second Law of Thermodynamics to explain the pervasive functionality of cultures viewed as ecological systems. Most Boasians loathed his unapologetic declaration that his theory of progressive cultural evolution was "substantially the same as that advanced by Lewis H. Morgan and E. B. Tylor many decades ago" (White 1943, 351). Those nineteenth century anthropologists made some mistakes about race, White conceded, but "a mistake made in the use of a tool does not render the tool worthless" (White 1938, 387). Kroeber, suspicious of disclaimers like this, deployed his stock arguments about anthropology's proper level of description to dismantle what he saw as a Trojan horse.

Boas himself greeted White's initial articulation of his views at a 1929 AAA gathering with stony silence. Henceforward White's mode of addressing him and his students was polemical. He attacked the Boasians as a cabal of worshipers at the shrine of a teacher who enforced a naïve historicist orthodoxy on anthropology. In response, Boasians tended to regard White's views as the ravings of a slightly deranged crank (Peace 2004, 106–116). This reaction was not entirely groundless. White asserted that their supposedly high scientific standards meant that in practice Boasians turned up nothing but an endless chaotic welter of particulars with little or no general significance. He alleged that this had led them to deny biological evolution, which actually exhibits, he said, plenty of law-governed pattern and direction.[8] With even less plausibility he accused them of repudiating science itself, which he said had already discovered the relevant generalizations governing cultural evolution whose very possibility Boas and his school irrationally refused to accept. "The anti-evolutionist outlook and not a little of the spirit of Wm. Jennings Bryan," White claimed, "lives on among the disciples of Franz Boas" (White 1945, 247).

Kroeber was spared White's fiercest criticisms because in White's eyes he had helped him press his own charge that Boas's overly lofty empiricist standards made it impossible to connect up the flotsam and jetsam of cultural history. Both were firmly committed to culture as a freestanding object of anthropological inquiry that must not be undermined by psychology or sociology. White objected to little in Kroeber and Kluckhohn's definition of culture. His concern was with *explaining* culture as they defined it. This he did in evolutionary terms. Kroeber, for his part, accepted the notion of "cultural evolution," but was at odds with the meanings White assigned to both "evolution" and "history."

White made it easy on himself by characterizing the Boasians as radical historicists according to whom history is constituted of unique, unrepeatable spatiotemporal events that resist any subsuming under generalizations. He interpreted Boas's skepticism about how much sense empirical methods can make of such a buzzing booming confusion as a virtual confession of historicism so construed. He was even more dismissive about whether the problem of historical generalization can ever be solved by historicists. "No amount of mere accumulation of facts will ever produce understanding," he wrote, "at least in the form of basic principles or generalizations of science" (White 1946, 84). Nor did he think Kroeber's interpretive variant on the theme was any more promising, a judgment he passed on to his student Marvin Harris, who says that Boas and Kroeber only imagined themselves as "at opposite sides of the history-science dichotomy" (Harris 1968, 321). White's point was not to contest what he took to be Boas's and Kroeber's view of historical events, whether natural or cultural, as a welter of particulars. It was to contrast history with evolution, which is "concerned with *classes* of events independent of a specific time and place," not with the *unclassified* particulars on which narratives impose fictive meanings (White 1938, 237; 1945, 230, our italics). He claimed that by consigning culture to the sphere of history Boas and the Boasians, including Kroeber, failed to see culture as evolutionary and *a fortiori* to see that cultural evolution is governed by knowable, even known, scientific laws.

White asserted that anthropologists actually know "more about cultural evolution than the biologist, even today, knows about biological evolution.... We know not only how culture evolves, but why, as well." We know that "the urge, inherent in all living species to live, to make life more secure, more rich, more full, to insure the perpetuation of the species" expresses

> the law of cultural evolution: Culture develops when the amount of energy harnessed by man per capita per year is increased; or as the efficiency of the technological means of putting this energy to work is increased; or, as both factors are simultaneously increased.
>
> (White 1943, 338–339)

This sweeping claim evaded the emptiness of functionalist teleology without denying temporality by substituting the teleonomic or end-oriented working out of natural laws such as the Second Law of Thermodynamics (White 1943, 339; on "teleonomy," Pittendrigh 1958, 394). Accordingly, White downplayed turning points in the history of culture that were important to Kroeber because they enhanced the cultural transmission of knowledge, such as the invention and improvement of writing and the emergence of ethical religions. Instead, he restricted the decisive moments in cultural evolution to innovations in harnessing energy to perform work more efficiently, such as the agricultural revolution of 10,000 years ago and the fossil-fuel driven mechanization of modern times (White in Tax and Callendar 1960, III, 229).

The issues raised by White pervade and perhaps distort Kroeber's *magnum opus*, *Configurations of Cultural Growth* (1944). After describing dozens and dozens of cultures he concludes anticlimactically that his book is no more than "a mass of data … that is essentially descriptive." There is, he added,

> no evidence of any true law in the phenomena dealt with by those convinced that human culture evolves: cyclical, regularly repetitive, or necessary. There is nothing to show either that every culture must develop patterns

> within which a fluorescence of quality is possible, or that, having once so flowered, it must wither without chance of revival.
>
> (Kroeber 1944, 761)

Graduate students who waded through all this data only to arrive at this downbeat ending were sometimes told that his book was a "magnificent failure" (Darnell 2001, 86). But if Kroeber had anticipated or arrived at a more satisfying result he would in his own estimation have betrayed his cultural "depictions" and his calling as an anthropologist. Repeating a point that had served him as a pole star from the beginning of his career, he declared at the outset of *Configurations*:

> I deliberately refrain from any ultimate explanation. An adequate description must precede any attempt at explanation because the phenomena are cultural and their first understanding must be in cultural terms: how they actually behave culturally; how far they are historically alike or unlike; whether there is a type of events underlying the several cultural physiognomies. *At present*, any explanation can hardly be more than descriptive.
>
> (Kroeber 1944, 19, our italics)

These remarks were probably directed at White. Kroeber drew a distinction within the notion of history with which Boas may not have been comfortable, but over which Kroeber thought White was running roughshod. Cultural history provides stabilizing, knowledge-enhancing background conditions for the narratological work of historians in the ordinary sense. This distinction reflects the proper division of labor and demarcation between anthropologists and historians. Cultural history can be described as evolutionary insofar as it deals with innovations such as the spread of writing and literacy, which affect the conditions for cultural transmission that ensued from them. But this, Kroeber insisted, does not mean that cultural evolution is not historical. Kroeber's theory of cultural history was predicated, like Boas's, on an analogy with Darwinian natural history, according to which cultural functions arise from beginnings just as accidental as those from which organic adaptations are subsequently shaped by natural selection (Chapter 2 of this book). As they enter into the work of culture-building, accidental beginnings make possible the contingency-cum-intelligibility of history narrowly considered:

> All history – whatever the field – worth its salt does deal with relations, with functions, with meanings. It certainly is not a tracing of the wanderings of detached and unrelated items through time and space, or … an arid roster of names, dates, and places.
>
> (Kroeber 1946, 2)

In contrast to the radical, anti-epistemic, flotsam-and-jetsam sort of historicism that White tried to pin on Boas and the Boasians, the historicism of cultural history as Kroeber conceived it retains the inquirer's ability to make empirical sense of contingent practices whose functions and meanings are as dependent on the circumstances from they came to be as biological adaptations are in Darwinian natural history. In this way his historicism contrasts with functionalism, which, like aprioristic adaptation*ism* in evolutionary biology, presupposes some sort of magical and timeless guarantee that cultures will always have the traits they need. Despite their differences, this principle put both Boas and Kroeber at odds with White's view of cultural evolution as the necessary working out of natural laws. Kroeber's painstaking interpretive and comparative studies of cultural centers and cultural diffusion were conducted in the spirit of contingent events that sometime get meaningfully connected – and sometimes don't.

The underlying weakness of White's conception of evolution is that it failed to honor the contingencies of both natural and cultural history because it remained transformational rather than variational, to use a contrast drawn by Levins and Lewontin (1985, 85–86; Fracchia and Lewontin 1999; Chapter 7 of this book). Like Spencer's, Haeckel's, and Lewis Morgan's, White's notion of evolution was modeled on the law-like unfolding of embryos. Kroeber correctly noted that White's idea of cultural evolution was "an unfolding of imminences" in accord with the old nineteenth century conception of evolution as development or ontogeny writ large (Kroeber 1946, 9). In this sense, Kroeber judged White's thinking pre-Darwinian, pre-Weismannian, and out of touch with developments in the evolutionary science in whose flag he so ostentatiously wrapped himself (Ingold 1986, 82–83).

An example of Kroeber's approach is his account of the history of writing, which he presented as cumulating in a way that shows its initial appearance and subsequent evolution as contingent, culturally pluralistic, and distinctively historical. In a revision of his textbook that had White in its gunsight, Kroeber argued that writing spread not because it was an inevitable unfolding stage of human development, but because of its amplifying importance for the transmission of cultural knowledge (Kroeber 1948a). The very diversity of writing systems shows this. This degree of diversity would not exist if writing systems were "explainable … by showing that they represent and express certain stages in a common developmental process," as White claimed (White 1945, 240). Why should the same thing not be true of the agricultural and industrial revolutions? They, too, are events that would be misdescribed if they were portrayed as having to happen necessarily rather than for contingent, but consequential and widely diffused reasons.

In 1959, the year before his death in Paris, where he was attending a conference, Kroeber found himself on a discussion panel with White at the University of Chicago's Darwin Centennial Celebration. Chaired by Kluckhohn, the panel's task was to discuss a series of propositions on "social and cultural evolution" that the organizers had assembled. The propositions were precipitates of a decade-long effort by evolutionary biologists and anthropologists to assess the implications of the Modern Synthesis for anthropology (Chapter 5 of this book). Kroeber followed these efforts with growing enthusiasm. They were, he told members of the panel and the audience at the University of Chicago,

> building a new through-highway [freeway] bridge between biology and the psychology-anthropology-humanities inquiries which deal with culture. It is gratifying, and an assurance of permanence of the new connection, that this pushing out forward has come mainly from biology.... The older biology has been rescued from shipwreck by the new science of genetics that hit its stride two decades or so after the turn of the century, and began to realize its potential. A rapprochement was effected between it and evolutionary biology, which grew more intimate and drew in many other lines of biological research, from cytology and ecology to paleontology and taxonomy.
>
> (Kroeber 1955, 294)

Kroeber's hopes for aligning anthropology with the Modern Synthesis were grounded in his correct perception that advocates of the new population genetic evolutionary biology were not violating his demarcational strictures against biologizing anthropology any more than the anthropologists with whom they were interacting were anthropologizing biology. Instead, the propositions under discussion at Chicago in 1959 resonated with his idea of a superorganic relationship between the two sciences. This very idea had first been sketched by the geneticist Theodosius Dobzhansky and the anthropologist Ashley Montagu in a 1947 article proposing that culture arose among some hominids because of selection pressure for flexible behavior (Dobzhansky and Montagu 1947; Chapter 4 of this book). The theorems presented to the panel at the Darwin Centennial Celebration painted the human capacity for culture as a population-genetically based, species-wide biological adaptation that once it was up and running provided a platform for "much more rapid adaptation and more rapid evolutionary advances" through "cultural innovation and transmission." These "advances," "breakthroughs," or "revolutions," the propositions stated, are products of "cultural variation, selection, and retention." They are "essentially independent of genetic differences between human subgroups and races" (Tax and Callendar 1960, III, 210; page numbers in the following paragraphs refer to this volume).

The discussion of these propositions at the Chicago Darwin Centennial served as an occasion for renewing the old quarrel between Kroeber and White by raising the question of whose view is better supported by the new population-genetic evolutionary biology. Kluckhohn asked Kroeber to begin the discussion. The propositions at hand, Kroeber said, support the idea that "except for certain overtones of connotation" history and evolution mean the same thing, namely, "long term change." The new biology's Darwinian model of variation, selection, and retention puts the accent on "history" by redefining the concept of evolution in ways that stress its contingency and historicity (212). The purging of transformational models from biology makes it more difficult to retain them in cultural evolution. Expelling them is a good thing.

When White's turn came he countered by claiming that the propositions put the accent on the term "evolution." "Things that have been called history," he remarked, "are now appearing in the clothing and phraseology of cultural evolutionism" (234). White was attempting to take credit for this shift without asking himself whether his transformational conception of cultural evolution was consistent with the variational model that the thematic propositions endorsed. Kroeber took a degree of malicious pleasure in maneuvering him into repudiating his former view by imputing to him the new, historically contingent view of evolution that the texts under discussion promoted. "It turns out," he said, "that … Leslie White and I have been sleeping in the same bed for thirty years without knowing it" (235). Recognizing the trap, White shot back, "I do not think 'history 'and 'evolution' are synonyms" (236). Continuing his needling, Kroeber said, "I only just discovered that White and I were sleeping in the same bed for thirty years and now he says there were two beds" (236). The exchange ended when Kluckhohn called attention to its vulgarity. "Watch these metaphors, gentlemen," he said (237).

This was not a complete triumph for Kroeber. Even if anthropologists saw the light about cultural evolution, evolutionary biologists in 1959 did not. Throughout the 1950s and 1960s most devotees of the Modern Synthesis adopted interpretations of cultural evolution like that of Julian Huxley, who was also on the panel on social and cultural evolution at the Chicago celebration. The author of *Evolution: The Modern Synthesis* spoke with a certain authority when he said that there are higher grades in cultural as well as in biological evolution and that cultures, too, have ontogenies (224–225). He also said that he agreed with "Kroeber's idea that cumulative transmission of experience is a second method of inheritance acting like the inheritance of acquired characters" (213). Kroeber had said no such thing either on this or any other occasion. Nonetheless, so widespread became the view that cultural transmission is a second kind of heredity, even a superior kind in virtue of its greater fidelity, speed, and cumulativity, that it became commonplace to see in cultural evolution a more rapid, more adaptive, more progressive, more Lamarckian continuation of biological evolution, which in turn is a continuation of cosmological evolution. "Dual inheritance," as it came to be called, points to a succession of cultures that as they arise confer on humans increasing degrees of independence from environmental dependency (Ruse 1996, 450–455). It was not until the conceptual structure of the Modern Synthesis had been clarified by bitter debates over Behavior Genetics and Sociobiology in the 1970s and 1980s that Kroeber's resolute confining of heredity to biology and his demarcation of separate spheres for biology, anthropology, and psychology could be fully appreciated. Had he lived longer he would have seen in the "dual inheritance" of Sociobiology a violation of his superorganic line of demarcation. He might even have seen that a shadow of what he so consistently opposed also fell over an idea that he accepted, cultural evolution (Fracchia and Lewontin 1999).

Notes

1 Much later Kroeber's second wife Theodora cast Ishi in a rather James Fenimore Cooperish light as "the Last Wild Indian in North America" (T. Kroeber 1961).

2 G refers to numbers assigned to letters in the collected correspondence of Kroeber and Sapir in Golla 1984. They do not refer to the page numbers of Golla's volume.

3 This is not entirely true. Kroeber gave expert testimony to several commissions on land theft from California Indians (Stewart 1961).

4 Kroeber and Lowie were reading the shorter, more popular of Rickert's books on demarcation, *Kulturwissenschaft und Naturwissenschaft* (1899). His longer treatise was *Die Grenzen der naturwissenschaftlichen Begriffsbildung* (*The Boundaries of Natural Scientific Concept Formation*) (1896–1902).

5 The Wikipedia entry for "superorganism" confuses Kroeber's idea of the superorgan*ic* as the trans-biological with the superorgan*ism* concept developed by students of social insects (accessed August 20, 2014).

6 Kroeber probably had in mind the Protestant William Jennings Bryan and the Catholic hierarchy as opposing eugenics; "the professedly skeptical" probably refers to the newspaperman H. L. Mencken.

7 Kluckhohn may have thrown in with Human Relations because he believed that "courses of behavior … [are] determined [only] in part by culture" (Kluckhohn 1949, 170; Gilkeson 2010, 284). A little psychologizing snuck in.

8 The widely circulated view that Boas was anti-evolutionist originated in this accusation.

References

Alsberg, Carl L. 1936. "Alfred L. Kroeber: Personal Reminiscences." In *Essays in Anthropology: Presented to A. L. Kroeber in Celebration of His Sixtieth Birthday, June 11, 1936*, edited by Robert H. Lowie, xiii–xviii. Berkeley: University of California Press.

Barkow, Jerome H., Leda Cosmides and John Tooby, eds. 1992. *The Adapted Mind: Evolutionary Psychology and the Generation of Culture*. New York: Oxford University Press.

Barrett, Deborah and Charles Kurzman. 2004. "Globalizing Social Movement Theory: The Case of Eugenics." *Theory and Society* 33 (5): 487–527.

Bidney, David. *Theoretical Anthropology*. New York: Columbia University Press, 1953.

Boas, Franz. 1927. *Primitive Art*. Oslo: H. Aschehoug.

Boas, Franz. 1932. "The Aims of Anthropological Research." *Science* 76 (1983): 605–613.

Boas, Franz. 1936. "History and Science in Anthropology: A Reply." *American Anthropologist* 38 (1): 137–141.

Boas, Franz. 1972. *The Professional Correspondence of Franz Boas* [Microfilm edition]. Wilmington: Scholarly Resources.

Brock, Adrian. 1992. "Was Wundt a 'Nazi'? *Völkerpsychologie*, Racism and Anti-Semitism." *Theory & Psychology* 2 (2): 205–223.

Buckley, Thomas. 1996. " 'The Little History of Pitiful Events:' The Epistemological and Moral Contexts of Kroeber's Californian Ethnology." In *Volkgeist as Method and Ethic: Essays on Boasian Ethnography and the German Anthropological Tradition*, edited by George W. Stocking, 257–297. Madison: University of Wisconsin Press.

Daly, Martin and Margo Wilson. 1988. *Homicide*. New York: A. de Gruyter.

Danziger, Kurt. 1983. "Origins and Basic Principles of Wundt's *Völkerpsychologie*." *British Journal of Social Psychology* 22 (4): 303–313.

Darnell, Regna. 2001. *Invisible Genealogies: A History of Americanist Anthropology*. Lincoln: University of Nebraska Press.

Dikötter, Frank. 1998. "Race Culture: Recent Perspectives on the History of Eugenics." *American Historical Review* 103: 467–478.

Diriwächter, Rainer. 2004. "*Völkerpsychologie*: The Synthesis that Never Was." *Culture & Psychology* 10 (1): 85–109.
Dobzhansky, Theodosius and Ashley Montagu. 1947. "Natural Selection and the Mental Capacities of Mankind." *Science* 105*:* 587–590
Ellis, Lee. 1996. "A Discipline in Peril: Sociology's Future Hinges on Curing Its Biophobia." *American Sociologist* 27 (2): 21–41.
"Eugenics Called Snare and Joke." 27 February 1914. *New York Tribune* 27 February 1914.
Fracchia, J. and R. Lewontin. 1999. "Does Culture Evolve?" *History and Theory* 38(4): 52–78.
Geertz, Clifford. 1973. *The Interpretation of Cultures: Selected Essays*. New York: Basic.
Geertz, Clifford. 2000. *Available Light: Anthropological Reflections on Philosophical Topics*. Princeton: Princeton University Press.
Gieryn, Thomas F. 1999. *Cultural Boundaries of Science: Credibility on the Line*. Chicago: University of Chicago Press.
Gilkeson, John S. 2010. *Anthropologists and the Rediscovery of America*, 1886–1965. New York: Cambridge University Press.
Goldenweiser, Alexander. 1917. "The Autonomy of the Social." *American Anthropologist* 19 (3): 447–449.
Goldenweiser, Alexander. 1920. "A New Approach to History." *American Anthropologist* 22 (1): 26–47.
Golla, Victor. 1984. *The Sapir-Kroeber Correspondence: Letters between Edward Sapir and A. L. Kroeber, 1905–15*. Berkeley: Survey of California and Other Indian Languages, University of California, Berkeley.
Grant, Madison. 1916. *The Passing of the Great Race: Or the Racial Basis of European History*. New York: Scribner's.
Haeberlin, Herman K. 1915. "Anti-Professions: A Reply to Dr A. L. Kroeber." *American Anthropologist* 17 (4): 756–759.
Haeberlin, Herman K. 1916a. *The Idea of Fertilization in the Culture of the Pueblo Indians*. Lancaster: New Era Print.
Haeberlin, Herman K. 1916b. "The Theoretical Foundations of Wundt's Folk-Psychology." *Psychological Review* 23 (4): 279–302.
Harris, Marvin. 1968. *The Rise of Anthropological Theory: A History of Theories of Culture*. New York: Thomas Crowell.
Hoernle, A. W. 1933. "New Aims and Methods in Social Anthropology." *South African Journal of Science* 30: 74–92.
Ingold, Tim. 1986. *Evolution and Social Life*. Cambridge: Cambridge University Press.
Jackson Jr., John P. 2010. "Definitional Argument in Evolutionary Psychology and Cultural Anthropology." *Science in Context* 23(1), 121–150.
Kevles, Daniel J. 1985. *In the Name of Eugenics: Genetics and the Uses of Human Heredity*. New York: Knopf.
Klein, Julie Thompson. 1990. *Interdisciplinarity: History, Theory, and Practice*. Detroit: Wayne State University Press.
Kroeber, Alfred L. 1901. "Decorative Symbolism of the Arapaho." *American Anthropologist* 3: 308–336.
Kroeber, Alfred L. 1915. "Eighteen Professions." *American Anthropologist* 17: 283–288.
Kroeber, Alfred L. 1916 "Inheritance by Magic." *American Anthropologist* 18: 19–40.
Kroeber, Alfred L. 1917. "The Superorganic." *American Anthropologist* 19: 163–213.

Kroeber, Alfred L. 1918. "The Possibility of a Social Psychology." *The American Journal of Sociology* 23: 633–650.
Kroeber, Alfred L. 1923. *Anthropology*. New York: Harcourt, Brace.
Kroeber, Alfred L. 1928a. "The Anthropological Attitude." *American Mercury* 13: 490–496.
Kroeber, Alfred L. 1928b. "Sub-Human Culture Beginnings." *The Quarterly Review of Biology* 3 (3): 325–342.
Kroeber, Alfred L. 1929. "Review of *Primitive Art* by Franz Boas." *American Anthropologist* 31 (1): 138–140.
Kroeber, Alfred L. 1935. "History and Science in Anthropology." *American Anthropologist* 37 (4): 539–569.
Kroeber, Alfred L. 1939. *Cultural and Natural Areas of Native North America*. Berkeley: University of California Press.
Kroeber, Alfred L. 1944. *Configurations of Culture Growth*. Berkeley: University of California Press.
Kroeber, Alfred L. 1946. "History and Evolution." *Southwestern Journal of Anthropology* 2 (1): 1–15.
Kroeber, Alfred L. 1948a. *Anthropology: Race, Language, Culture, Psychology, Prehistory*. New York: Harcourt, Brace.
Kroeber, Alfred L. 1948b. "White's View of Culture." *American Anthropologist* 50 (3): 405–415.
Kroeber, Alfred L. 1949a. "An Authoritarian Panacea." *American Anthropologist* 51 (2): 318–320.
Kroeber, Alfred L. 1949b. "The Concept of Culture in Science." *The Journal of General Education* 3 (3): 182–196.
Kroeber, Alfred L. 1952. *The Nature of Culture*. Chicago: University of Chicago Press.
Kroeber, Alfred L. 1955. "History of Anthropological Thought." In *Yearbook of Anthropology, 1955*, edited by William L. Thomas, Jr., 293–311. New York: Wenner-Gren Foundation for Anthropological Research.
Kroeber, Alfred L. 1956. "The Place of Boas in Anthropology." *American Anthropologist* 58 (1): 151–159.
Kroeber, Alfred L. 1960. "Evolution, History, and Culture." In *The Evolution of Man*, edited by Sol Tax and Charles Callendar. Vol. 3, 1–16. Chicago: University of Chicago Press.
Kroeber, Alfred L. 1963. *Culture: A Critical Review of Concepts and Definitions*. New York: Vintage.
Kroeber, Alfred L. and Clyde Kluckhohn. 1952. *Culture: A Critical Review of Concepts and Definitions*. New York: Vintage.
Kroeber, T. 1970. *Alfred Kroeber: A Personal Configuration*. Berkeley: University of California Press.
Kronfeldner, Maria E. 2009. " 'If There is Nothing Beyond the Organic …': Heredity and Culture at the Boundaries of Anthropology." *N. T. M.* 17: 107–127.
Kronfeldner, Maria E. 2010. "Won't You Please Unite? Cultural Evolution and Kinds of Synthesis." In *The Hereditary Hourglass. Genetics and Epigenetics, 1868–2000*, edited by Ana Barahona, Edna Suarez-Diaz and Hans-Jorg Rheinberger, 111–125. Berlin: Max Plank Institute for the History of Science.
Le Bon, Gustave. 1912. *The Psychology of Peoples*. New York: G.E. Stechert.
Levins, Richard and Richard C. Lewontin. 1985. *The Dialectical Biologist*. Cambridge, MA: Harvard University Press.

Lindsay, Brendan C. 2012. *Murder State: California's Native American Genocide, 1846–1873*. Lincoln: University of Nebraska Press.

Lowie, Robert H. 1936. "Professional Appreciation." In *Essays in Anthropology: Presented to A. L. Kroeber in Celebration of His Sixtieth Birthday, June 11, 1936*, edited by Robert H. Lowie, xix–xxiii. Berkeley: University of California Press.

Lynd, Robert Staughton and Helen Merrell Lynd. 1929. *Middletown, a Study in Contemporary American Culture*. New York: Harcourt, Brace.

Lynd, Robert Staughton and Helen Merrell Lynd. 1937. *Middletown in Transition; a Study in Cultural Conflicts*. New York: Harcourt, Brace.

Madley, Benjamin. 2016. *An American Genocide: The United States and the California Indian Catastrophe, 1846–1873*. New Haven: Yale University Press.

Malinowski, Bronislaw. 1922. *Argonauts of the Western Pacific*. London: Routledge and Kegan Paul.

Mead, Margaret and Ruth Leah Bunzel. 1960. *The Golden Age of American Anthropology*. New York: Braziller.

Megill, Allan. 1989. "Recounting the Past: 'Description,' Explanation, and Narrative in Historiography." *American Historical Review* 94: 627–653.

Meloni, Maurizio. 2016. *Political Biology: Science and Social Values in Human Heredity from Eugenics to Epigenetics*. New York: Palgrave Macmillan.

Miller, Jay. 2007. *Regaining Dr. Herman Haeberlin: Early Anthropology and Museology in Puget Sound, 1916–17*. Seattle: Lushootseed.

Peace, William J. 2004. *Leslie A. White: Evolution and Revolution in Anthropology*. Lincoln: University of Nebraska Press.

Pittendrigh, C. 1958. "Adaptation, Natural Selection, and Behavior." In *Evolution and Behavior*, edited by Anne Roe and George Gaylord Simpson, 390–416. New Haven: Yale University Press.

Rickert, Heinrich. 1889. *Kulturwissenschaft und Naturwissenschaft*. Freiburg: B. Mohr.

Rickert, Heinrich. 1962. *Science and History: A Critique of Positivist Epistemology*. Translated by George Reisman. Princeton: Van Nostrand.

Ruse, Michael. 1996. *Monad to Man: The Concept of Progress in Evolutionary Biology*. Cambridge: Harvard University Press.

Sapir, Edward. 1917. "Do We Need a 'Superorganic'?" *American Anthropologist* 19 (3): 441–447.

Smith, Woodruff D. 1991. *Politics and the Sciences of Culture in Germany, 1840–1920*. New York: Oxford University Press.

Stewart, Omer. 1961. "Kroeber and the Indian Claims Commission Cases." *Kroeber Anthropological Society Papers* 25: 181–190.

Stocking, George W. 1968. *Race, Culture, and Evolution: Essays in the History of Anthropology*. Chicago: University of Chicago Press.

Tax, Sol and Charles Callendar. 1960. "Panel Three: Man as an Organism." In *Issues in Evolution* (vol. 3 of *Evolution after Darwin*, edited by Sol Tax and Charles Callendar), 145–174. Chicago: University of Chicago Press.

Taylor, Charles Alan. 1996. *Defining Science: A Rhetoric of Demarcation*. Madison: University of Wisconsin Press.

Ward, Lester Frank. 1891a. *Neo-Darwinism and Neo-Lamarckism*. Washington: Gedney & Roberts.

Ward, Lester Frank. 1891b. "Transmission of Culture." *The Forum* 11: 312–319.

Ward, Lester Frank. 1893. *The Psychic Factors of Civilization*. Boston: Ginn.

White, Leslie A. 1938. "Science is Sciencing." *Philosophy of Science* 5 (4): 369–389.

White, Leslie A. 1943. "Energy and the Evolution of Culture." *American Anthropologist* 45 (3): 335–356.

White, Leslie A. 1945. "History, Evolutionism, and Functionalism: Three Types of Interpretation of Culture." *Southwestern Journal of Anthropology* 1 (2): 221–248.

White, Leslie A. 1946. "Kroeber's *Configurations of Culture Growth*." *American Anthropologist* 48 (1): 78–93.

White, Leslie A. 1949. *The Science of Culture: A Study of Man and Civilization*. New York: Grove.

Wolf, Eric R. 2004. "Alfred L. Kroeber." In *Totems and Teachers: Key Figures in the History of Anthropology*, edited by Sydel Silverman, 27–47. 2nd edition. Walnut Creek: Alta Mira.

Wong, Wan-chi. 2009. "Retracing the Footsteps of Wilhelm Wundt: Explorations in the Disciplinary Frontiers of Psychology and in *Völkerpsychologie*." *History of Psychology* 12 (4): 229–265.

Wundt, Wilhelm. 1916. *Elements of Folk Psychology: Outlines of a Psychological History of the Development of Mankind*. Translated by Edward L. Schaub. London: Allen & Unwin.

Wundt, Wilhelm. 1900–1920. *Völkerpsychologie*, vols 1–10. Leipzig: Engelmann.

4 Theodosius Dobzhansky and the argument from definition

Dobzhansky, the Modern Synthesis, and human evolution

In 1927, Theodosius Dobzhansky, a twenty-seven-year-old Ukrainian-born Russian naturalist and geneticist, arrived in New York to apprentice himself to T. H. Morgan's fruit fly laboratory at Columbia University. He wanted to learn more about Morgan's method of mapping genes on chromosomes.[1] A year later Morgan moved his laboratory, including his ingenious research assistants Alfred Sturtevant and Calvin Bridges, to California Institute of Technology in Pasadena, where they founded its Division of Biology (Kohler 1994). Dobzhansky and his wife Natasha went too, probably having resolved already not to return to the Soviet Union if they could help it.

The only naturalist in a group of experimentalists, Dobzhansky formulated a research program that allowed him to explore evolutionary processes directly. He trapped samples of local fruit fly populations on the sides of California's mountains and subsequently further afield. Back in the laboratory and later on site, he used microscopy to correlate differences in specimens taken from different places, altitudes, or seasons with changes in the arrangement of light and dark bands on their chromosomes. With assists from Sturtevant, who made the first genetic map in 1913, and soon by mail from the mathematically adept geneticist Sewall Wright at the University of Chicago, he then inferred differences in the genetic composition of these local populations and occasionally their ecological causes. By combining field, laboratory, and statistical analysis in ways pioneered by the Russian geneticists who were his first mentors – his American colleagues saw him as an ambassador from "the Russian school" of genetics[2] – Dobzhansky established the geographic ranges of various species and subspecies of fruit flies. In a few cases he was able to catch evolution in the act of turning "geographical races" into genetically isolated species.

Dobzhansky focused on speciation. Of particular importance was his discovery that two strains of fruit fly *Drosophila pseudoobscura*, so morphologically identical that museum specialists classified them as the same species, are physiologically, because genetically, incapable of sustained interbreeding (Dobzhansky 1941, 1962–1963, 356; Dobzhansky to Ernst Mayr, December 15, 1970, Dobzhansky Papers). His discovery supported ongoing efforts to reform systematics by liberating it from the combined influence of amateur naturalists and museum-based systematists, the amateurs classifying species by recognizable traits, the professionals by static traits they presumed to be shielded from natural selection. The "New Systematists" of the 1930s and 1940s opposed both approaches. They demanded that classification be grounded in the real biological processes that constantly occur in spatially and temporally distributed populations of interbreeding organisms, preeminently adaptive natural selection.

"What is to be done with forms like the two 'races' of *D. pseudoobscura,*" asked Julian Huxley, Thomas Henry Huxley's grandson, specialist (to the extent that he specialized) in avian ethology, and advocate of the New Systematics in Great Britain? In spite of their physical similarity "their inter-sterility is of the same order of magnitude" that separates good species like *D. melanogaster* and *D. simulans* (Huxley 1940, 24, citing Muller 1939). His question was rhetorical. Like Dobzhansky, Huxley already knew what to do: Define species in a way that turns one of these look-alike races of *D. pseudoobscura* into the sibling species *D. persimilis* (Dobzhansky and Epling 1944). Such a definition looks upward at species from the perspective of geographical races that are becoming reproductively isolated, not down on them from the static perspective of the genera they partition. It does not use observed similarities and dissimilarities as decisive criteria for classification. After all, these exist only at an arbitrary moment in the churning evolutionary history of lineages. Trait markers are tools for tracking biological lineages undergoing racial dispersion and species integration.

Dobzhansky delivered a series of lectures at Columbia in 1936 in which he used the Russian-school's signature insight that a great deal of potentially selectable genetic variation exists in the recessives of diploid chromosomes to explain the evolution of adapted populations, geographical races, and full-blown, reproductively isolated species. Thanks to being laid up with a broken kneecap – he fell from his horse – he was quickly able to turn these lectures into one of the most influential books in the history of evolutionary biology, *Genetics and the Origin of Species* (Dobzhansky 1937; Dobzhansky to Dunn, September 24, 1937, Dobzhansky Papers). With this success under this belt, Dobzhansky returned to Columbia in 1940 as a full professor.

Like Darwin, whose *Origin of Species* served as a model for his own book, Dobzhansky had a genius for turning allies into friends. Among these were the avian systematist Ernst Mayr and the vertebrate paleontologist George Gaylord Simpson. In New York Dobzhansky began promoting the New Systematics in league with these curators at the American Museum of Natural History. As an expert in the biogeographical distribution and classification of birds, Mayr added fuel to Dobzhansky's fire by "demonstrat[ing] to him the magnificent geographic variation in South Sea Island birds" (Mayr 1980b, 420). Mayr was "delighted" with Dobzhansky's approach because it illuminated, among other things, a phenomenon that fascinated him: "ring species," a linked series of slightly genetically divergent populations that interbreed with their immediate neighbors until, when the ring is closed, two morphologically indiscernible species may be living in the same location but remain reproductively isolated. If the new approach was to be as revolutionary as Dobzhansky and Mayr believed it to be it would have to explain temporally successive and not just spatially dispersed, taxa, most of them extinct – a difference Dobzhansky compared to making movies rather than taking snapshots (Dobzhansky 1944, 255). This is where Simpson came in. He provided evidence for the hypothesis that higher taxa are long-term consequences of the same evolutionary factors that are at work below and at the species level (Simpson 1947; Dobzhansky 1937, 12). If Dobzhansky's key concept was race and Mayr's species, Simpson's was evolutionary grade.

In 1942, Columbia University Press published Mayr's Jesup lectures, *Systematics and the Origin of Species*, as a follow up to *Genetics and the Origin of Species.*[3] Simpson's *Tempo and Mode in Evolution* appeared in 1944. In 1950, Dobzhansky's protégé in California, W. Ledyard Stebbins, added the evolution of plants (Stebbins 1950; 1995, 11; Dobzhansky 1962–63; Smocovitis 2006, 26–27). Julian Huxley had already named the international movement of which these naturalists were part in his 1942 *Evolution: The Modern Synthesis.* By viewing Mendelian genetics and Darwinian natural selection as processes in dynamic populations, Huxley hoped to fulfill his grandfather's aspiration to synthesize biology's disparate fields around an evolutionary core (Huxley 1942, 1; Smocovitis 1996; Cain 2009). His New York colleagues would have preferred more on speciation and phylogenetic diversification and fewer proofs that species-marking traits previously thought have no adaptive significance are indeed adaptations.[4] Still, Dobzhansky, Simpson, and Mayr joined Huxley in championing the unification of biology around the banner of "population thinking" (Chapter 6 of this book). When Mayr complained to Dobzhansky about his (temporary) inability to secure funds for their proposed journal, *Evolution*, his friend rallied him: "I feel very strongly that the whole future of evolutionary biology not only in the USA but everywhere depends on the success or failure of this undertaking" (Dobzhansky to Mayr, July 15, 1946, Dobzhansky Papers).[5]

Scholars treat *Genetics and the Origin of Species* as the founding text of the Modern Synthesis and Dobzhansky as *primus inter pares* among its founders. In the evening of his life – he died of leukemia in 1975 – Mayr told him that as a "member of the Russian school" he had "brought true population thinking to genetics" and so had brought true genetics to the study of natural (not just laboratory) populations (Mayr to Dobzhansky, February 25, 1973, Dobzhansky Papers). With some false modesty, Dobzhansky recorded in his *Reminiscences* that all he had done was catalyze what was "already in the air" (Dobzhansky 1962–1963, 397–398). Truth be told, if anyone catalyzed the Modern Synthesis it was Dobzhansky's patron, colleague, chairman, co-author, close friend, and fellow geneticist Leslie C. Dunn, who took Morgan's place at Columbia after the latter decamped to California. It was Dunn who commissioned Dobzhansky's and the other Jesup lectures, arranged for Columbia University Press to publish them, secured Dobzhansky's appointment in his own department, and founded with him an Institute for the Study of Human Variation. Dobzhansky called Dunn, who died in 1974, "one of the best human beings whom I had the luck of meeting in my life" (Dobzhansky 1962–1963, 290; on Dunn's career, Gormley 2006).

From the start, Dunn and Dobzhansky shared a concern with the problems of human evolution. Dobzhansky discussed "the species problem" with Mayr, but wondered how to apply a population-genetic re-conception of race to our species with Dunn and the anthropologist Ashley Montagu. Human evolution had been animating Dobzhansky since his student days. He had hoped by studying ladybugs and fruit flies to say something of scientific and ethical value about "specifically human problems:" "If you wish a justification of biological science it is in what it says about man" (Dobzhansky 1962–1963, 244, 634; 1973b, ix). Without Dunn it is unlikely that Dobzhansky would have been allowed to indulge this ambition. Without Montagu, the most effective opponent of scientific racism since Boas, he would probably not have viewed racial issues as he did.

This chapter addresses how Dobzhansky's approach to evolutionary biology undergirded his ideas about human evolution, including race. The topic was already on the table when Dunn brought him back to Columbia. Dunn wanted to enlist his help revising the persistently best-selling, but dated and eugenics-pockmarked genetics textbook he had co-authored with Edmund Sinnott (Sinnott and Dunn 1925; Dunn to Dobzhansky, April 5, 1937, Dobzhansky Papers). Even before finishing the job, Dunn and Dobzhansky co-authored *Heredity, Race, and Society* (1946), a little book addressed to the postwar public with a view to showing that since traits segregate independently one cannot treat physical traits as signs of psychological abilities. On the contrary, Mendelian genetics supports cultural diversity and equality of the Boasian sort, with which Dunn, a prominent figure in faculty affairs at Columbia, had long been familiar and in which he, Montagu, and others instructed Dobzhansky. When the book failed to make an impression, Dobzhansky turned to Montagu to help him get out the word. A largely British-trained physical anthropologist, he finished his education with a Ph.D. at Columbia under Boas and his second-in-command, Ruth Benedict. He said that Boas's were

> by far the most important courses I had ever taken.... The benefits of Boas's teachings have been immense.... Boas's book, *The Mind of Primitive Man* ... and his course on 'race' that I attended in 1935 both had considerable influence on my own thinking.
>
> (Undated manuscript, Montagu Papers)

Dobzhansky turned to Montagu because his wartime tract *Man's Most Dangerous Myth: The Fallacy of Race* revealed a talent for communicating with the public that the geneticist was eager to imitate and enlist (Montagu 1942b; Dobzhansky to Montagu, January 4, 1943, Montagu Papers). In 1947, the pair offered readers of *Science* a scenario showing how from the rich pool of genetic diversity contained in wild populations natural selection evolves genetically based capacities for responding flexibly to environmental change. Such flexibility is exhibited by our species' capacity for cultural life; our genetically based openness to education and attendant ability to pass learned experience across generations makes us more rapidly responsive to environmental challenges than even the quickest shift in gene frequencies (Dobzhansky and Montagu 1947). Dunn, an accomplished scientific diplomat, helped ensure that this conjecture – much about it needed to be proved – retained its central place in UNESCO's *Second Statement on Race* after Montagu's *First Statement* ran into opposition from physical anthropologists (UNESCO 1951, Section 6; Gayon 2003; Brattain 2007). In the context of global decolonization, the revelation of the horrors of the Holocaust, and the stirrings of the Civil Rights Movement in America, the anti-racist consensus that Dunn, Dobzhansky, and their anthropological allies hoped to catalyze on biological grounds began to influence public discourse.

In 1962, after interacting closely with American anthropologists for over a decade, Dobzhansky set out his own interpretation of the UNESCO Statements in *Mankind Evolving* (see also Dobzhansky 1955a). His authorial voice positioned itself as occupying common-sense middle ground, like nature and the judicious democratic public itself, between two perverse extremes: genetic determinism, which confuses the heritability of genetic factors with the immutability of gene expressions, and an excessively culturalist view that "points to the utter insignificance of biological factors in any consideration of behavior variation" (Dobzhansky 1962a, 54). Dobzhansky used Kroeber's "superorganic" to find the vital center (Dobzhansky 1962a, 8, 20). "Culture," he wrote, "is man's most potent means of adaptation to his environment and genetically conditioned educability his most potent biological adaptation to his culture" (Dobzhansky 1962a, 264). There was, however, a difference. Kroeber did not deny that heritable differences between individuals and groups exist, only that it is anthropology's business to find them. Those who take themselves to have found these differences, he argued, invariably misdescribe the cultural context in which they are expressed (Chapter 3 of this book). Dobzhansky claimed that heritable differences between individuals and groups do exist. It's just that, since our capacity for culture is our most salient adaptation, these differences are too modulated by culture and too independent from other traits to carry the implications advocates of racial inequality and eugenics ascribe to them. His larger claim was that with their permeable, meritocratic social structures, tolerant values, and egalitarian cultural practices, not least free choice of marriage partners, liberal democratic regimes more or less automatically favor optimal matches between individual genetic propensities, the varying circumstances under which they are expressed, and socio-economic roles (Dobzhansky 1962a, 247–248).

Dobzhansky's aim in writing *Mankind Evolving* was to support social, political, and cultural practices that let nature do its work and avoid those that hinder it. Framing the issue around the unhappy political history of his homeland, he divided unhelpful institutions into "aristocratic," which confuse social standing with inherited superiority, and "utopian," which level social distinctions either on the incorrect assumption that humans do not differ at all by nature or by thinking that only in uniform social environments can genetic advantages and disadvantages reveal themselves. He argued that in liberal democratic societies only genetic diseases arising from double-dose lethal recessives and chromosomal disorders that cannot be treated by environmental modification should be objects of eugenic intervention. Everything else should be left as nearly as possible to informed individual choice.

Scholars differ about why Dobzhansky used the term "eugenics" at all in connection with practices that are close to present-day genetic counseling. Did his decision to do so reflect continuity with or departure from geneticists' pre-war "eugenicist consensus" (Dobzhansky 1962a, 332–333; Paul 1994; Beatty 1994)? We will add our own view. Still, close study of Dobzhansky's evolutionary theory, style of argumentation, and the postwar context is needed even to properly frame, let alone answer, this and related questions. So is refusal to separate the history of biology and anthropology. Our approach is to interpret *Mankind Evolving* as re-enacting Dobzhansky's debates with Montagu and the geneticist Hermann Muller. He cast both as avoiding extremes of genetic and cultural determinism, but as still failing to hit his happy medium, Montagu falling toward side of culture, Muller toward genetics.

Dobzhansky, it must be said, did not quite hit his happy medium. He left a hostage to fortune by intimating, and in his last decade asserting, that as equality of opportunity is achieved genetic causes and effects that are masked in aristocratic and utopian societies will become visible (Dobzhansky 1973a). To assess why he said this, and why his own theory precludes it, we must first appreciate how well he did in the rhetorical situations, scientific and political, with which he was confronted.

Definition in Dobzhansky

Scholars agree that *Genetics and the Origin of Species* drew hitherto contending field naturalists, laboratory experimentalists, and mathematical population geneticists toward what became the Modern Synthesis. It appealed to all three sources of evidence simultaneously and so made practitioners of each sub-discipline sense that its author was on their side (Allen 1994; Cecarrelli 2001). But in this seminal work and other professional writings, even the most technical, we find something more than finessing meanings that different communities of inquiry assign to key terms. We find a dialectical strategy of argumentation that leads readers, professional and lay alike, from the inadequacies of common sense "thingism" about individual organisms to the elevated perspective of dynamically changing populations.[6] From this perspective Dobzhansky offered definitions of key terms in evolutionary biology – population, race, species, and evolution itself – that to his mind resolve the contradictions, inadequacies, and antinomies of the overly concrete lowlands and enable readers to see and interpret otherwise obscure phenomena. His definitions are wings on which even lay students of biology could ascend from a world of fixed individual substances to one of dynamic processes in populations – and to a new appreciation of their democratic freedoms.

The higher perspective of population thinking first came into view when mathematically-minded geneticists working earlier in the century showed how

Mendel's laws of segregation, independent assortment, and dominance at each locus of each diploid chromosome look when they are expanded to an entire population of freely interbreeding organisms. Until they are perturbed by mutation, natural selection, migration, or genetic drift, the distribution of recessive and dominant genetic variants (alleles) in Mendelian populations will remain in a state of equilibrium. There will be twice as many chromosomal loci that contain a recessive and a dominant allele (heterozygotes, 2Aa) as loci with two dominants (AA homozygotes) or two recessives (aa homozygotes). Factors, agents, or forces disturbing this "Hardy-Weinberg" equilibrium (named after the mathematicians who derived it) are, of course, always in play. Still, this insight undercut Francis Galton's worries about a built-in tendency of good-making traits to regress to an undistinguished norm, thereby taking away a key assumption of early eugenicists.

Dobzhansky's own preoccupation was with how much genetic variation exists in populations. He argued that most "gene pools" – his term – contain any number of alleles that can slot into chromosomal loci and that the frequency of particular alleles waxes or wanes as natural selection keeps populations tuned to changing environments. To fix this picture in readers' minds, he defined populations that reproduce in accord with Mendel's laws – "Mendelian populations" – as "reproductive communities of individuals which share in a common gene pool"; evolution as "change in the genetic composition of Mendelian populations"; races as "Mendelian populations that differ in the frequencies of some gene or genes"; and species as "Mendelian populations that have become integrated into complexes within which interbreeding is possible, but between which it is limited or eliminated entirely" (Dobzhansky 1955b; for variant definitions, see Dobzhansky 1937, 11, 1951, 16, 138, 261; Dunn and Dobzhansky 1946, 118).[7]

Dobzhansky's characteristic figure of argumentation is the argument *to* and *from* definition. We might imagine that his markedly genocentric definitions commend themselves solely for reasons of theorectical consistency, reductionistic simplicity, and scientific unity. Science is always engaging in redefinition in this sense as it keeps facts in tune with lagging explanations and aging theories. If scientists did not explicitly assign meanings to the abstract terms named in their theories they could not move from sound premises to valid conclusions or allow other scientists to follow and assess their arguments. Keeping definitional gaps of this sort from becoming yawning chasms entangles scientists and their philosophical handmaidens in persistent disputes about whether the sometimes unobservable entities defined in scientific theories (genes, for example) identify the real natures of things that common sense language misses or instead have only instrumental, "pragmatic," or "operational" value in helping scientists keep track of and predict future configurations of their data. In this respect, Dobzhansky was as instinctively a scientific realist as, say, the physicist Ludwig Boltzmann, who killed himself when he could not get his colleagues to believe in the reality of atoms. From the heights toward which he led his readers he showed that biology must really be about large-scale, long-term temporal processes, not (only) about the seemingly stable and mostly visible entities millennia of pre-evolutionary thinking have deposited in ordinary language. We can see from the perspective of population thinking that organisms develop but don't evolve and, conversely, that populations evolve but don't develop, an insight that liberated the new evolutionary biology from the parallel between ontogeny and phylogeny on which nineteenth century stadial evolutionism was based.

Still, if we come to issues about definition solely from the perspective of the philosopher's insistence on necessary and sufficient conditions, we are likely to find Dobzhansky wanting. If Mendelian genetics reveals that racial essentialists bundle together too many separately varying traits to pick out entities that really exist in nature – as Dunn and Dobzhansky showed in their 1946 book – can following the track of a single genotype as it changes frequencies in this but not that nook or cranny of a Mendelian population pick out anything more than an investigator's interest in whether a "race" so defined might lead to speciation? Is change in gene frequencies really all there is to evolution? How, too, can Dobzhansky's stress on evolution as a population process be reconciled with his resolve to make species discrete entities (Gannett 2013)?

Dobzhansky's definitions acquire a different look, however, if we approach questions like these from the perspective of his passion for illuminating human affairs by viewing them in the light of evolution and viewing evolution from the perspective of population-level expansions of Mendel's laws. When human affairs are placed at the center of his concerns, his definitions appear not as efforts to reduce biology to genetics but as figures of argument in which genes stand in (as metonymies) for larger biological, social, and ideological processes that would otherwise be missed or misconstrued (Lewontin 2000, 7).

This point can be made more perspicuously by looking at definition from the perspective of rhetorical theory. In studying definition less as a component of an argument than as a strategy of argumentation in its own right, the rhetorical scholar Edward Schiappa contrasts ordinary definitional gaps with "definitional ruptures," in which in the course of offering and testing definitions "the process of defining itself become[s] an issue" (Schiappa 2003a, 8). In rhetorical situations in which definitional ruptures arise, going around and around about realism and anti-realism is a symptom of failure to appreciate the ethical, normative, and existential stakes of using a term in this or that way. Plato, for example, on whom Schiappa has also written, took himself to be living at a time when Athenian democracy was degenerating into a might-makes-right cult of force that was affecting language as much as it was corrupting politics (Schiappa 2003b). It is not a coincidence that, like his hero, role model, and mouthpiece Socrates, Plato raised questions about what is at stake in defining words like piety, justice, and knowledge in the course of vividly portraying a society on the verge of or already in the throes of definitional rupture. Anyone whose encounter with Plato's *Dialogues* is restricted to textbook philosophical questions about Platonic Forms will miss the urgency that attends his worries about the corruption of the Athens that his Socrates says gave him his very identity (Plato, *Crito*, 50d). The phrase "defining moment" retains something of the kind of urgency Schiappa ascribes to definitional ruptures.

A similar urgency is hard to miss in the published and archived texts that emerged from Dobzhansky's interactions with Dunn and Montagu. All three were aware that in the rhetorical situation posed by World War II failing to ask questions such as, "What is the purpose of defining or not defining evolution as change in gene frequencies?" or, "What, if anything, should we count as a race?" might well result in the revival after the war of the racism, imperialism, and eugenics that in their opinion led to the war in the first place. "Race" is a loaded term. Montagu followed Huxley, who in the 1930s worried about Hitler's rants about "the races of Europe," in proposing that scientists use "ethnic group" instead of "race" in public discussions of human diversity (Haddon and Huxley 1936; Huxley 1941; Montagu 1942b). "'Ethnic group,'" Montagu told Dobzhansky, "eliminates obfuscating emotional implications" and "acknowledges cultural factors in our species that 'race' obscures" (Montagu to Dobzhansky, May 23, 1944, Montagu Papers). But rather than seeing in this proposal a shrewd grasp of persuasion in mass society – a subject in which Montagu had instructed himself by studying communication-oriented semanticists such as Alfred Korzybski, I. A. Richards, and C. K. Ogden (Montagu to Dobzhansky, May 23, 1944, Montagu Papers; Montagu 1945) – Dobzhansky rebuffed it. "The propagandist trick of making people swallow something under a different name," he told Montagu, "will do nothing to deter racists like [Madison] Grant, who ever since Hitler blackened their name have been laying low and waiting for their day, which may come as soon as the war ends or even sooner." What was to keep a postwar racist from complaining that his or her ethnic group was being persecuted by another (Dobzhansky to Montagu, May 22, 1944, Montagu Papers)? Like Plato, Dobzhansky prescribed definitional cures for definitional ruptures. By re-defining "race" and intertwined terms in genetic terms he hoped to induce in readers a wholesale "frame-shifting" or "seeing as" that would lead them to understand biological processes as population genetics-informed evolutionary scientists do (Zarefsky 2006). They would then recognize their prejudices and repent of them.

Dobzhansky called for frame shifting in response to the threat of revived racism in postwar America, but also because of the rising influence in the Soviet Union of Lysenkoism, a version of Lamarck's theory of the heritability of acquired characteristics. If new traits could be induced and inherited directly, the Communist Party could at one and the same time burnish its scientific pretensions and increase agricultural yields, which the Soviet Union badly needed to do, without having to wait around for slow genetic changes. In 1948, Lysenkoism became official party and state doctrine. Stalin began persecuting Russia's highly developed community of geneticists, including some of Dobzhansky's friends and mentors. To be a Darwinian, especially a genetic Darwinian, was now to be an enemy of the Revolution and an agent of its Western foes. Accordingly, Dobzhansky and Dunn's efforts to discredit pseudo-scientific racism were complemented by their effort to stiffen the spine of left-leaning American colleagues who, in rightly stressing the role of culture in human differences, might fall prey to traces of Lamarckism. For this reason they insisted *as a matter of definition* that biological evolution occurs at the level of genes; it is *by definition* change in gene frequencies in Mendelian populations. To be sure, the phenotypic plasticity that Lysenko misinterpreted, in which the same genes express themselves differently in different environments, is a real and important phenomenon. But wide diversity in the way genes are expressed does not mean that gene expression is not rooted in strictly inherited germ line factors.[8]

To bring out how engaged Dobzhansky's research on fruit flies was with the ethical and political issues of the middle decades of the twentieth century we will observe him in professional controversies in which he showed himself quick to see even slight departures from the population-genetic definitions he assigned to key terms as standing in the way of the liberal post-war order he was convinced up-to-date evolutionary biology supports. Seen in connection with the pseudo-science that oppressed his country of origin and the racism that dishonored his adopted land, his definitions were aimed at elevating democratic citizens (including scientists) to a cognitive position that would at one and the same time inoculate them against moral evils, justify their democratic social practices and political institutions, and liberate them from false views of nature's ways.

Dobzhansky was unwilling to enter into dialogue with just anybody. He was, for example, simply contemptuous of Cyril Darlington, an adept in chromosomal mechanics, and R. Ruggles Gates, a British botanist. Their belief that all human traits are under direct genetic control, with enculturation having little effect on gene expression no matter what the environment, was for Dobzhansky inseparable from the fact that they were out-and-out racists and not coincidentally knew next to nothing about population biology. The "excrement" of a recent article by Gates, Dobzhansky told Montagu in 1948, is nothing compared to the "evil smell" of Darlington's claim that the "values of races and classes," even the sounds of their languages, are coded in their genes. "I told you we are in for a big comeback of racialism in a virulent form" (Dobzhansky to Montagu, May 1, 1948, Montagu Papers; Dobzhansky 1962, 13). Thus in *Mankind Evolving* Darlington is placed at one unreasonable end of a spectrum at the other end of which we find the anthropologist Leslie White's equally excessive claim that human differences are determined entirely by cultural variables (Dobzhansky 1962; White 1949, 140; Chapter 3 of this book). It was different, however, with Montagu and Muller. In Dobzhansky's scheme, Montagu was closer to White's cultural determinism and Muller to Darlington's hyperbolic genetic determinism, but both were close enough to the vital center identified by Kroeber's superorganic to make dialectically engaging with them worthwhile.

In the next section we will follow the ascending and descending arc of Dobzhansky's interactions with Montagu, in which Dobzhansky opened himself to the culture concept and Montagu to instruction in population genetic evolutionary theory. Dobzhansky was always grateful to the anthropologist for offering to do "whatever bulldogging may be necessary" to help enlighten the public on how genetics bears on race (Montagu to Dobzhansky, October 9, 1944, alluding to T. H. Huxley's reputation as Darwin's "bulldog," Montagu Papers; Dobzhansky to Dunn, January 23, 1947, Dunn Papers). We think it probable, however, that he saw Montagu's persistence about "ethnic groups" as symptomatic of a deeper failure to appreciate the transformative power of population genetic definitions. This meant for him that Montagu missed an opportunity to give the American public the best possible biological argument for the racially pluralistic democracy that he and Dobzhansky both prized. *Mankind Evolving* expresses Dobzhansky's decision to do his own bulldogging.

Dobzhansky's dispute with Muller, a geneticist more technically accomplished than he, arose from the conflict between Muller's Morgan-influenced "classical" view of how genetic variation is distributed in populations and Dobzhansky's "balance" view of population structure and natural selection's mode(s) of operation. For Muller, selection turns only the best alleles into adaptations and eliminates the others. For Dobzhansky, selection is more "creative" than that. It preserves, indeed amplifies, the supply of potentially adaptive variations (polymorphisms), especially by favoring heterozygotes, which bank recessive alleles that might prove adaptive in new environments (Beatty, 1994). In addition, Dobzhansky believed that heterozygotes are generally adaptively superior because their phenotypic expressions are more environmentally flexible; the well-documented phenomenon of hybrid vigor depends on heterozygote superiority or heterosis. Since his and Montagu's scenario about how sapient humans acquired their capacity for culture rests on the claim that wherever it can natural selection will promote the evolution of flexible agency, preeminently by preserving variation and favoring heterozygotes, Dobzhansky threw everything he could against Muller's conviction that selection eliminates all but the best alleles from Mendelian populations.

The disagreement bore on eugenics. For Muller, the need for overt eugenic measures arises from the "genetic load" of harmful variations that a long history of benevolent human practices and recently the effects of atomic radiation have allowed to collect in human gene pools (Muller 1950). Commendably, Muller insisted that the ethical right to do something about genetic loads cannot be exercised unless social conditions are in place that do not mask the true effects of genes by confusing high social or racial status with good genes and low status with bad. But Dobzhansky disagreed that the best social condition for identifying the effects of genotypes is the communist flattening of the social structure from which he had fled and toward which Muller gravitated. Communism, Dobzhansky argued, would eliminate the social diversity that is required to realize the distinctive qualities, both inherited and acquired, of individuals. He also rejected Muller's (and Huxley's) assumption that high IQ elites are the preferred agents of eugenic change. Dobzhansky saw in Montagu remnants of old ideas about classification that prevented him from fully grasping the fundamental concepts of population genetic Darwinism. Muller, by contrast, knew his population genetics, but his political prejudices and eugenic preoccupations led him to distort what he knew in ways that failed to capture what John Beatty calls "the biology of democracy" (Beatty 1994).

In using rhetorical-critical methods to examine how fact and value entwine in Dobzhansky's view of evolutionary biology our aim is suggest that, although it may be paradoxical that his abstract, gene-centered definitions of evolutionary biology's key terms facilitated good judgment about the biological and political controversies of his time, there is something to it, even if he left it to his scientific heirs more closely to approach his happy mean.

Race and biology: Dobzhansky, Montagu, and the specter of Hitler

On April 5, 1937, as Dobzhansky, still in Pasadena, was putting the finishing touches on *Genetics and the Origin of Species*, Dunn wrote, "[Edmund] Sinnott and I are in the throes of a new edition of the genetics textbook and your material will come just in time for the new treatment of genetics and evolution that we hope to make" (Dunn to Dobzhansky, April 5, 1937, Dobzhansky Papers). The text certainly needed revising. Subsequent editions had slowly been softening but not eliminating its proclamations that hereditary genius runs in families, that members of these families should intermarry and with help from state subsidies raise large broods, and that carriers of hereditary degeneracy and criminality should be legally prevented from reproducing (Sinnott and Dunn 1925, 402–414). In rewriting the book Dobzhansky raised Dunn's consciousness, with the result that when it was finally published in 1950 eugenics was gone. Its focus was now on "the special interest of the two authors chiefly responsible for the revision" in "population genetics and speciation … and genic effects on development" (Sinnott *et al.* 1950, ix; Sinnott, having been named Chair of Botany at Yale, played little role in the new edition). In this way, Dobzhansky's population genetics liberated Dunn, a man of the democratic left, from having his name still peddling in the era of Hitler remnants of the eugenics that until the emergence of the Modern Synthesis, and in some cases after, geneticists shared to one degree or another (Paul 1984).

At the same time, Dunn raised Dobzhansky's consciousness about race and culture. "Dunn can be seen as carrying out a Columbia University campaign," writes his biographer, to support the equality of races that had been "initiated by Franz Boas and picked up by Ruth Benedict" (Gormley 2006, 414–415; 184–185; 390–393; Benedict 1940; Benedict and Weltfish 1943). When the tide of war turned, Dobzhansky and Dunn began to fear that "the racialists who were so overwhelmingly strong in the USA before 1932 and before Hitler" would "come back in force" as soon as hostilities ended (Dobzhansky to Montagu, May 22, 1944, May 1, 1948, Montagu Papers). To deflect this possibility, they wrote a short trade book on how Mendelian genetics supports the main themes of Boas's views about race. *Heredity, Race, and Society* is a geneticists' complement to Benedict's *Race: Science and Politics* (1940), Benedict and Weltfish's *The Races of Mankind* (1943), and Montagu's *Man's Most Dangerous Myth: Race* (1942b). It reaffirmed Boas's claims that there never were any pure races, that attributions of race membership are context-relative, that variation within races is higher than between them, and that race mixture not only does no harm but as hybrid vigor can be "biologically desirable" (Dunn and Dobzhansky 1946, 114–115, 121; Chapter 2 of this book).

To undermine the categorizing impulse blocking recognition of these perceptions Dunn and Dobzhansky invited readers to rise to the population level. From there it is easier to see that Mendel's law of independent assortment ensures that, "The distribution of genes that determine human 'racial' traits … vary independently.… The distribution of stature … agrees neither with that of blood type nor that of head shape or [skin] coloration" (Dunn and Dobzhansky 1946, 124). Keeping in mind the sheer number of independently varying traits in interbreeding populations will discredit scientific racism's essentialism. Even if a statistical correlation between two traits is found there will be many exceptions and what at first glance look like clear cut differences will turn out to be graded continua. From this perspective, races reappear as populations "marked off only by the relative frequency of [certain] genotypes" in a part of an interbreeding population whose trajectory a biologist has an interest in tracking. It's not that races don't exist. It's that "there are too many races of too many different kinds" to make the inferences "racialists" do (Dunn and Dobzhansky 1946, 124).

Dobzhansky and Dunn were disappointed when Simon and Schuster declined to publish their book on the ground that "[i]t is not popular enough" (Dobzhansky to Dunn, July 31, 1945, Dunn Papers). It appeared a year later, but didn't make a stir. "Our Penguin book has disappeared from the market," Dobzhansky told Dunn, who was in Europe at the time. "No reviews, no nothing" (Dobzhansky to Dunn, March 10, 1947, Dunn Papers).[9] Dobzhansky hoped for better luck working with Montagu, with whom he initiated a correspondence in 1943 by thanking him for bringing the interpretive framework of *Genetics and the Origin of Species* to the attention of his anthropological colleagues. In a 1942 article in *American Anthropologist* Montagu had quoted Dobzhansky's remarks that, "Geographical distributions of the separate genes compassing a racial difference are very frequently independent" and, "Blood group distributions are independent of skin color, cephalic index distributions, and so on" (Montagu 1942a, 374; Dobzhansky 1937, 77; Brattain 2007, 1394). But Dobzhansky also wrote to ask Montagu's advice about writing science for the public as he began drafting his book with Dunn. "I have strong ideas about what popular writing should be like," he told him, but "I have never done it" (Dobzhansky to Montagu, January 4, 1943, Montagu Papers).

Montagu jumped at the chance. By return mail he told Dobzhansky how grateful he was to him for attuning him to population genetics and offered to help him address the public in any way that might be required (Montagu to Dobzhansky January 4, 1943, Montagu Papers). They soon began using each other as editors and conduits through which to gain access to and credibility in each other's fields, delicacy about which bothered Dobzhansky more than Montagu (Dobzhansky to Montagu, April 27, 1944; Montagu Papers). Dobzhansky always admitted he was "an amateur in anthropology," but, his back now covered, he began writing for anthropology journals and addressing the AAA (Dobzhansky 1962–1963, 634, 1944; 1963a, c).

For his part, Montagu was eager to learn how population genetic arguments could bolster his anti-racist crusade. In their early interactions, Dobzhansky patiently corrected his technical misunderstandings. In his 1942 article in *American Anthropologist*, Montagu ascribed most racial differences to accumulations of chance mutations, migration, sexual selection, and the caste-making practices he called "social selection" (Montagu 1942a, 371, 375). He was silent on natural selection. This astigmatism recurred in a draft revision of *Man's Most Dangerous Myth* on which he asked Dobzhansky to comment in spring, 1944. Dobzhansky told Montagu that his book was "admirable," but in handwritten notes and a follow-up letter pressed him on basic points (Dobzhansky to Montagu, May 22, 1944, Montagu Papers). When he spoke of "the inherent variability of genes" as a cause of race formation, Dobzhansky asked, was he referring to mutation or genetic drift? It turned out that Montagu meant drift: the spread of alleles through small isolated populations by chance of the sort that allows a roulette ball to land on red six or eight times in a row without offending against the principle that in the long run black and red landings will even out. In that case, Dobzhansky wrote, Montagu should also introduce natural selection, since "without it no races, or 'divisions' or 'ethnic groups' of the actually existing kind would be formed." Drift played a larger role in Dobzhansky's thinking before the mid 1940s, but even then he invoked it as an igniter of adaptive natural selection of the sort that leads to speciation (Provine 1986; Beatty 1987).

Nor, Dobzhansky told Montagu, are adaptation and natural selection alternatives. To think so was simultaneously to conflate adaptation with environmental accommodation and draw too large a contrast between natural and sexual selection, which Dobzhansky conceived as a variant of adaptive natural selection in which mating preferences shift gene frequencies. Finally, in contrast to his 1942 article, Montagu claimed that some differences between human races, such as skin color, are adaptations. It was unclear to Dobzhansky how this could be if Montagu persisted in contrasting adaptation with natural selection. He asked him why he was so sure that skin color differences diffused *because of* their reproduction-enhancing effects, as adaptation by natural selection demands. It's possible, Dobzhansky said, but hard to show because the way skin color changes in human populations is so graded and continuous that "[w]e can't assume the classes and names we would use" in assigning differences to races are nature's (Dobzhansky to Montagu, May 3, 1944, April 15, 1947, Montagu Papers).[10]

In responding to this barrage, Montagu, who was so well known for his obstreperousness that Dunn cast a cold eye on Dobzhansky's engaging with him at all, was surprisingly docile.[11] He thanked Dobzhansky for his "exceedingly valuable comments on the manuscript" and assured him that he would be happy "to substitute genetic drift and enlarge a little more on the meaning of that." Dobzhansky should "write [him] a definition of genetic drift that I can use in the book" so that I don't have to "send it to you for your approval" (Montagu to Dobzhansky, May 23, 1944, Montagu Papers).

What lay at the root of Montagu's tendency to slight natural selection? The answer is that Dunn's skepticism about Dobzhansky's engagement with Montagu was based on more than his reputation as a difficult person. Dunn knew that Montagu was funded by a group of scientifically outdated anti-Malthusians who under the influence of the nineteenth century Russian ecologist Petr Kropotkin stressed cooperation to the exclusion of competition in part because they took natural selection to be purely eliminative (Kropotkin 1902). By contrast, the Modern Synthesis sees adaptation as arising from a process in which reproductively more successful alleles gradually amplify their representation in Mendelian populations. "The modern concept of natural selection," Simpson wrote,

> is ... different from Darwin's. Darwin emphasized the survival of favored (or the early death of unfavored) individuals. The survivors were, for the most part, the "fittest," in the sense of qualification for success in competition, the "struggle for existence." That is still the usual non-scientific understanding of the process, but to specialists in the study of evolution "natural selection" now means the average production of more offspring by such organisms in a population as are distinguished by any particular heredity factors. "Fitness" is now defined solely as relative success in reproduction.
>
> (Simpson, 1959, xi; see also Huxley 1955, 275)

The grip of the eliminative mindset goes some way to explaining Montagu's early resistance to natural selection and his mistaken contrast between it and adaptation. Nor did his core convictions entirely disappear after he had seen the light. If by 1947 he had become a recovering anti-Darwinian it was only because under Dobzhansky's tutelage he now saw the cooperative tendencies in nature to which he remained devoted as arising from natural selection's preference for genotypes that express themselves flexibly. Under his cooperation-fixated interpretation of their hypothesis about the evolutionary origins of our educability, Montagu pursued a career of public appearances in which he maintained that human beings are cooperative by nature, women are the superior sex, and the male-dominated American family is a neurotic, authoritarian mess (Radick, unpublished).

Dobzhansky was pleased with their 1947 paper. He told Montagu that, "The [natural] selection of genotypes permitting a phenotypic plasticity of mental traits is the most important idea that has ever occurred to me" (Dobzhansky to Montagu, June 23, 1947, Montagu Papers). He informed Dunn that, "Montagu is very excited about my idea ... and intends to take upon himself propaganda for [it]" (Dobzhansky to Dunn, February 15, 1947, Dunn Papers). But their interpretations differed. Inspecting the paper when it appeared in print, Dobzhansky told his co-author that in hindsight he wished they had made it clear that "[w]hen a trait becomes plastic it exhibits more and more 'Lamarckian' modifications without thereby altering the [inherited] trait in the offspring." He worried, too, that they may have left a false impression that, having evolved the educability that leads to something like Lamarckian transmission of culturally acquired knowledge, natural selection has had no further effects on human evolution. Finally, he told Montagu that he wished they had said that "[a]daptation within human groups is achieved by cooperation *as well as* conflict, not *rather than* [conflict]" (Dobzhansky to Montagu, July 20, 1947, Montagu Papers, our italics).

That Montagu remained untroubled by these nuances does not mean that Dobzhansky did not give him plenty of support in the immediate postwar years, when intellectuals were articulating and anti-totalitarian political authorities were promoting a global vision of racial equality and cultural pluralism. In 1950, Montagu served as *rapporteur* of UNESCO's [First] *Statement on Race*. Asked to comment on his draft, Dobzhansky reported that he was "on the whole wholeheartedly in agreement" with it (Dobzhansky to Robert C. Angell, January 17, 1950, Dunn Papers; see Brattain 2007, 1397, n. 34). Why not? At its heart was the proposition that

> The one trait which above all others has been at a premium in the evolution of men's mental characters has been educability, plasticity. This is a trait which all human beings possess. It is indeed a species character of *Homo sapiens*.
>
> (UNESCO 1950, Section 11)

A month later, Dobzhansky encouraged Montagu "as one of the leaders of American anthropology in the fight for a modern race concept" to take his anti-racist message to the world (Dobzhansky to Montagu, March 29, 1950, Montagu Papers). When he got wind that a group of physical anthropologists was seeking to derail the *Statement on Race* by demanding that UNESCO publish a new one, he secretly warned Montagu that Huxley was recommending that Darlington serve on a reshuffled drafting committee (for no better reason, Dobzhansky told him, than Huxley's "senility") (Dobzhansky to Montagu, February 24, 1951, marked confidential, Montagu Papers; Alfred Metraux to Dobzhansky, February 21, 1951, UNESCO Papers, cited in Brattain 2007, 1400, n. 45). He feared that the new statement would be "pretty bad" if, as seemed probable, neither he nor Dunn was able attend its scheduled meeting in Paris.

In the end, schedules were shuffled and the diplomatic Dunn became *rapporteur* of a *Second Statement on Race* written by a committee that excluded Darlington and retained Montagu. The result was a hard-fought, but uncertain victory. The *Second Statement* still supported such Boasian theorems as the grading nature of human races, the higher degree of variation within races than between them, and the harmlessness of interracial marriage, but it did so only by asserting that there was no evidence *against* these claims. The text omitted Montagu's peroration about human nature as inherently cooperative, dropped the "ethnic group" idea, and came close to opposing racial and cultural rank ordering purely on ethical grounds – human rights – instead of evolutionary arguments (Gayon 2003; Müller-Wille 2005; Brattain 2007). Even so, some members of the committee and consultants refused to endorse it. So uncertain was the victory that Provine cited the *Second Statement's* negatively phased locutions and influential dissenters as evidence for his contention that the racial and eugenic views of biologists shifted after the war for political, not biological reasons (Provine 1973; Provine and Russell 1986). In point of fact, even the political views of the dissenters remained unreformed.[12]

This situation frustrated Dobzhansky. In spite of Dunn's effort to stem the tide, Montagu's overreaching had led to a backlash. Accordingly, in *Mankind Evolving* he set out biological arguments for the unity of the species, the equal capacities of all human groups, and the benign nature of racially mixed marriages based on the very points he wished he and Montagu had made in 1947: (1) the flexible agency conferred by our shared capacity for cultural learning offers no support for Lamarckian inheritance as a biological phenomenon (Dobzhansky 1962a, 8, 20, 52); (2) the phenotypic plasticity of our genetic inheritance means that, "Genes create the setting for cultural traits but do not compel the development of any particular ones" (Dobzhansky 1962a, 322); (3) natural selection is still at work in human evolution, maintaining adapted gene pools in the face of new selection pressures created by our cultural way of life (Dobzhansky 1962a, 150–154); and (4) the evolution of adaptedness requires competition as well as cooperation:

> Ashley Montagu, a very able modern exponent of the theory of the innate goodness of man, has stated, "It is not evil babies who grow up into evil human beings, but an evil society that turns good babies into disordered adults."

But, Dobzhansky asked, "If human nature is really good why does it not resist the disordering influence of evil social environments" (Dobzhansky 1962a, 52)?

A few pages later, Dobzhansky identified taking separate traits as the fundamental *explananda* of evolutionary theory as a root cause of misunderstanding the dynamic interrelatedness of environments, genotypes, and phenotypes (Dobzhansky 1962a, 55–56). To find adaptationist rationales for species-marking traits such as Darwin's finches it may be enough to map the mean distribution of characters onto resources afforded by or dangers presented by particular environments. But if this model is generalized it results in describing the interaction of genes, traits, and environments in static, thing-like ways that make too much of the analogy between organisms and artifacts, which really are assemblies of separate components rather than developmental processes. This makes it difficult to see why, or even that, natural selection favors genotypes with phenotypic plasticity or how races at the extremes of a distribution, not its mean, become species.

As early as 1937, Dobzhansky pointed out that genetic and environmental factors are constantly changing under the impact of the activities of organisms themselves as they degrade resources in the course of making a living (Dobzhansky 1937, 126, 150, 179). Changing environments, developing organisms, and genomes are even more dynamically intertwined for Dobzhansky than distance, time, and velocity for Galileo. The genetic contribution to an observed population-level difference goes to 100 percent by definition in uniform environments and to zero when genotypic variation has been eliminated. Outside of this dynamical framework parsing traits into genetic and environmental causes makes no sense. It is possible to construct experimental set-ups in which genetic and environmental factors are plotted against each other in a linear way: so much more fertilizer, so many more peas or apples per inch, foot, or acre of soil. But in such cases one is exploring nature by holding it at bay. In the wild, genotypes depend on their shifting relation to other genotypes to evolve a range of responses to no less dynamic environments at various stages in the ontogenetic process. Variables that at first seem to change in lockstep with others suddenly cease doing so. Fertilizer inputs go up, for example, but apple production goes down.[13]

Following Richard Woltereck, Dobzhansky identifies a genotype's range of responses to different environments as its "norm of reaction" (Dobzhansky 1937, 170, 1951, 21–22; 1955a, 369; 1962a, 42, 81; Dunn and Dobzhansky 1946, 36). Strictly speaking, it is these norms that evolve, not traits. So viewing an organism as a sum of traits goes against the grain of evolutionary dynamics. This is not to say that traits are not useful. But a trait, Dobzhansky says, is "an abstraction, a semantic device" that gives us a handle by which to compare populations as we go about answering questions about the processes of adaptation, racial separation, and speciation. "There are as many traits as we see fit to use" as probes in inquiring into how genes express themselves in the development of organisms in inherently slippery environments (Dobzhansky 1962a, 56; 1973, 68). For Dobzhansky, in an evolving world pragmatism about the use of concepts leads to realism about evolutionary processes. This point is especially relevant to race: "An individual or a population might belong to one 'race' as far as the gene A is concerned, to a different 'race' with respect to the gene B, [and] to a still different 'race' with respect to C" (Dobzhansky 1962a, 266; Dobzhansky and Epling 1944b, 138; Montagu 1950, 317). Soon after the publication of *Mankind Evolving*, the physical anthropologist Frank Livingston argued that by Dobzhansky's population-genetic standards there are plenty of races of other species, but no human races, just character gradiants (clines). This was Huxley's view, too, and it remains influential among physical anthropologists. But Dobzhansky rejected it for the same dialectical reason he rejected Huxley and Montagu's "ethnic group" proposal: It leaves the old race concept intact in the background, where it "plays into the hands of bigots." Livingstone replied, "The fact that some crank may make political hay out of a biological fact" is "incompetent, irrelevant, and immaterial" (Livingstone and Dobzhansky 1962, 280).

On Dobzhansky's view, extrapolating the effect of genes in different environments is a tricky business. In most cases, the genetics we need is not, as with Mendel's peas or eye color, qualitative but quantitative. "What we want to know are the relative magnitudes of genetic and environmental components in the variance observed in a given trait [in] a certain population at a particular time" (Dobzhansky 1962a, 56). Norms of reaction can be determined only by rising to the level of population-wide distributions over time to see what phenotypic expressions correlate with what genotypes under what environmental conditions. To find these effects, experimental crosses of the sort Dobzhansky learned in Morgan's lab must be undertaken. Inquiry of this sort is especially complicated in the human case. Cultural variables are in a feedback relationship with biological variables that are already feeding back on each other. "The human personality always functions within the framework of a certain social and cultural setting and cannot be understood apart from that framework" (Dunn and Dobzhansky 1946, 36). This being so, humans cannot be turned into experimental subjects like fruit flies. That racist and eugenicist Nazis thought so makes it all the more incumbent on liberal democracies to avoid anything of the sort, preferably by grounding this prohibition as deeply as possible in biological science instead of relying only on ethical objections.

Montagu had absorbed most of this by 1950, when Washburn and Dobzhansky invited him to give a high-profile address to a joint conference of anthropologists and geneticists at the biological research station at Cold Spring Harbor on Long Island (Chapter 5 of this book). Backpedalling, Montagu assured his audience that in advising them to use "ethnic group" rather than "race" when addressing the public about human evolution his aim was to only to "open the door … to understanding" by leading "laymen" away from old preconceptions and toward the new population genetic approach. "The term 'race,'" he said, "constitutes the greatest impediment to getting the layman to understand what the scientist means when he uses it" (Montagu 1950, 319, 335). Among themselves, experts are free to talk about races if they restrict themselves to genotype-phenotype relations in temporally and spatially shifting environments. In this way Montagu hoped to cure biological anthropologists of the bad habit of establishing races and species merely by averaging collections of expressed traits, which, like Dobzhansky, he saw as a pseudo-statistical approach held over from the typological classificatory practices of the old systematics (Montagu 1950, 323; Chapter 6 of this book).

Montagu used the occasion to say that his reason for recommending "ethnic group" had always differed from Huxley's (Huxley and Hadden 1936, Huxley 1941). Huxley's proposal was motivated by a well-placed worry about Hitler's racist eugenics. The cephalically indexed "races of Europe" really are, at most, ethnic groups. It now seemed to Montagu, however, that Huxley was assuming that biological races are Johannsen's and Boas's "pure types." How otherwise explain why Huxley took the way human traits grade, blur, and blend to imply that there are no human races – or his devotion to eugenics without racism (Montagu 1950, 320)? Had Montagu not said that human races exist in *Man's Most Dangerous Myth* (Montagu 1950, 318, 1945, 2)? What he opposed, he informed the assembled worthies at Cold Spring Harbor, and was now using Dobzhansky's definitions to show, is the static essentialism about races that fosters both eugenics and racism by linking many physical and psychological traits. When biologists and anthropologists "average the characters of a group, knock individuals together, give them a good stirring, and serve the resulting omelette as a 'race'" they are still making a semi-essentialist mistake (Montagu 1950, 318).

Montagu also used Dobzhansky's authority to say that against a background of genetic harmony a single genetic difference can take a geographical population off onto a trajectory that evolves a new race and in some cases a new species. After all, this view is closer to the conception of races as lineages that prevailed before the hypertrophy of trait-based racial (and racist) characterizations. It was, for example, the view that figured in the subtitle of the *Origin of Species*, in which Darwin refers to "favored races." The new, Dobzhansky-tutored Montagu was especially intent on warning that in the case of so polytypic and polymorphic a species as *H. sapiens* the presumption must be that what look like racial markers are more likely to be environmentally induced expressions of the same plastic genotypes. "Andes man," he says, is not a race unless and until it has been shown that a genetic mutation or a heritable chromosomal rearrangement underwrites the ability of Andes-dwelling humans to adjust to life in high altitudes. Until then we should call Andes-man a "habitat type" and presume that the ability to thrive in thin air expresses a widely shared norm of reaction (Montagu 1950, 322, 336).[14]

All this is close to orthodox Dobzhansky. The difficulty comes when we imagine what Montagu would say when his conditions for positing a human race *have* been met, that is, when phenotypic differences have been traced to genetic differences. "Where there is *no doubt*," he says, "we [scientists] should continue to use the term 'race'" (Montagu 1950, 322). But even then more remains to be done:

> Ideally, *all* variable genes and chromosome structures would have to be taken into account to *describe* a given race. At the present level of knowledge this ideal is unattainable. The *description* of races may become more and more exact as knowledge grows.... Differences in the frequency of a single gene must be interpreted with caution. Such a difference may be significant or not.
>
> (Montagu 1950, 317, our italics)

This passage suggests that Montagu's revised conception of race was aimed at describing the objective existence of "real" biological races as co-adapted (cooperative) collections of genotypes that either validate or invalidate conventionally recognized racial differences. When enough of these correlations have been found we are entitled to say that we have identified a race.

Was this as close to Dobzhansky's thinking as Montagu assumed? For Dobzhansky, genotypes are difference makers, and so their effects are relative to what they are compared to (Waters 1990). It is difficult for this reason to imagine a complete description of a race or a species. Dobzhansky was not above listing the races of *H. sapiens* any more than of fruit flies (Dobzhansky 1962a 263). "Classifying and systematizing are devices used to make diversity intelligible and manageable," he wrote (Dobzhansky 1962a, 266, 1973a, 68). How many races one posits depends on what question one is trying to answer. But it does not follow from the fact that "the number of races we choose to recognize is a matter of convenience" that what gets picked out does not exist. "Pragmatic and theoretic race studies should be complementary and not rival" (Dobzhansky 1962a, 266).

For purposes of his 'speciationist' research program, Dobzhansky identified races of fruit flies by using one or two genotypes to follow a population that is, or may be, becoming isolated enough to become a new species. There is no suggestion that we fall short if our resolution of this diachronically inflected question never adds up to a complete synchronic description of a race in the sense that Montagu countenanced in the opening pages of *Man's Most Dangerous Myth* and reaffirmed at Cold Spring Harbor. This being so, Montagu's approach to population genetics was not as processive as Dobzhansky's. At one point in his address he even came close to rejecting Dobzhansky's definition of races as "Mendelian populations that differ in the frequencies of one or more genes." "Differences in the frequency of a single gene must be interpreted with caution," he wrote (Montagu 1950, 317). Here he was echoing, if faintly, the classification-oriented essentialism to which he feared the public would attach itself if the language of race were not censored in public and severely disciplined in the technical sphere.

One might also see a similar tendency to fall back toward what he was opposing in Montagu's way of belatedly embracing natural selection as a "creative force" in evolving adaptations (Montagu 1950, 325). In spite of Dobzhansky's cautions, his newfound affection for adaptive natural selection encouraged him confidently to assert that, "The broad nose of the Negro and the narrow nose of whites represent adaptive characters" and that, "Populations living in regions of extreme cold … tend to be relatively short and well padded with fat" because these present to the environment less surface area than "populations which have been long resident in regions of high temperature" (Montagu 1950, 325, 331). In *Mankind Evolving*, Dobzhansky reiterated his cautions. "Shocking as it may be," he wrote, "solid and conclusive evidence concerning the adaptive significance of racial traits in man is scant in the extreme" (Dobzhansky 1962a, 271). Treating genetically based traits as presumptively adaptive *differentia* strengthens the perception that conventionally identified races such as Negroes or Eskimos might after all be races in something close to the old typological sense. This is not full-fledged population thinking.

These are not the only traces of older styles of thinking in Montagu. He never fully disconnected his stress on cooperation as the distinctively human trait from his Kropotkin-inspired suspicion of Darwinism. At Cold Spring Harbor, he reworked Dobzhansky's stress in the late 1940s on harmonious co-adapted genomes to support an idiosyncratic claim that the cooperative tendencies in nature that culminate in naturally cooperative human beings are prefigured in the cooperative genome (Montagu 1950, 326–328). In 1952, he published a book (dedicated to Kropotkin, and really a defense of his original *Statement on Race*) in which he saw Darwinism as still distorted by dog-eat-dog Malthusianism, especially in the popular mind (Montagu 1952). Dobzhansky wrote to say, "I have read [an early draft of] *The Darwinian Fallacy*" – a title later changed to *Darwinism: Competition and Cooperation* – "and have major objections impossible to describe in a letter" (Dobzhansky to Montagu, December 17, 1950, Montagu Papers). He loyally wrote a non-committal jacket blurb, but it was omitted when the book was published (Radick, pers. comm.). Darwinism, it read "has repeatedly been perverted to serve prejudice and malevolence.... Professor Ashley Montagu deserves gratitude for his attempt to winnow the sound biological grain from the chaff of sociological misuse" (Montagu Papers). The old pattern persisted: on-stage alliance, nitpicking in public, off-stage disagreement.

Montagu and Dobzhansky came together because they were both acutely aware that they were living in a time of definitional rupture. Montagu's objection that Dobzhansky could not possibly hope to drive out the old ideologically crippled pseudo-biology simply by teaching children "the ABCs of genetics" – Mendel's laws and the Hardy-Weinberg equilibrium – is well taken (Montagu to Dobzhansky, May 22, 1944, Montagu Papers). Dobzhansky's problem-centered, pragmatic race concept is so flexible that, except among cognoscenti, it cannot serve as an effective firewall against ingrained habits of using trait-markers as racially marked signs of co-varying capabilities and incapacities and so of retaining commonplace conceptions of race. To Montagu, Dobzhansky's belief that this result could be circumvented by re-educating whole societies in theoretical biology was no less utopian than, say, Plato's proposals in the *Republic*. Responding to Mayr's Dobzhansky-echoing assertion at Cold Spring Harbor that, "The obliteration of racism depends on popular acceptance of the new race concept," Montagu said, "I'm afraid that the elimination of racism will depend on much more than popular acceptance of the statistical conception of race" (Montagu 1950, 336; see also Gannett 2001, 2013). His worry about Dobzhansky's retention of the term "race" arose from more than the inability of re-definition to exorcise typology. It expressed the political realism of one who as a Jew – he was born Israel Ehrenberg – had personally experienced anti-Semitism and witnessed the genocidal power of misused language. In bad times, only good rhetoric can drive out bad, as the semanticists Korzybski, Ogden, Richards and, he might have added, Kenneth Burke realized (Burke 1939). Wholesale dialectical reframing must await better days.

Dobzhansky thought those days had come. He was as quick as Montagu to pick up the scent of racism *redivivus*, but he took the emerging world order of the postwar years as a timely opportunity to cut the ground from beneath it by shifting paradigms. His sense of the new rhetorical situation created by the defeat of Hitler was more serenely Olympian than Montagu's. Montagu fought his enemies by nipping at the heels of biology's old classificatory paradigm. It is not surprising that working at such close quarters he fell back toward what he opposed. Dobzhansky wanted to replace the old paradigm lock, stock, and barrel in the public no less than the technical sphere. Taking advantage of post-Sputnik calls for reform in science education, he worked with Muller, Simpson, Mayr, and other biologists to advocate teaching evolutionary biology in the schools (Nelkin 1982). When he repeated in *American Biology Teacher* in 1973 his 1964 remark that, "Nothing in biology makes sense except in the light of evolution," he had in mind not just the population genetic account of evolution, but his own democratically inflected, anti-racist version of it (Dobzhansky 1964, 1973a).

Dobzhansky and Montagu remained allies into the early 1960s. Responding to a polite inquiry about a fly collecting expedition abroad, Dobzhansky told him good naturedly, "The ethnic groups of local *Drosophila* have proved … interesting" (Dobzhansky to Montagu, October 7, 1960, Montagu Papers). Nonetheless, in the rhetorically charged atmosphere surrounding the Civil Rights Act of 1964 and a year later the Voting Rights Act Montagu repudiated even the technical use of "race" he had endorsed at Cold Spring Harbor and moved *avant la lettre* toward social constructionism. "While it is usually true," he wrote, "that populations differ in one or more genes from one another, it serves no useful purpose to call that fact a matter of 'race,' especially in the case of man" (Montagu 1965, 91). This may have disappointed Dobzhansky, but it probably didn't surprise him.

Eugenics and democracy: Dobzhansky, Muller, and the specter of Stalin

Hermann Muller's life was an epitome of his troubled times. Born in New York of a German father and a Jewish mother – he was Kroeber's cousin on her side – he was educated at Columbia, receiving his B.A. in 1910 and his Ph.D. in 1916. He trained as a geneticist under Morgan, working in the famous "fly room" when Sturtevant was making the first genetic maps. At the invitation of Huxley, who founded its Department of Biology, Muller taught at Rice Institute (later Rice University) in Houston, Texas between 1914 and 1918. His research at Rice, Columbia, the University of Texas, Amherst, and finally the University of Indiana focused on the mutagenic effects of X-radiation, which he proved in papers published beginning in the mid-1920s. In the anxious atmosphere of the atomic bombing of Japan and the gathering Cold War, he won the Nobel Prize in Medicine for this body of work in 1946.

Muller was a full-blown eugenicist and man of the internationalist left. Participating in the genetic revolution as it unfolded, he was as aware as Dobzhansky of the intimate interaction between genes and environments. Unlike Dobzhansky, however, who held that eugenics and racism are difficult to separate, he believed with Huxley that eugenics, positive and negative, could realize its potential only when it was freed from racism. He also believed that the genetically fit – he nominated Lenin – would begin revealing themselves only when the socio-economic environment is leveled.

In 1932 Muller went to Berlin to work with the eminent Russian geneticist Nikolai Timofeev-Ressovsky. There his contempt for American pop eugenics, in which Babe Ruth and Rudolf Valentino figured as eugenically fit, was eclipsed by his horror at Hitler's racist version. In 1933 he moved to Leningrad, where he completed *Out of the Night* (Muller 1935). In it he predicted a socialist future whose realization would be hastened by the willingness of men whose biological superiority revealed itself in the effort to build a new social order to allow their sperm to be repeatedly used to artificially inseminate the eggs of women whose own superiority would be shown by their willingness to put most of these children up for adoption. Naively, he sent a copy of his book to Stalin, who promptly decided that genetics was decadent bourgeois science and declared Muller *persona non grata*. It was the beginning of the turn to Lysenkoism. Muller could not simply leave the Soviet Union without exposing his Russian colleagues to danger. So he cleverly, or perhaps desperately, volunteered to serve the Republican cause in the Spanish Civil War. He soon left the field to take up a position at the University of Edinburgh before returning to the United States in 1940, going first to Amherst and in 1945 to Indiana.

By that time Muller's claim was that even without a socialist revolution superior men and women could voluntarily do democracy a big favor by mating and producing superior children, thereby raising the average IQ and preventing further degeneration of the population due to dysgenic social practices, economically unjust institutions, and radiation-induced mutations. In 1950 he used the bully pulpit afforded by his Nobel Prize to warn that atomic testing and *a fortiori* nuclear war would further compromise human fitness by magnifying the already large "genetic load" of deleterious recessive alleles that millennia of bad social practices had failed to flush from human populations (Muller 1950). This warning made him as suspect to Lewis Strauss, the powerful head of the Atomic Energy Commission, as he had been to Stalin (Beatty 1987b).

Dobzhansky and Muller were on good terms, exchanging fruit fly stocks and visiting each other's labs. Dobzhansky signed the *Geneticists' Manifesto* that Muller drafted when, just as war was breaking out in Europe, he and his Edinburgh colleagues hosted the Seventh International Genetics Conference. The statement combined support for fact-based eugenics with a timely assertion that no nation or race can be said to have a monopoly on biological fitness. Racism and social inequality, it said, would have to be ended before the true effects of genes could be assessed. Dobzhansky did not object. The high water mark of their cooperation was an effort in the later 1940s to free Soviet geneticists in German prisoner of war camps and, later, sidelined or exiled in the USSR by the rise of Lysenkoism. Dobzhansky congratulated Muller when he won the Nobel Prize (Dobzhansky to Muller, November 1, 1946, Muller Papers). Muller returned the compliment by telling Dobzhansky "how highly I value your own work and my personal regard for you" (Muller to Dobzhansky, November 26, 1946, December 2, 1946, Muller Papers). Dobzhansky saw Muller as one of the two greatest geneticists of the century (Lewontin 1995, 93; the other was Sewall Wright). Their mutual respect managed to survive a protracted disagreement in the 1950s about the structure and dynamics of Mendelian populations.

Dobzhansky and Muller's ideas about the retention of harmful mutations in gene pools were as different as their ideas of democracy. Dobzhansky did not deny that mutation increases what Muller called "our load of mutations" (Muller 1950; Dobzhansky 1959, 158, 1962a, 295). But his encounter with the Boasian legacy in anthropology sensitized him to the powerful role of culture in human development and evolution. He spent the 1950s laying the scientific groundwork for his assertion in *Mankind Evolving* that people "do nicely in their natural habitats despite the fact that they bear enormous genetic loads" (Dobzhansky 1962a, 295, 288–290). These habitats are cultures.

One might think otherwise if one were to assume that the ethical protections of civilization have made our cultural habitat unnatural by cossetting the unfit. But few things annoyed Dobzhansky more than the presumption that we are "hastily made over apes" who are not at home in our cultural world (Dobzhansky 1962a, 330, quoting Muller 1960, 458). That is why he was so delighted to get a glimpse of how natural selection could have evolved our cultural niche in his 1947 paper with Montagu. The claim implies not only that enculturation is natural to our development, but also that more than any other adaptation it enhances our agency in the face of environmental contingencies. Like Kroeber, Dobzhansky had found a way to reject the presumption of eugenicists, who assume that if human life is to be set right it will have to be made over by experts. With the ironic edge of a refugee from Bolshevism lecturing a sometime enthusiast, he called Muller's scheme in *Out of the Night* "utopian" in the dystopic sense of *Brave New World* (Dobzhansky 1962a, 327). The most eugenically fit society, he argued, will leave it to individuals to choose their mates, manage their reproduction, and pursue their callings. The only exceptions are posed by circumstances in which the prospect of great and predictable suffering from diseases caused by double-dose recessives cannot be alleviated by environmental modifications such as dietary change, pharmaceuticals, or eyeglasses and when, in addition, communicative conditions for informed consent do not obtain. In such cases responsibility for what Dobzhansky called "this much of eugenics" devolves onto medical personnel whose decisions are to be legally protected in a science-friendly democratic state whose citizens, legislators, and judges will be, indeed must be, knowledgeable enough about genetics and evolution to see why preventing reproduction in a particular case, preeminently their own, is the best policy (Dobzhansky 1962a, 330–333). The last of these conditions helps explains why Dobzhansky was so intent on instructing the entire population in "the ABCs of genetics."

It has been claimed that Dobzhansky's concession to "this much of eugenics," together with his signature on the *Geneticists' Manifesto*, made him party to a "eugenic consensus" that stretched from early in the twentieth century into the postwar period (Paul 1984). Admittedly, Diane Paul places Dobzhansky with Muller and Huxley on the anti-racist "reform" side of that consensus. Still, he shares less with them than they do with each other. Dobzhansky's "this much of eugenics" is a matter of preventing "inborn errors of metabolism," from one of which Dunn's son Stephen suffered, from causing more suffering in offspring and families (Gormley 2006; Dobzhansky 1962a, 331; 1962–1963, 464–466).[15] In contrast to the brands of eugenics preached by Muller and Huxley, the interests of the individual are focal, not the state.

Why did Dobzhansky use a word that invited lumping him into a "eugenic consensus" to which these figures certainly belonged? Whenever he referred in print or correspondence to views as diverse as those of Darlington, Gates, Fisher, Huxley, and Muller, he spoke derisively of eugenics. That was his stance in

Mankind Evolving, too, except in the passage in which he accepts "this much of eugenics." It will not do to guess that Dobzhansky brought his penchant for redefining to terms like "eugenics." He confines defining to science and uses the language of political movements pretty much as he finds it. In the background of "this much of eugenics," we suspect, is that Muller's eugenics got a second wind when molecular geneticists like Linus Pauling, James Watson, James Neel, and their allies in genetic medicine accurately predicted that the identification of genetic diseases with mutations in DNA sequences would make genetic risks known and in the foreseeable future allow manipulation of the genetic material, including the germ line. This idea came burdened with the clinician's distinction between normal and abnormal functioning. That mapped well enough onto Muller's conviction that natural selection tries to eliminate all but the fittest allele in a given environment, but flew in the face of Dobzhansky's conviction that natural selection banks the variation on which the future of the species depends in the recessives of heterozygotes. When a new breed of molecular eugenicists recommended that all heterozygote carriers of the recessive for sickle cell anemia, retinoblastoma, and other "genetic diseases should be asked (or made) to refrain from having children," Dobzhansky realized that this policy would affect the main mechanism that preserves genetic diversity (Zuckerkandl and Pauling 1962; Suarez-Diaz, 2017). His recommendation that only carriers of double recessives of genetic diseases, such as Stephen Dunn, refrain from having children was combined with hope that in the future "eugenics would come into its own" when we learn how to engineer genes and modify environments without compromising evolvability (Dobzhansky 1973b, 49).

Political ideals were also in contention. The issue of eugenics comes up in *Mankind Evolving* as part of Dobzhansky's dialectic with Muller about what society and state evolutionary science requires for ethical and medically appropriate intervention in the reproductive life of its citizens. Dobzhansky was aware that power granted to genetic experts will be abused under any form of government except the personal-freedom-loving, science-respecting, individual-ability-facilitating liberal democratic institutions that, like Columbia's Dewey, Boas, Dunn, Kroeber, Benedict, Mead, Washburn and Dobzhansky himself endorsed. Foremost in his mind, however, was his conviction that the dynamics of natural selection themselves require liberal institutions if the best possible distribution of genes is to be achieved. To understand this claim we must enter into technical issues that divided Muller and Dobzhansky.

Selection pressure can favor the disproportionate spread of the dominant, recessive, or heterozygote. Heterozygote superiority, variously called hybrid vigor, over-dominance, or heterosis, means that "neither A nor a is best by itself because the fittest, the heterozyote, Aa, has them both" (Dobzhansky *et al.* 1977, 109, slightly amended). Fisher showed mathematically that this situation will obtain whenever persistent low-intensity mutation and weak selection pressure coincide (Fisher 1930, 56–59). The issues between Muller and Dobzhansky concerned the long-term fate of recessive alleles hidden in heterozygotes and the extent of heterosis in natural populations. Muller believed that where the heterozygote is selectively favored, its recessive component will slowly be flushed from the population – in some cases very slowly, since recessive alleles, including mutations caused by radiation, can lurk in heterozygotes for many generations. He thought so because he had inherited from Morgan a set of assumptions that Dobzhansky called "classical," in part because they hark back to Mendelism's classic scenario in which populations contain at each locus only one optimally fit allele, the dominant or wild type; recessive alleles, including those hidden in heterozygotes, are less fit than wild types because they impose costs on reproductive success; and reproductive success or fitness at one locus is not linked with fitness at others (Dobzhansky 1959, 257; Morgan 1932). Fisher viewed heterosis more favorably than Muller. At least heterozygotes have the effect of retaining variation for a time as fuel for natural selection. Dobzhansky viewed it even more favorably. Environments are so changeable that if variation-conserving mechanisms had failed to take hold the evolution of complex organisms would have ground to a halt long ago. The long-lived lineages we see around us survived because through trial and error, not any sort of directedness, the structures and mechanisms that make the spread of these lineages possible – diploid or polyploid chromosome structure, for example – not only have the effect, but something close to the function of conserving variation for future contingencies.

According to Dobzhansky, the evolution of mechanisms that make heterosis possible tells us that wherever it can natural selection hedges its bets by maintaining a balance between alleles that serve current adaptedness and those kept in reserve for future adaptability. Several points bearing on his dispute with Muller follow from Dobzhanky's "balance" theory of natural selection. One is that, while heterosis increases genetic loads, so important is future evolvability that "[t]he possibility must be considered that sexual species may have become adapted to their genetic loads and even may have used these loads to some advantage" (Dobzhansky 1959, 258). Another is that the price of evolutionary advance is the predictable uncovering of harmful and lethal recessives, the ineluctability of which gave moral poignancy to Dobzhansky's suffering-acknowledging and suffering-alleviating concession to "this much of eugenics," and occasion for him to fret about theodicy in the last, anxiety-ridden decade of his life. By then, Dobzhansky, a professed theist, was deep into dialogue with theologians about how a good God could have come up with such a seemingly perverse, if also creative, mechanism for evolutionary progress as heterosis (Krimbas 1994, 188–191; Greene and Ruse 1996).

In the 1950s a variant hemoglobin allele was discovered in human populations that in its heterotic configuration confers some protection against sickle cell anemia and so spreads in affected populations. Dobzhansky took this case as a harbinger of the prevalence of heterozygotes in the adapted genome. Muller admitted the case, but thought it atypical. His "classical" assumptions and heightened sensitivity to the destructive effects of radiation led him to believe that "almost all mutants … are unconditionally deleterious," that "no mutants are truly recessive," and that "their [deleterious] effects in heterozygous condition are qualitatively similar to, although often weaker than, in [recessive] homozygotes" (summarized in Dobzhansky 1959, 257). Dobzhansky believed this was wrong because Muller embraced the shadow, if not quite the substance, of Morgan's claim that mutation is the creative factor in every form of evolutionary advance and natural selection merely a pruning mechanism. To Dobzhansky, Muller remained too faithful to the "eugenic consensus" to embrace the "creativity" of natural selection that is the hallmark of the Modern Synthesis in all of its forms (Dobzhansky 1962a, 430–431; Huxley 1942, 28; Mayr 1980, 18, 22). Montagu failed fully to grasp Dobzhansky's definitions of evolutionary theory's key terms. Muller grasped them, but his biases led him to interpret them incorrectly. Mayr, Dobzhansky, and Simpson never regarded him as party to the Synthesis at all, let alone as a founder.

Were these diverging interpretations doomed to remain just interpretations, expressing little more than how the same facts look to people with different views about eugenics and politics, or was it possible to experimentally test them? Dobzhansky and Muller took the issue to be empirically resolvable. In fact, they had been working on it for a long time. In 1943, Dobzhansky and his co-workers discovered that sections of the chromosomes of *D. pseudoobscura* invert their order on a seasonal basis. Inversions protect blocks of alleles from being broken up by sexual recombination. According to Dobzhansky, they are common because they keep populations adaptively tuned to seasonal fluctuations in their environments (Dobzhansky 1943). Here was evidence, first, that recombination of existing genes is a source of variation at least as important as mutation and, second, that natural selection favors inversion polymorphisms as ways of adapting populations to environmental cycles.

Dobzhansky's embrace of adaptationism of this distinctive sort was intensified by his discovery in 1947 that heterozygote superiority goes *down* when geographically but not yet fully reproductively isolated races of flies are crossed (Dobzhansky 1950). In a state of excitement intensified by his intuition a few days earlier of the flexibility-enhancing trait of educability as decisive in the adaptive evolution of *H. sapiens*, Dobzhansky informed Dunn that this discovery offered "one of the most elegant proofs of natural selection known," since it showed that natural selection *itself* and not haphazard migration and drift contributes to reproductive isolation by insuring hybrid sterility between races-turning-into-species even as it maintains hybrid vigor within them (Dobzhansky to Dunn, April 26, 1947, Dunn Papers). The insights kept coming. A selection-based mechanism of species splitting by disruptive natural selection became clear. He told Dunn that the intermediate territories in which this process begins are "zones of tension" in which heterozygotes are no longer reproductively superior in the way we observe in hybrid vigor. It is significant, though, that heterozygote superiority returns when splitting has turned races into reproductively isolated species (Dobzhansky and Levene 1951).

Dobzhansky was conscious of how these facts, if they held up, provided a way of settling his disagreement with Muller. "This is the end of the classical theory of heterosis," he told Dunn, "because here it is crystal clear that heterosis is a [natural, adaptive] tendency of the heterozygote and not a beneficial dominant covering up a deleterious recessive" (Dobzhansky to Dunn, April 26, 1947, Dunn Papers). On the basis of his "classical" assumptions, Muller had predicted that in crosses of the sort Dobzhansky had made the frequency of heterozygotes would go up as populations, deprived of hitherto optimally adaptive circumstances, began shedding deleterious mutations into heterozygote halfway houses. When that didn't happen, Dobzhansky triumphantly wrote Muller, "You may recall that you and I have a bet for ten cents about the outcome of [these] experiments. I can tell you that the first results suggest my income will be increased by ten cents" (Dobzhansky to Muller, October 13, 1947, Muller Papers). Muller was a gentleman about it. As soon as you are sure, he replied, "I shall be giving you a dime. Let me know so I can adjust my resources for the occasion" (Muller to Dobzhansky, November 5, 1947, Muller Papers). Dobzhansky proudly displayed the dime in his lab (Lewontin, "Notes on Th. Dobzhansky." 1989, 29, Lewontin Papers).

The balance view of population structure and its companion doctrine that natural selection maintains a balanced supply of variation is the scientific ground of *Mankind Evolving*'s politics. To prove that liberal democracy, and neither communist leveling nor aristocratic caste-formation, affords the most favorable human environment Dobzhansky described how basic evolutionary principles disrupt eugenic aspirations. Only where environments have been made rigidly uniform, such as laboratories, are we entitled to say that differences have genetic causes. If each chromosomal locus contains only one optimally adapted genotype for each environment, accordingly, leveling the social environment might well reveal the genetic potentials of individuals, families, and sub-populations, as Muller claimed. But if heterozygotes, with their built in phenotypic flexibility, are generally superior the inference is invalid. In that case, the eugenic practices of both aristocratic and communistic societies alike will lump people into groups on the basis of criteria that inevitably run roughshod over environmental and genetic differences and underestimate the width of norms of reaction. "Inequality of opportunity acts ... to hide, distort, and falsify genetic diversity," Dobzhansky concluded (Dobzhansky 1962a, 260). It is tolerant democracies that afford the best conditions for fostering optimal interaction between genetic endowment and environmental situations (Beatty 1994). In societies that prize equality of opportunity, nature does well what eugenics is doomed to do badly everywhere else.

This line of argument meant that the contest between Dobzhansky and Muller boiled down to whether heterozygotes are generally superior and how common they are in natural populations. Dobzhansky maintained with increasing tenacity that heterozygotes predominate in most populations in most species, Muller that heterosis is more exception than rule. Natural selection, he believed, tends to eliminate, not preserve, heterozygotes. In 1951, Dobzhansky and Dunn's experimentally talented former student Bruce Wallace offered spectacular experimental support for the superiority of heterozygotes. Turning Muller's authority on radiation against him, he found that repeatedly crossing conspecific flies dosed with X-radiation actually enhances heterozygote superiority if the radiation is not so high that it physically damages the chromosomes and sterilizes the flies (Wallace 1951). In 1958, he discovered that the same thing happens when X-ray induced mutations are bred into lines of fruit flies previously made homozygous by removing their latent genetic diversity (Wallace 1958). Wallace's data intensified Dobzhansky's belief that the heritable variation on which selection draws is largely supplied by recombination and depends so remotely on mutation that the force of Muller's genetic load argument was blunted. At the 1959 Darwin Centennial celebration in Chicago, he remarked, "Suppression of the mutation process … would probably have little effect on the evolutionary plasticity of a population for some time to come" (Tax and Callendar 1960, III, 115).

Things did not, however, stay on this satisfying trajectory. Wallace's results decreased with attempts to replicate them, enabling Muller and his students to fight back. Throughout the 1950s, the classical and balance schools went at each other with such ferocity that Mayr, observing one of their exchanges, remarked on what "prima donnas" these geneticists are, including his friend Dobzhansky (Mayr to Lerner, June 27, 1960, quoted in Beatty 1987b, 293). It got personal among the new generation and, to the extent that they instinctively sided with their protégés, the principals. Wallace became convinced that when Muller's lab claimed to be unable to replicate his results they were intentionally setting radiation doses so high that they couldn't help damaging chromosomes (Wallace 1963; Beatty 1987b; Burian and Siegel, unpublished). Dobzhansky backed him. "Muller and [Raphael] Falk … have claimed to invalidate Wallace's results," he wrote in *Mankind Evolving.* "Their experiments … are in no way repetitions of those of Wallace" (Dobzhansky 1962a, 298).

Arguments ensued about how many loci on how many chromosomes in how many species must conform to the predictions of each model if the balancing or the classical view was to prevail. "A substantial number?" "More than half?" "Forty to sixty percent?" "More than twenty five percent?" (Dobzhansky, Lewontin, James Neel, and James Crow in discussion in 1963, our paraphrase from Beatty 1987b, 297–298). The grand vision of nature and society Dobzhansky embedded in his definitions of evolutionary biology's key terms shriveled up in this way into stipulative definitions expressing little more than operationalist agreements among contentious professionals. Having chosen to meet Muller on the field of experiment, Dobzhansky and his school lost credibility when they failed to prove that "genotypes with adaptively flexible manifestations" are more often than not heterotic (Dobzhansky to Mayr, December 15, 1970, Dobzhansky Papers). Characteristically, Dobzhansky and Wallace moved to higher ground by insisting that interactions between genetic loci are spread across entire co-adapted genomes. In consequence, fitness can't be predicated of individual genotypes, let alone of a uniquely fittest one (Wallace 1991, 90; 1994). Muller dismissed this notion as "mystical" (Muller 1958, 152; Muller to J. C. King, May 19, 1960, in Beatty 1987b, 313).

On this note, the classical-balance controversy petered out into a "stalemate" in the early 1960s (Beatty 1987b). It lived on in a new form, however, when in 1966 Lewontin experimentally validated two lynchpins of "the [research] program ... my professor ... initiated": the great extent of genetic variation in natural populations and the pervasiveness of epistatic interactions in genomes (Lewontin and Hubby 1966; Lewontin 1974; Lewontin *et al.* 2001, 29). These results had wide-ranging implications. The first backed Dobzhansky's support for Boas's theorem that there is more diversity within conventional races than between them (Lewontin 1972). The second put Lewontin on a collision course with Sociobiologists, whose "gene for" rationales for human psychological traits clashed with his school's genetic holism. To protect these core points of Dobzhansky's research program, Lewontin jettisoned the very idea of adaptation, whether of particular traits or whole co-adapted genomes, as a pseudo-scientific remnant of natural theology (Lewontin 1978; Gould and Lewontin 1979). This renunciation burnished his growing authority as an austere experimentalist, especially among molecular geneticists whose techniques he had used to show the extent of polymorphism at the level of protein evolution. But he also had bad news for them and their medical allies. The neo-eugenicist ambition to cure genetic diseases by monkeying with gene sequences is an illusion springing from Muller's classical view of population structure and evolutionary dynamics and thin conception of culture. Lewontin was not being captious when he objected to a sentence in a letter his colleagues were drafting nominating Dobzhansky for a Nobel Prize in which he was said to have believed that "[h]uman welfare depends on ... human genetic *health.*" This, Lewontin wrote, "is completely at variance with everything [Dobzhansky] stands for" (Lewontin to I. M. Lerner, November 22, 1974, Lewontin Papers, our italics).

Narrow criteria for assessing scientific theories might suggest that it was Muller, not Dobzhansky, who remained within the bounds of experimental evidence and so showed himself the better scientist. If we tend to believe Dobzhansky was on the side of the angels in his quarrel with Muller it might be because we think we already know who the angels are: the better angels of our democratic culture, not the testimony of hard-nosed scientists. Still, Muller's eugenic enthusiasms have been abandoned, but most of the evolutionary mechanisms Dobzhansky postulated have been entered into the inventory from which evolutionary biologists draw explanations: the extent of genetic variation in natural populations, the balance view of how variation is maintained, the importance of inversion polymorphisms, Dobzhansky's model of speciation, his success in explaining the etiology of sickle cell anemia and evolution's way of reducing its pathology, and his holistic recognition that organisms are environmentally sensitive, phenotypically plastic developmental systems. Physical anthropologists have widely adopted his definition of race (Farber 2015). It is probably true that Dobzhansky's biological arguments against scientific racism and eugenic interference in nature's ways could not be appreciated until liberal democracy of the sort he championed consolidated itself. But in light of the more socially context-sensitive views of theory choice now prevailing, we might justifiably conclude that, far from standing in the way of new scientific knowledge, the postwar shift in the political wind made it possible to see weaknesses in Muller's science that to his credit Dobzhansky spotted from the start.

Viewed in this historical light, it is Montagu who poses a bigger challenge to Dobzhansky's legacy. His growing social constructionism about race could not make the best case for itself until the 1970s, when attention shifted from the psychological biases of individuals and groups to the pervasive, largely anonymous discursive practices of societies and communities of inquiry. This linguistic turn has made it possible for anthropologists to assert with confidence that race talk is indeed a social construction; it biologizes a discursive practice (Durham 1991; American Anthropological Association 2008; Fuentes 2012; Goodman 2013). In this respect Montagu's skepticism seems more prescient than Dobzhansky's effort to redeem the term "race" by redefining it in terms he hoped the citizens of science-respecting countries would happily and readily learn.

This is especially so because in his last decade Dobzhansky undercut himself by trying empirically to prove his claim that, *contra* Muller, "[t]he closer the approach to equality of opportunity in a society the more observed differences between its members are likely to reflect their genetic differences" (Dobzhansky 1962a, 260). When he accepted the presidency of the newly founded Behavior Genetics Association (BGA) in 1971, it was probably with the aim of elevating its members to the perspective of full-fledged population thinking and engaging them to look for the genetic effects of heterozygotes, which would make themselves visible in liberal democracies in the telltale genetic diseases that inevitably attend them. The BGA, however, would have none of it. They blithely partitioned variance into "genes for" psychological traits and environments that they too quickly took to be the same. They confused heritability with fixity and projected population-level facts onto individuals.[16] Rejecting Dobzhansky's cautions about "lunatic fringes" on both sides of the debate about Jensen's claim that Great Society programs would not affect the fifteen point average difference in black and white IQ scores, the BGA's leaders backed Jensen. They wanted to protect their own freedom of inquiry, but they also thought he might be right (Jensen 1969; Osborne to Dobzhansky, May 11, 1973, Loehlin to Dobzhansky, May 11, 1973, Dobzhansky Papers; Dobzhansky 1973b, 1976).

The BGA exasperated Lewontin. "You have been taken in by a false issue," he told one of its leaders (Lewontin to S. Scarr-Salpatek, May 8, 1973, Dobzhansky Papers). But it also led him to intensify his effort to protect Dobzhansky's legacy from his liabilities. On a fly-hunting outing in the Anza Borrego desert not long before Dobzhansky died (and not far from the San Bernardino mountains in which he had done his early field work on fruit flies), Lewontin tried to persuade his professor that because it disrupts the work of enculturation any effort to identify the same genetic effects across human environments implies that heterozygotes have lower norms of reaction than Dobzhansky preached. This, he said, made him a genetic determinist (Lewontin *et al.* 2001, 30; Lewontin, "Notes on Th. Dobzhansky." 1989, Lewontin Papers). We read this as a rhetorical *reductio* aimed at bringing Dobzhansky back to his core commitments: the dynamic interaction among gene, organism, and environment, and Kroeber's superorganic demarcation of how culture and biology are entwined. It probably didn't succeed. By then Dobzhansky's diachronic bent had metastasized into a construction of evolution itself, not just natural selection, as a creative process in which physical evolution gives rise to biological evolution, which in turn gives rise to cultural evolution (Dobzhansky 1967, 1973b). This is just the kind of thinking that worried Kroeber as corrupting both biological and anthropological inquiry (Chapter 3 of this book). It worried Lewontin too. Unlike Kroeber, he eventually gave up on the very idea of cultural evolution (Fracchia and Lewontin 1999). Dobzhansky, by contrast, imagined that manipulating the germ line would be an important feature of the more evolved culture life of the future (Dobzhansky 1973b, 105). In this respect he was indeed party to the "eugenic consensus." What this amounts to, however, depends on closely following the contours of the controversies in and through which he did his scientific work.

Notes

1 Genetic maps are based on the assumption that the further apart genes are on chromosomes the higher the probability they will be broken up and recombined by meiotic division.
2 On the Russian school of genetics, Adams 1994; Burian 1994.
3 Dobzhansky's lectures were retroactively baptized "Jesup lectures" (Cain 2002). The name recalled an earlier set of Jesup lectures. Between 1897–1902 industrialist Morris Jesup bankrolled an American Museum of Natural History anthropological expedition to investigate the genealogical and cultural relationships of peoples flanking the Bering Sea. Boas directed it, linking him obliquely to the new lecture series. Dunn's editorship of the classics of the Modern Synthesis revived the Columbia University Press Biological Series in which the original Jesup lectures were published, which was discontinued in 1910.
4 The Modern Synthesis embraced Dobzhansky, Mayr, and Simpson's "speciation project," as we might call it, and the "adaptationist" effort to find the functional significance of traits pursued in Great Britain by E. B. Ford, Bernard Kettlewell, A. J. Cain, Philip Shepherd, and others. Both sprang from the New Systematics. Huxley's *Evolution: The Modern Synthesis* integrated them, but "speciationists" objected to his formulating genetic drift as an evolutionary factor working independently of natural selection and a tendency to match traits to genes by "beanbag genetics" (Huxley 1942, 59; Dobzhansky 1937, 184; Dobzhansky to Montagu, May 22, 1944, Montagu Papers; Mayr to Dobzhansky, June 11, 1943; Dobzhansky to Mayr, October 18, 1954, Dobzhansky Papers; Mayr 1963, 263; Provine 1986). These differences waned in the 1950s and 1960s, only to resurface in the 1980s (Gould 1983).
5 On the founding of *Evolution* and its parent organization, The Society for the Study of Evolution, see Cain 1994.
6 We choose the term "dialectic" carefully. His former student Bruce Wallace recalled how Dobzhansky "enjoyed the interplay of ideas – pushing and pulling at the logic of any idea to see how far its consequences could be carried" (Wallace 1983, ix).
7 Dobzhansky's diachronic bent was evident in his early definition of a species as a "stage" in evolution characterized by complete genetic isolation (Dobzhansky 1935, 354; 1937, 312). He saw races similarly: "What is essential about races is not their state of being but of becoming. When the separation of races is complete we are dealing with races no longer, for what have emerged are separate species" (Dobzhansky, 1937, 62–63). As a spatially oriented biogeographer, Mayr protested

that Dobzhansky was describing the process of speciation, not defining species, the results of a process (Mayr 1942, 119). Dobzhansky retreated in the second edition of *Genetics and the Origin of Species* (Dobzhansky 1951, 262) and even more in 1955b. Nonetheless, his temporal approach to the "genetics of the evolutionary process" – the title of his 1970 treatise – constantly resurfaced.

8 Collopy suggests that Dobzhansky's worries about Lysenkoism led him to police uses of the term 'race' in America (Collopy 2015). We think his concerns about American racism were stimulus enough, and might have fueled his opposition to Lysenkoism.

9 Dobzhansky told Dunn that he had

> found out from Penguin [that] … they can't get paper [due to wartime shortages] for a second run.… Odd, they had enough paper to publish Huxley's *Man Stands Alone*, which as you know is absolute trash.… But not even Dunn is President of UNESCO.
>
> (Dobzhansky to Dunn, April 26, 1947, Dunn Papers)

Their disappointment was premature. By the early 1960s the book had sold almost half a million copies (Gormley 2006, 398, n. 20).

10 On the evolutionary biology of skin color, Jablonski 2012.

11 Dobzhansky defended collaborating with Montagu by telling Dunn, "Though I know that anthropologists dislike him he is the only anthropologist I know who is interested in things that interest me. So our association may still be profitable for both of us" (Dobzhansky to Dunn, January 23, 1947, Dunn Papers).

12 The majority report was so contested that Dunn felt it necessary to publish the dissenting minority's arguments (UNESCO 1952). Some comments, including Darlington's and Sturtevant's, were overtly racist (UNESCO 1952, 63; see also Darlington 1978). Others stressed eugenics (UNESCO 1952, 33). The minority report also divided advocates of classical and balance views of population structure: Muller dissented because he saw eugenics as still being asked in the revised Statement to shoulder the burden of proof.

13 Lancelot Hogben had already spotted this problem in R. A. Fisher's experiments and his seminal tract on population genetics, *The Genetical Theory of Natural Selection*, where it appears in the claim that only "additive" genetic variation, that is, variation that is correlated in a linear way with environmental change, is relevant to adaptive evolution by natural selection (Fisher 1930; Hogben 1933; Tabery 2014).

14 This condition seems to have been met in the case of Tibetan adaptation to high environments (Huerta-Sanchez *et al.* 2014).

15 In his *Reminiscences* Dobzhansky observes that Stephen Dunn "is now an anthropologist, does all kinds of things due to good support, married another spastic, and did better than his normal brother" (Dobzhansky 1962–1963, 466; Gormley 2006).

16 Aaron Panofsky has argued that fault lines within the BGA were so deep that, in spite of the large data bases cited in its members' papers, the "*ethos* of moderation" with which it discussed claims like Jensen's, and its repetition of the mantra that "differences are not defects," Behavior Genetics never had a prayer of becoming the "normal science" that its founders envisioned (Panofsky 2014, 31). Instead, it "turned into a kind of sanctuary" for the likes of Jensen, Hans Eysenck, William Shockley, and J. Philippe Rushton (Panofsky 2014, 95; Chapter 7 of this book).

References

Archival sources

L. C. Dunn Papers, American Philosophical Society, Philadelphia, PA.
Theodosius Dobzhansky Papers, American Philosophical Society, Philadelphia, PA.

Richard C. Lewontin Papers, American Philosophical Society, Philadelphia, PA.
Ashley Montagu Papers, American Philosophical Society, Philadelphia, PA.
Hermann Muller Papers, Lilly Library, University of Indiana.

Secondary sources

Adams, Mark, ed. 1994. *The Evolution of Theodosius Dobzhansky: Essays on His Life and Thought in Russia and America*. Princeton: Princeton University Press.

Allen, Garland E. 1994. "Theodosius Dobzhansky, the Morgan lab, and the Background of the Naturalist-Experimentalist Dichotomy 1927–1947." In *The Evolution of Theodosius Dobzhansky: Essays on His Life and Thought in Russia and America*, edited by Mark B. Adams, 87–97. Princeton: Princeton University Press.

American Anthropological Association. 2008. *Race: Are We So Different?* Arlington, VA: A Project of the American Anthropology Association. DVD.

Beatty, John. 1987a. "Dobzhansky and Drift: Facts, Values, and Chance in Evolutionary Biology." In *Probabilistic Revolution*, edited by Lorenz Kruger, Gerd Gigerenzer and Mary S. Morgan, 2:271–311. Cambridge: MIT Press.

Beatty, John. 1987b. "Weighing the Risks: Stalemate in the Classical/Balance Controversy." *Journal of the History of Biology* 20: 289–319.

Beatty, John. 1994. "Dobzhansky and the Biology of Democracy: The Moral and Political Significance of Genetic Variation." In *The Evolution of Theodosius Dobzhansky: Essays on His Life and Thought in Russia and America*, edited by Mark B. Adams, 195–218. Princeton: Princeton University Press.

Benedict, Ruth. 1940. *Race: Science and Politics*. New York: Modern Age.

Benedict, Ruth and Gene Weltfish. 1943. *The Races of Mankind*. New York: Public Affairs Committee.

Brattain, Michelle. 2007. "Race, Racism, and Anti-Racism: UNESCO and the Politics of Representing Science to the Postwar Public." *American Historical Review* 112: 1386–1413.

Burian, Richard. "Dobzhansky on Evolutionary Dynamics: Some Questions about his Russian Background." In *The Evolution of Theodosius Dobzhansky: Essays on His Life and Thought in Russia and America*, edited by Mark Adams, 129–140. Princeton: Princeton University Press.

Burian, Richard and Paul Siegel. "Interview with Bruce Wallace." Unpublished transcript.

Cain, Joe. 1994. "Ernst Mayr as Community Architect: Launching the Society for the Study of Evolution and the Journal Evolution." *Biology and Philosophy* 9: 387–427.

Cain, Joe. 2002. "Co-opting Colleagues: Appropriating Dobzhansky's 1936 Lectures at Columbia." *Journal of the History of Biology* 35: 207–219.

Cain, Joe. 2009. "Rethinking the Synthesis Period in Evolutionary Studies." *Journal of the History of Biology* 42: 621–648.

Ceccarelli, Leah. 2001. *Shaping Science with Rhetoric: The Cases of Dobzhansky, Schrödinger, and Wilson*. Chicago: University of Chicago Press.

Collopy, Peter Sachs. 2015. "Race Relationships: Collegiality and Demarcation in Physical Anthropology." *Journal of the History of the Behavioral Sciences* 51 (3): 237–260.

Darlington, C. D. 1978. *The Little Universe of Man*. London Allen and Unwin.

Dobzhansky Theodosius. 1935. A Critique of the Species Concept in Biology. *Philosophy of Science* 2: 344–355.

Dobzhansky Theodosius. 1937. *Genetics and the Origin of Species*. 1st edition. New York: Columbia University Press.

Dobzhansky Theodosius. 1941. "The Race Concept in Biology." *Scientific Monthly* 52: 161–165.

Dobzhansky Theodosius. 1943. "Temporal Changes in the Composition of Populations of *Drosophila Pseudoobscura*." *Genetics* 28: 162–186.

Dobzhansky Theodosius. 1944. "On Species and Races of Living and Fossil Man." *American Journal of Physical Anthropology* 2: 251–265.

Dobzhansky Theodosius. 1951. *Genetics and the Origin of the Species*. 3rd edition. New York: Columbia University Press.

Dobzhansky Theodosius. 1955a. *Evolution, Genetics and Man*. New York: Wiley.

Dobzhansky Theodosius. 1955b. "A Review of Some Fundamental Concepts and Problems of Population Genetics." *Cold Spring Harbor Symposium on Quantitative Biology* 20: 1–15.

Dobzhansky Theodosius. 1959. "Variation and Evolution." *Proceedings of the American Philosophical Society* 103 (2): 252–263.

Dobzhansky Theodosius. 1962a. *Mankind Evolving*. New Haven: Yale University Press.

Dobzhansky Theodosius. 1962–63. "The Reminiscences of Theodosius Dobzhansky, conducted by B. Land for the Oral History Research Office of Columbia University." Dobzhansky papers.

Dobzhansky Theodosius. 1963a. "Possibility that *Homo Sapiens* Evolved Independently Five Times is Vanishingly Small." *Current Anthropology* 4: 360.

Dobzhansky Theodosius. 1963b. "A Debatable Account of the Origin of Species." *Scientific American* 208: 169–172.

Dobzhansky Theodosius. 1963c. "Anthropology and the Natural Sciences: The Problem of Human Evolution." *Current Anthropology* 4: 360–366.

Dobzhansky Theodosius. 1964. "Biology: Molecular and Organismic." *American Zoologist* 4: 443–452.

Dobzhansky Theodosius. 1967. *The Biology of Ultimate Concern*. New York: New American Library.

Dobzhansky Theodosius. 1973a. "Nothing in Biology Makes Sense Except in the Light of Evolution." *The American Biology Teacher*. 35 (3): 125–129.

Dobzhansky Theodosius. 1973b. *Genetic Diversity and Human Equality*. New York: Basic.

Dobzhansky Theodosius. 1976. "The Myths of Genetic Predestination and of *Tabula Rasa*." *Perspectives in Biology and Medicine* 19 (2): 156–170.

Dobzhansky, Theodosius and C. Epling. 1944. "Contributions to the Genetics, Taxonomy and Ecology of *Drosophila pseudoobscura* and its Relatives." *Carnegie Institute of Washington Publications* 554: 1–183.

Dobzhansky, Theodosius and Ashley Montagu. 1947. "Natural Selection and the Mental Capacities of Mankind." *Science* 105: 587–590.

Dobzhansky, Theodosius, Francisco Ayala, G. Ledyard Stebbins and James Valentine. 1977. *Evolution*. San Francisco: Freeman.

Dunn, Leslie C. and Theodosius Dobzhansky. 1946. *Heredity, Race, and Society*. New York: New American Library.

Durham, William. 1991. *Coevolution: Genes, Culture, and Human Diversity*. Stanford: Stanford University Press.

Farber, Paul L. 2015. "Dobzhansky's and Montagu's Debate on Race: The Aftermath." *Journal of the History of Biology*. DOI: 10.1007/s10739-015-9428-1.

Fisher, R. A. 1930. *The Genetical Theory of Natural Selection*. Oxford: Oxford University Press.

Fracchia, Joseph and Richard C. Lewontin. 1999. "Does Culture Evolve?" *History & Theory* 38 (4): 52–78.

Fuentes, Agustin. 2012. *Race, Monogamy, and Other Lies They Told You: Busting Myths about Human Nature*. Berkeley: University of California Press.

Gannett, Lisa. 2001. "Racism and Human Genome Diversity Research: The Ethical Limits of 'Population Thinking.'" *Philosophy of Science* 68: 479–492.

Gannett, Lisa. 2013. "Theodosius Dobzhansky and the Genetic Race Concept." *Studies in History and Philosophy of Biological and Biomedical Sciences* 44: 250–261.

Gayon, J. 2003. "Do Biologists Need the Expression 'Human Races'? UNESCO 1950–1951." In *Bioethical and Ethical Issues Surrounding the Trials and Code of Nuremberg. Nuremberg Revisited*, edited by J. Rozenberg. Lewiston: Edwin Mellen.

Goodman, Alan. 2013. "Bringing Culture into Human Biology and Biology Back into Anthropology." *American Anthropologist* 115: 359–373.

Gormley, Melinda. 2006. *Geneticist L. C. Dunn: Politics, Activism, and Community*. Unpublished Ph.D. dissertation. Oregon State University.

Gould, Stephen Jay. 1983. "The Hardening of the Modern Synthesis." In *Dimensions of Darwinism*, edited by M. Grene, 71–93. Cambridge: Cambridge University Press.

Gould, Stephen Jay and Richard C. Lewontin. 1979. "The Spandrels of San Marco and the Panglossian Paradigm: A Critique of the Adaptationist Programme." *Proceedings of the Royal Society of London B* 205 (1161): 581–598.

Greene, John C. and Michael Ruse. 1996. "On the Nature of the Evolutionary Process: The Correspondence Between Theodosius Dobzhansky and John C. Greene." *Biology and Philosophy* 11: 445–491.

Hogben, Lancelot. 1933: "The Limits of Applicability of Correlation Techniques in Human Genetics." *Journal of Genetics* 27: 379–406.

Huerta-Sánchez, Emilia, Xin Jin, Asan, Zhuoma Bianba, Benjamin M. Peter, Nicolas Vinckenbosch, Yu Liang, *et al.* 2014. "Altitude Adaptation in Tibetans Caused by Introgression of Denisovan-like DNA." *Nature* 512 (7513): 194–197.

Huxley, Julian 1940. "Toward the New Systematics." In *The New Systematics*, edited by Julian Huxley, 1–46. Oxford: Oxford University Press.

Huxley, Julian. 1941. *Man Stands Alone*. New York: Harper.

Huxley, Julian. 1942. *Evolution: The Modern Synthesis*. London: Allen & Unwin.

Huxley, Julian. 1955. "Evolution and Genetics." In *What is Science?*, edited by James Newman, 256–293. New York: Simon and Schuster.

Huxley, Julian and Haddon, A. C. 1936. *We Europeans: A Survey of Racial Problem*. New York and London: Harper.

Jablonski, Nina G. 2012. *Living Color: The Biological and Social Meaning of Skin Color*. Berkeley: University of California Press.

Jensen, Arthur. 1969. "How Much Can We Boost IQ and Scholastic Achievement?" *Harvard Educational Review* 39 (1): 1–123.

Kohler, Robert E. 1994. *Lords of the Fly: Drosophila Genetics and the Experimental Life*. Chicago: University of Chicago Press.

Krimbas, Costas. 1994. "The Evolutionary Worldview of Theodosius Dobzhansky." In *The Evolution of Theodosius Dobzhansky: Essays on His Life and Thought in Russia and America*, edited by Mark B. Adams, 179–193. Princeton: Princeton University Press.

Kropotkin, Petr. 1902. *Mutual Aid: A Factor of Evolution*. London: Heinemann.

Lewontin, Richard C. 1972. "The Apportionment of Human Diversity." *Evolutionary Biology* 6: 381–398.

Lewontin, Richard C. 1974. *The Genetic Basis of Evolutionary Change*. New York: Columbia University Press.

Lewontin, Richard C. 1978. "Adaptation." *Scientific American* 239: 212–228.

Lewontin, Richard C. 1995. "Dobzhansky – Theoretician without Tools." In *Genetics of Natural Populations: The Continuing Importance of Theodosius Dobzhansky*, edited by L. Levine, 87–101. New York: Columbia University Press.

Lewontin, Richard C. 2000. "Natural History and Formalism in Evolutionary Genetics." In *Evolutionary Genetics: From Molecules to Morphology,* edited by Costas B. Krimbas and Rama S. Singh, 7–20. Cambridge: Cambridge University Press.

Lewontin, Richard C. and J. L. Hubby. 1966. "A Molecular Approach to the Study of Genic Heterozygosity in Natural Populations. II. Amount of Variation and Degree of Heterozygosity in Natural Populations of *Drosophila Pseudoobscura*." *Genetics* 54: 595–609.

Lewontin, Richard C., Diane Paul, John Beatty and Costas Krimbas. 2001. "Interview of R. C. Lewontin." In *Thinking about Evolution: Historical, Philosophical and Political Perspectives: A Festschrift for Richard C. Lewontin*, edited by Rama Singh, Costas Krimbas, Diane Paul and John Beatty, 22–46. Cambridge: Cambridge University Press.

Livingstone, Frank and Theodosius Dobzhansky. 1962. "On the Non-Existence of Human Races." *Current Anthropology* 3: 279–281.

Mayr, Ernst. 1942. *Systematics and the Origin of Species*. New York: Columbia University Press.

Mayr, Ernst. 1963. *Animal Species and Evolution.* Cambridge: Belknap Press of Harvard University Press.

Mayr, Ernst. 1980. "How I Became a Darwinian." In *The Evolutionary Synthesis: Perspectives on the Unification of Biology*, edited by Ernst Mayr and William B Provine, 413–429. Cambridge: Harvard University Press.

Montagu, Ashley. 1942a. "The Genetical Theory of Race and Anthropological Method." *American Anthropologist* 44: 369–375.

Montagu, Ashley. 1942b. *Man's Most Dangerous Myth: The Fallacy of Race.* New York: Harper, 1942.

Montagu, Ashley. 1945. "Some Anthropological Terms. A Study in the Systematics of Confusion." *American Anthropologist* 47 (1): 119–133.

Montagu, Ashley. 1950. "A Consideration of the Concept of Race." In *Origin and Evolution of Man: Cold Spring Harbor Symposia on Quantitative Biology*, edited by Milislav Demerec, 15, 315–334. Cold Spring Harbor: Long Island Biological Association.

Montagu, Ashley. 1965. *The Idea of Race.* Lincoln: University of Nebraska Press.

Morgan, Thomas Hunt. 1932. *The Scientific Basis of Evolution.* New York: Norton.

Muller, Hermann J. 1935. *Out of the Night: A Biologist's View of the Future*. New York: Vanguard.

Muller, Hermann J. 1950. "Our Load of Mutations." *American Journal of Human Genetics* 2: 111–176.

Muller, Hermann J. 1958. "Evolution by Mutation." *Bulletin of the American Mathematical* Society 64: 137–160.

Müller-Wille, Staffan. 2005. "Race et Appartenance Ethnique: La Diversite Humaine et l'UNESCO Declarations sur la Race (1950 et 1951)." In *60 Ans d'Histoire de l'UNESCO. Actes du Colloque International, Paris, 16–18* Novembre 2005, 211–220. Paris: l'UNESCO.

Nelkin, Dorothy. 1982. *The Creation Controversy: Science or Scripture in the Schools.* New York: Norton.

Panofsky, Aaron. 2014. *Misbehaving Science: Controversy and the Development of Behavior Genetics.* Chicago: University of Chicago Press.

Paul, Diane B. 1984. "Eugenics and the Left." *Journal of the History of Ideas* 45: 567–590.

Paul, Diane B. 1994. "Dobzhansky in the 'Nature-Nurture' Debate." In *The Evolution of Theodosius Dobzhansky: Essays on His Life and Thought in Russia and America,* edited by Mark B. Adams, 219–232. Princeton: Princeton University Press.

Provine, William B. 1973. "Geneticists and the Biology of Race Crossing." *Science* 182: 790–796.

Provine, William B. 1986. *Sewall Wright and Evolutionary Biology.* Chicago: University of Chicago Press.

Provine, William B. and Elizabeth S. Russell. 1986. "Geneticists and Race." *American Zoologist* 26: 857–887.

Radick, Gregory. 2009. "Ashley Montagu: A Darwin Critic at Rutgers in the Age of McCarthy." Unpublished lecture at Rutgers University, with permission of the author.

Schiappa, Edward. 2003a. *Defining Reality: Definitions and the Meaning of Politics.* Carbondale: Southern Illinois University Press.

Schiappa, Edward. 2003b. *Protagoras and Logos: A Study in Greek Philosophy and Rhetoric.* Columbia: University of South Carolina Press.

Selcer, Perrin. 2012. "Beyond the Cephalic Index: Negotiating Politics to Produce UNESCO's Scientific Statements on Race." *Current Anthropology* 53 (S5): S173–184.

Simpson, George Gaylord. 1944. *Tempo and Mode in Evolution.* New York: Columbia University Press.

Simpson, George Gaylord. 1959. "Foreward" to *The Life and Letters of Charles Darwin.* New York: Basic Books. v-xvi.

Sinnott, E. and L. C. Dunn. 1925. *Principles of Genetics.* NewYork: McGraw-Hill.

Sinnott, E., L. C. Dunn and T. Dobzhansky. 1950. *Principles of Genetics.* New York: McGraw-Hill.

Smocovitis, Vassiliki Betty. 2006. "Keeping up with Dobzhansky: G. Ledyard, Jr. Stebbins, Plant Evolution, and the Evolutionary Synthesis." *History and Philosophy of the Life Sciences* 28: 9–48.

Stebbins, G. L. 1950. *Variation and Evolution in Plants.* Columbia University Press, New York.

Stebbins, G. L. 1995. "Recollections of a Coauthor and Close Friend." In *Genetics of Natural Populations: The Continuing Importance of Theodosius Dobzhansky*, edited by L. Levine. 7–13. New York: Columbia University Press.

Suárez-Díaz, Edna. 2017. "The Molecular Basis of Evolution and Disease: A Cold War Alliance." *Journal of the History of Biology*, March, 1–28.

Tabery, James. 2014. *Beyond Versus: The Struggle to Understand the Interaction of Nature and Nurture.* Cambridge: MIT Press.

Tax, Sol and Charles Callendar. 1960. "Panel Three: Man as an Organism." In *Issues in Evolution* (vol. 3 of *Evolution after Darwin*, edited by Sol Tax and Charles Callendar), 145–174. Chicago: University of Chicago Press.

UNESCO. 1951. Statement on the Nature of Race and Race Differences. In *Four Statements on the Race Question*, UNESCO 1969. Available online at http://unesdoc.unesco.org/images/0012/001229/122962eo.pdf

UNESCO. 1952. *The Race Question.* Paris: UNESCO.

Wallace, Bruce. 1951. "Genetic Changes within Populations after X-irradiation." *Genetics* 36: 612–628.

Wallace, Bruce. 1958. "The Average Effect of Radiation-Induced Mutations on Viability in Drosophila Melanogaster." *Evolution* 12: 532–552.

Wallace, Bruce. 1963. "Further Data on the Overdominance of Induced Mutations." *Genetics* 48: 633–651.

Wallace, Bruce. 1991. "Coadaptation Revisited." *Journal of Heredity* 82: 89–95.

Waters, C. Kenneth. 1990. "Why the Anti-reductionist Consensus Won't Survive: The Case of Classical Mendelian Genetics." *PSA* 1990, 1: 125–139. Reprinted in *Conceptual Issues in Evolutionary Biology*, edited by Elliott Sober, 2nd edition (1993) and 3rd edition (2006). Cambridge: The MIT Press.

White, Leslie A. 1949. *The Science of Culture: A Study of Man and Civilization*. New York: Grove.

Zarefsky, David. 2006. "Strategic Maneuvering through Persuasive Definitions: Implications for Dialectic and Rhetoric." *Argumentation* 20: 399–416.

Zuckerkandl, Emile and Linus Pauling. 1962. "Molecular Disease, Evolution and Genetic Heterogeneity." In *Horizons in Biochemistry*, edited by Ibn Michael Kasha and Bernard Pullman, 189–225. New York: Academic Press.

5 Unifying science by creating community

The epideictic rhetoric of Sherwood Washburn

The New Physical Anthropology

In the previous chapter we saw how Theodosius Dobzhansky, L. C. Dunn, Ashley Montagu, and Ruth Benedict aligned anti-stadial, anti-racist, and anti-eugenicist themes in Boasian anthropology with population-genetic evolutionary theory. Their aim was to unite cultural anthropology with the politics of multicultural democracy. To succeed they needed physical anthropologists to adapt their field, too, to the Modern Evolutionary Synthesis. The problem was that even after World War II physical anthropology was riddled with race-based classifications, non-selectionist ideas about evolution's causes, and ignorance of genetics. This chapter is devoted to Sherwood Washburn's successful effort to put population genetic foundations under both physical and cultural anthropology, thereby integrating four-field anthropology in a new way and more effectively enlisting all of anthropology in the service of the postwar anti-racist cause.

Washburn was the leading voice in American physical anthropology from the 1940s to the mid-1970s (Haraway 1988; Marks 2000; Fuentes 2010). He knew from his association with Dobzhansky and the other makers of the Modern Synthesis that "geneticists and experimental zoologists … have created the 'New Systematics,' 'Modern Synthesis of evolutionary studies,' or whatever one wants to call it," that "the new evolutionary thinking differs from the old in its emphasis on process rather than classification and names," that "the emphasis on evolutionary process, including race, gives anthropology a unity it never had before," and that "archeology provides evidence on all these topics, as does linguistics and cultural anthropology." But Washburn also knew that so far "comparative anatomy has been included to a very minor extent." The physical anthropologist is needed "because his interests and training include these other absolutely necessary branches of knowledge," but add a missing dimension (Washburn to Hooton, August 29, 1951, Washburn Papers). Without the unification of physical anthropology with its sister fields through the Modern Evolutionary Synthesis, Washburn feared that four-field anthropology would falter, with large and unwelcome consequences for the issues about race that had stimulated it.

Washburn framed the problem clearly and suggested how to solve it. In a stream of ten- to twelve-page papers delivered at professional gatherings for over three decades, many soon published in professional venues, he argued that structural-functional analysis demonstrates how natural selection further modified the homologous anatomical structures that enable monkeys, great apes, and hominids to live lives adapted their niches in order to fit *H. sapiens* for cultural life. These analyses were to be based not only on fossil remains, but also on anatomical experimentation in the lab and observation of primate behavior in the wild.

Washburn opposed his favored position to a claim that he urged his audience to reject. The rejected side of his binaries always included typification and classification torn from the "functional complexes" that he took to be manifestations of adaptive gene frequency shifts in populations. "Traditional descriptions that substituted typological classification for real biological evolution" are at odds with real evolutionary theory, he said, by which he meant population-genetic natural selection (Washburn 1953, 725). When this strategy has been pursued far enough he expected physical anthropology to sing to the tune of integrated ways of life that can be read from the fragmentary fossil record, thereby preventing cultural anthropologists from going off on their own under the false impression that physical anthropologists are hopelessly mired in the typological thinking of the past.

Washburn arrived early at a number of claims about human evolution that, although he was not alone in adopting them, subsequent inquiry has treated so kindly that we regard them as commonsensical. Bipedalism, first seen in the *Australopithecines* of Southern and East Africa, facilitated tool making as well as tool use. Differences in the structures, functions, and uses of the hand between monkeys, great apes, and the more proximate ancestors of *H. sapiens* provide evidence that "the hand was freed by the assumption of bipedal locomotion" (Washburn 1959, 24). The primary use of handmade, handheld tools – sharpened rocks – was to kill and butcher prey that lived on the savannah alongside the immediate ancestors of our species. At first, prey were small, but as tools improved they became bigger even as the dentition of the hunters became smaller. "It is my belief," Washburn wrote, "that the decrease in the size of the anterior teeth and tripling of the size of the brain came after man was a tool user and were the result of selection pressures coming in with the use of tools" (Washburn 1959, 25).

The selection pressures in question were for enhanced cooperation in hunting, killing, and butchering dangerous animals. Throughout the Pleistocene, what Washburn called "proto-culture" (and, in our own species, simply "culture") led to an enlarged brain because "in hunting cooperation and the necessity of communication and language," including gendered role differentiation and attendant sexual dimorphism, requires ever more "memory, foresight and originality" (Washburn 1959, 25, 31).[1] Distinctive about Washburn's treatment of this theme was his insistence that, "It is more correct to think of much of our structure as the result of culture than it is to think of men anatomically like ourselves as slowly discovering culture" (Washburn 1959, 21). In this respect his research supported Dobzhansky's claim in *Mankind Evolving* that culture does not sit uncomfortably atop our biology, but instead emerges by a process in which its stirrings in great apes and hominids ("proto-culture") feed back onto bodies that have been increasingly refashioned for cultural life. Natural selection of this gene-soma-culture interactionist sort is still at work changing gene frequencies to fit populations to changed environments. But for Washburn, as for Dobzhansky and Montagu, this process occurs in such close alignment with our developmentally entrenched capacity for culture that it cannot possibly carry any implication that we are "hastily made over apes" who must pay a biological cost for leading a cultural form of life.[2]

It was not merely by dint of repetition or the clarity of his short papers that Washburn made his project known. The venues in which he did so were just as important. Washburn gained a hearing for his hope that anthropology would deepen its unity by aligning itself with the new evolutionary biology by staging, with help from hand-picked allies, a string of symposia and conferences in the 1950s that brought anthropologists and biologists together into a community of inquiry formed by the process of consultation and communication he fostered. In his presentations, Washburn typically praised what, in explicit reference to the New Systematics, he called the New Physical Anthropology and blamed those resisting it as obtuse. For this reason we propose that his effort to unify anthropology and beyond that the biological and human sciences generally is best understood as epideictic rhetoric.

The name comes from Aristotle's *Rhetoric*, which defines epideictic as the oratory of praise and blame on ceremonial occasions. Aristotle identified three kinds (*genê*) of artful (*technikos*) rhetoric (Aristotle *Rhetoric*, I.3.1358a36-b28). The two kinds with which he contrasts epideictic are also focused on praise and blame. But in the judicial rhetoric exercised in law courts praise and blame take the form of convicting or exonerating one who has been accused of a crime, while in the political rhetoric of the legislative assembly it takes the form of praising or blaming a proposed course of collective action, such as invading another *polis*. Epideictic's aim is to form or shore up a community that hears occasion-bound speeches such as Pericles's famous funeral oration or, in our own polity, Lincoln's striking riff on it at Gettysburg. We view Washburn's tireless organizational efforts in the two decades following World War II as arranging epideictic occasions for building a new community of anthropologists and biologists and, in his own contributions to these events, steering that community to praise what is good (true) in the new population-genetic biology and blame what is bad (false) in the old.

The titles, locations, and names of the editors of the proceedings of the seminars, symposia, colloquia, and commemorations on which this chapter is built are as follows:

1 Summer Seminars in Physical Anthropology, 1945–1952 (Kaplan 1946–1952).

2 Cold Spring Harbor Seminar on Quantitative Evolution XV, 1950. "Origin and Evolution of Man" (Demerec 1950).
3 Wenner-Gren Foundation International Symposium on Anthropology, 1952 (Kroeber 1953; Tax 1953).
4 Columbia University Bicentennial Conference, 1954 (Leary 1955b).
5 Behavior and Evolution, 1955, 1956 (Roe and Simpson 1958).
6 The Evolution of Man's Capacity for Culture, 1956 (Spuhler 1957).
7 Cold Spring Harbor Symposium on Quantitative Evolution XXIV "Genetics and Twentieth Century Darwinism" (Woolridge 1959; a Darwin Centennial Event).
8 University of Chicago, Darwin Centennial Celebration, 1959 (Tax 1960).
9 The Social Life of Early Man (Washburn 1961).
10 Classification and Human Evolution, 1961 (Washburn 1963a).
11 Symposium on Man the Hunter, 1966 (Lee and DeVore 1968).

It may seem odd that scientific conferences and their published proceedings should be considered epideictic rhetoric. Accordingly, we call attention to recent work in the rhetoric of science that shows how well matched are the characteristics of epideictic rhetoric and scientific texts that, in the words of the rhetorician of science Leah Ceccarelli, "catalyze community" within and between scientific disciplines because, as another rhetorical scholar of science, Celeste Condit, puts it, "the focus of the event is inevitably … on unity and sharing" (Ceccarelli 2001, 3; Condit 1985, 289; Sullivan 1991, 1994; Casper 2007).

Commemorative occasions (such as 4 and 8 in the list above) are obvious sites of scientific epideictic (Abir-Am 1999). There is nothing odd about the appearance of biologists and anthropologists as featured speakers at events such as the fiftieth anniversary of the University of Chicago (Redfield 1942), Columbia University's Bicentennial Conference (Leary 1955b), or celebrations marking the centennial of the publication of the *Origin of Species* at the Anthropological Society of Washington (1959), the American Philosophical Society in Philadelphia (1959), the Cold Spring Harbor biological research station on Long Island (Wooldridge 1959), and the University of Chicago (Tax 1960). An earlier generation of biologists orated in a similar vein at the fiftieth anniversary celebration of Darwin's book at Columbia University and elsewhere. A similar hubbub unfolded in 2009. The purpose of these celebrations is to strengthen bonds between professional communities of inquiry by recalling, reenacting, and thereby reanimating a foundational event that they all shared and, to various degrees, to expand this sense of community to the American public at large. The Darwin Centennial Celebration mounted by Washburn's colleague Sol Tax at the University of Chicago was specifically intended to assure the American public that in the wake of the Modern Evolutionary Synthesis it had nothing to fear from Darwinism (Tax 1960; Smocovitis 1999). These aims were also present, albeit less overtly, in the academic forums enumerated above: conferences, symposia, and proceedings that crossed disciplinary boundaries in searching for and creating scientific unity. Accordingly, we see epideictic community-formation in these occasions as well.

In all of these events, as in epideictic generally, correct opinion or belief – *ortho-doxy* in the original sense of the term – is praised and heterodoxy condemned (Vivian 2006).[3] It should not be forgotten, however, that in science as well as in discursive formations as different as religion or politics what counts as orthodox at a given time negotiates a shifting boundary between innovation and tradition. In the gatherings listed above, scientists tutored each other about what it is correct to believe in view of newly discovered facts and more penetrating concepts. In interventions of this sort, argument is aimed as much at securing a speaker's authority to address others on behalf of and in the name of his or her field (*ethos*) as it is at evidencing what a speaker is proposing as right to believe (*logos*). Authoritative mediations between innovation and tradition are achieved by deploying lines of argument that tend to create unity not only by reciting facts but by highlighting their significance. The characteristic argumentative strategy of epideictic is not reciting facts or wielding definitions, but amplifying and downplaying their importance. In scientific epideictic, commonplaces (*topoi*) – Washburn's appeal to the principle that form follows function, for example, or Dobzhansky's axiom that, "Nothing in biology makes sense except in the light of evolution" (Dobzhansky 1973) – are, as Aristotle says, "amplified by being invested with grandeur and beauty" (Aristotle *Rhetoric*, I.9.27–29). Recall in this connection the last sentence of Darwin's *Origin*: "There is grandeur in this view of life in which … from so simple a beginning endless forms most beautiful … have been, and are being, evolved" (Darwin 1859). Imbuing facts with significance imbues them with values.

Washburn, Dobzhansky, Mayr, Simpson and others did this by presenting new discoveries about evolution, including the evolution of humans, as amplifying Darwin's values more persuasively than Darwin. In spite of his insistence on the unity of the human species in and through our evolved moral sense, Darwin's picture of human evolution was marred by elements of Victorian imperialism, sexism, and racism that after his death led his name to be associated with views of evolution that favored stability by allowing a fixed pre-evolutionary hierarchy of types, including types of humans, merely to unfold in time (Darwin 1871; Richards 1987; Haller 1971; Rainger 1978; Lorimer 1988; Jackson and Weidman 2004; Beasley 2010). Peter Bowler has called this "The Non-Darwinian Revolution" (Bowler 1988). Boas both affirmed the existence of this tendency and questioned it by invoking Darwin's method against what in his day still passed even in his own mind as Darwinism (Chapter 2 of this book). Kroeber helped Boas's case by extirpating Lamarckian heredity and progress from cultural anthropology (Chapter 3 of this book). In this chapter we will see Washburn following suit by expelling colleagues in physical anthropology who retained remnants of late nineteenth century views of phylogeny and the racist messages they carried.

In bringing this charge against his own graduate adviser, Earnest A. Hooton (1887–1954), Washburn mediated innovation and tradition by insisting that the new anthropology values disciplinary unity as much as the old, but emphasizes biological processes and causes over "classification and names" in order to "give anthropology a unity it never had before" (Washburn to Hooton, August 28, 1951, Washburn Papers). His epideictic rhetoric, that is to say, amplified the shared value of scientific unanimity, but reversed the hierarchical importance his opponents assigned to methods of bringing it about.[4] Washburn did not eliminate classification, but, like Dobzhansky, demoted it from end to means. Nor did he claim to be the first physical anthropologist to do experiments or observe primates in the wild. But in amplifying the importance of these methods and downplaying morphological measurement he opened a path for those who were wavering between the old and the new physical anthropology to make their way toward the Modern Synthesis without breaking too severely with their past. In this way Washburn sought to form an inclusive community of inquiry. Still, when the unity he was attempting to create was threatened by complicity with or tolerance of racism he did not hesitate to remonstrate with offenders and to marginalize or denounce them if that failed. We will see him treating Hooton this way, as well as his fellow Hooton student, Carleton Coon (1904–1981).

Another aspect of epideictic rhetoric, scientific and otherwise, is visible in high wire acts like Washburn's. It been noted since antiquity that epideictic draws so much attention to the *ethos* of speakers that it is often called "display rhetoric" in the sense that in performing it an orator puts his or her character and skill as a performer on display and therefore on the line. It is facile to reduce this aspect of epideictic to hubris or chutzpah. It is more deeply rooted in the fact that misperforming speeches on ceremonial occasions carries risks of fracturing rather than uniting the addressed community. By the same token, performing well makes an epideictic speaker into a representative embodiment of the beliefs and values of a renewed community, thereby stimulating the wider dissemination and acceptance of its beliefs. The late Stephen Jay Gould was a virtuoso of this sort. In the generation before him so was Washburn.

He had more than enough confidence to perform this role. The son of a clergyman, Washburn was not in the least a show off. He was motivated by deep belief in the new biology and its relevance for American anthropology's stand against racism. We will trace his career in the next section, arguing that his early break with Hooton freed him for encounters with a variety of theoretical orientations whose useful features he absorbed in shaping his own views. Chief among these influences were the scientific anti-racism embodied by Boas, to which he was first attuned by an even more restive Hooton graduate student, Gabriel Lasker; Dobzhansky's genetics; and the brand of structural-functionalist anthropology that flourished at the University of Chicago when Washburn was a faculty member there. If everything hangs on theoretical consistency, these programs can easily be shown to contradict one another. Washburn recognized that functionalist anthropology is not Boasian (De Vore and Washburn 1992, 418). It was through Lasker and Dobzhansky that he came to the anti-racist convictions of Boasians of the strict observance. But having embraced them, he took the harder edges off competing theories by creating in himself the kind of "synergistic" community of "different perspectives" that he prized in anthropology departments (De Vore and Washburn 1992, 419).

In integrating aspects of the competing research programs he encountered in the course of his career, Washburn's New Physical Anthropology presupposed a robust enough conception of culture to prevent anthropology from being reduced to or replaced by various forms of evolutionary psychology.[5] He was not as opposed as Kroeber and other Boasians *stricto sensu* to pronouncing on supposed psychological differentia (Chapter 3 of this book). As we will see, however, he categorically rejected the "constitutional psychology" of William Sheldon, which Hooton endorsed, because it undermined the role of cultural life in shaping our minds as well as our bodies. Washburn insisted that, unlike other hominids, our tendencies to aggression and cooperation co-evolve with and are profoundly modified by cultural practices that exert selection pressures at points where material aspects of cultural life such as the use of weapons meet environments to be exploited, tamed, and transformed. As he pointed out in rejecting the Sociobiology of the 1970s, these are social facts. They cannot be reduced to or built up from the psychological tendencies of individuals (Washburn 1978a, b).

Unity within and among the sciences was in some ways the Holy Grail of postwar American thinking. It seemed necessary if scientists were to play a leading role in guiding society, thereby preventing recurrence of the irrationalism that culminated in National Socialism. From the perspective of the theory-oriented logical-empiricist view of scientific unity that became ascendant, Washburn's way of weaving together Dobzhansky's genetics, Chicago structural-functionalism, and Boas-inspired scientific anti-racism does not exhibit the epistemic virtues logical empiricists prize. This is not surprising. Unlike Kroeber, Washburn's professional life was conducted with only passing attention to meta-theoretical norms. It was filled instead with situation-specific argumentation addressed to the communities of inquiry whose influence he helped consolidate and project *extra muros* by encouraging his colleagues to thread their way through rhetorical situations that simultaneously threatened and, when effectively addressed, advanced their common interests. In this respect, Washburn's epideictic mode of scientific discourse serves as a useful vehicle for identifying a conception of the unity of science that is neither impersonally abstract nor personally abrasive, but focuses instead on argumentation that advances knowledge by building communities. In the last section of the chapter we will argue that this conception of unity was more widespread in Washburn's formative years than logical empiricist historiography suggests.[6]

Aligning anthropology with the Modern Synthesis: Washburn *contra* Hooton and Coon

Washburn (1911–2000) was a child of genteel, WASP privilege. He was raised in Cambridge, Massachusetts, where his father was Dean of the Episcopal Theological Seminary. As a boy, he frequented and later helped out at Harvard's Museum of Comparative Zoology. After his secondary education at the elite preparatory school Groton, he entered Harvard. He received his Ph.D. in anthropology there in 1940. His professional career was marked by westward migration. Having taken his first job as an Assistant and then Associate Professor of Anatomy at Columbia's medical school, the College of Physicians and Surgeons (1940–1947), he subsequently became Professor of Anthropology at the University of Chicago (1947–1958). His career reached its apex in Kroeber's department at the University of California, Berkeley, where he remained until he died in 2000. Given his background and record of success, it is no wonder that he was both self-confident and genial, with a touch of Yankee cussedness when he was sticking to a principle.

In forging his career, Washburn had to contend with his Ph.D. advisor, Hooton, who exerted as great an influence over physical anthropology's institutionalization in American universities as Kroeber did in securing an academic home for cultural anthropology (Wolpoff and Caspari 1997; Giles 2000; Brace 2005). He mentored no less than twenty-eight Ph.Ds in a forty-year career at Harvard (Giles 2010). His position on race was complicated. He embraced two parts of racialist anthropology. Anthropologists can reveal the reality of morphologically defined racial categories and people's physical characteristics do index their mental and moral characteristics. But he rejected "race propagandists" who "were not professional anthropologists" and attempted "to refer every manifestation of the psychological qualities assumed to be the exclusive property of this or that race to the physical type in question" (Hooton 1926, 76; Giles 2010). He disparaged the "ludicrous yet tragic history of the prostitution of the scientific conception of race to base political motives, religious intolerance, and economic advantage" (Hooton 1935, 27). But his opposition to the worst abuses of the traditional race concept had the effect of keeping the traditional, typological race concept alive in physical anthropology in the face of culturalist attacks on it and population biological re-conceptualizations of it, in consequence doing little to undermine the populist racism he despised.

His position was possible because Hooton never accepted the probative obligations Boas imposed on those who postulated fixed, morphologically defined races. In Chapter 2 we saw Boas marshaling evidence against the cephalic index as a way of shifting the burden of proof onto those who retain race as a fixed category by treating species-marking traits as adaptively neutral and hence permanent enough to serve as the basis of systematic classification. Hooton read Boas's challenge as assuming a crude nature-nurture binary and wrote him off as an extreme environmentalist. "Environmentalist onslaughts upon racial criteria," he wrote,

> have in no single instance terminated the usefulness of any standard measure of race differentiation.... They have merely applied a harsh and well-deserved castigation to rabid hereditarians who have assumed, without taking the least pains to investigate the matter, that physical features in man are solely the result of germinal combinations.
>
> (Hooton 1935, 28)

Having dissociated himself both from early genetic determinist views of racial inheritance and from what he wrongly took to be Boas's environmentalism, Hooton trained his students to classify specimens in terms of

> significant racial criteria ... based principally upon non-adaptive bodily characters.... The very insignificance of certain features, such as the form of the hair or the thickness of the lips, insures their hereditary transmission in the absence of selected adaptive modifications that have survival value.
>
> (Hooton 1926, 77)

His fixation on racial typologies led Audrey Smedley to conclude that Hooton projected "the legacy of racial and polygenist thought" into the twentieth-century by liberating it from the Darwinian adaptationism that in contrasting ways his students Coon and Washburn restored (Smedley 1998, 296; Wolpoff and Caspari 1997; Chapter 6 of this book).

Even as a student, Washburn was skeptical of Hooton's methods and preconceptions. When he served as his teaching assistant in a class on primate evolution, he recalled, "I stressed that all the major families of primates could be seen as adaptive radiations, but Hooton was sure that the families should be defined by non-adaptive characters" (Washburn 1983, 6). When he left Harvard in 1940 to take a job in the anatomy laboratory at Columbia he immediately put Hooton behind him. This allowed him to nurture his ambition to move physical anthropology beyond measuring and classifying skeletal specimens toward evolutionary explanation. In this he was encouraged by two men: S. R. Detwiler, his supervisor at Columbia, and Dobzhansky, whose return to Columbia from his ten-year stint in California coincided with Washburn's own arrival in New York (Chapter 4 of this book).

Detwiler denied that merely dissecting and comparing specimens could produce satisfactory explanations in anatomy. Experimentation was necessary. "Whereas the comparative method has rendered great service in the past," he wrote, "it has done little to reveal causal relations.... The dynamics underlying development and the production of organic form were left practically uninvestigated, and they present us today with problems which can be approached only by the experimental method" (Detwiler 1929, 565). Washburn agreed. In a co-authored 1943 paper he and Detwiler argued that so far physical anthropology has been "a descriptive science concerned with the evolution of mankind and with the variation among the groups of living men" (Washburn and Detwiler 1943, 171). Its descriptive bent produced a mass of data, but unfortunately prevented physical anthropologists from making virtually any progress in over a century in correctly framing and solving the problems anatomical comparison raised. What, for example, determines the shape of the brain case? The size of the brain? The muscles of the jaw? Both? Something else? These questions had been around since the nineteenth century, but even the best twentieth century physical anthropologists, such as Franz Weidenreich and Aleš Hrdlička, were no closer to solving them. The reason, Washburn and Detweiler alleged, is that anthropologists neglected the most relevant kind of data: experimental. "Even with abundant descriptive data which no one questions, explanation and interpretation still cause difficulties" because the data on hand are always capable of supporting rival interpretations, but offer no guidance in choosing among them. The traditional remedy for scientific deficiencies of this sort is controlled experimentation to get the right facts and the facts right. Washburn and Detwiler recommended it. Having done so, they urged those who had attempted to perform experiments in comparative anatomy to try harder by reporting on an experiment they themselves conducted on the growth of amphibian eye occipital bones. It showed that previous experimentation, having been compromised by manipulation of soft tissues, was worthless and should be done over (Washburn and Detwiler 1943).

Washburn's characteristic style of argument is already evident in this 1943 paper. He recalled that his and Detwiler's publication in the *Journal of Physical Anthropology* was accepted only after a considerable number of objections and significant revisions. "I had learned my lesson," he later told an interlocutor. "Anyone trying to use experimental methods in anthropology was going to have a hard time" (Washburn 1983, 7). Henceforth, whenever he opposed a claim he tried to obey the first rule of rhetoric: Seek the good will (*benevolentia*) of the audience. Washburn typically did so by recommending that all hearers or readers had to do to come to his conclusion was shift their methodological emphasis: "Experiments do not replace any of the usual anthropological procedures but serve to check the biological soundness of theory" (Washburn and Detwiler 1943, 176). The strategy of amplifying and downplaying aspects of a case characteristic of epideictic is already apparent. Physical anthropology needn't abandon what it was doing; it merely needed to reorder the importance it attached to its various methods. The deficiencies of one approach will be remedied by stressing, but not discrediting, others.

While he was at Harvard, Washburn became friends with Lasker, a younger Hooton student who disconcerted his adviser by writing a dissertation on hybridity and plasticity that was chock full of Boas's data and arguments (Bogin 2003). Lasker had grown up in the culturally pluralist ambiance of New York City and Columbia-style anti-racist anthropology. He introduced Washburn to the New Systematics and raised his consciousness about racist distortions in science. "Lasker and I discussed anthropology constantly," Washburn recalled. "He helped me to understand why the genetics of populations replaced typology and the nature of Boas's contributions to biometry.... He clearly saw the social importance of anthropology" (Washburn 1983, 5–6).[7] Accordingly, when Washburn arrived at Columbia it was natural that he would pay a call on Dobzhansky. He got nowhere with the geneticist, however, until he assured him that although he was Hooton's advisee he did "not believe in types.... It is populations that should be compared." Dobzhansky, who was just turning to anthropological issues, "beamed and shook my hand.... There began a very pleasant friendship" – and Washburn's initiation into population genetics (Washburn 1983, 20). He joined Montagu in encouraging Dobzhansky to publish in anthropology journals and began collaborating with him in organizing meetings, notably the 1950 Cold Spring Harbor Symposium on the "Origin and Evolution of Man," which drew physical anthropologists and biologists of the Modern Evolutionary Synthesis into conversation.

In his 1951 manifesto "The New Physical Anthropology," Washburn wrote that physical anthropologists who, unlike Hooton, favor marking off species by adaptive characters had not quite faced up to the question of how to "determine the precise nature of a particular adaptation." "Suggestions of adaptations," he said, "are not enough" (Washburn 1951, 300). Fortunately, he reported, "Recently, evolutionary studies have been revitalized and revolutionized by an infusion of genetics into paleontology and systematics.... Population genetics presents the anthropologist with a clearly formulated, experimentally verified, conceptual scheme" (Washburn 1951, 298). In that scheme, adaptations are effects of natural selection, usually working in conjunction with other evolutionary factors. But the conditions that must be met to prove that a trait is really an adaptation are onerous. They must be grounded in genotypic changes that have spread through a population precisely because they have made that population more reproductively successful than its temporal or spatial neighbors. In this sense, adaptation is an inherently historical phenomenon, "an effect ... meaningful only as it has contributed to population success" (Haraway 1989, 213). Washburn acknowledged complications in applying this already burdensome criterion. An adaptation must also be distinguishable from an environmentally induced accommodation due to the phenotypic plasticity of a genotype and from "modifications of [gene] frequencies result[ing from] mutations, [genetic] drift, and migrations" (Washburn 1951, 298–299). In remarking that adaptation of this natural-historical sort "is essentially a return to Darwinism," Washburn encouraged his readers to embrace the population-genetic style of evolutionary reasoning on Darwin's authority (Washburn 1951, 299). More proximally, he was following Dobzhansky (Chapter 4 of this book).

What Washburn added was deep knowledge of comparative anatomy. The guiding principle of physical anthropology, he asserted, and the key to his own research program, "must be that the major force in evolution is selection of functional complexes" (Washburn 1951, 300). Morphological and physiological modifications properly underwritten by gene frequency changes are adaptive insofar as they realize biological functions. Strictly speaking, however, it is the adapted activities that physical modifications serve that confer adaptedness on them. To show how this proposal worked in anthropology, Washburn divided the primate body into units of functional behavior such as bipedal locomotion, mastication, and communication. These activities are key to identifying the physical structures that carry them out. Accordingly, what Washburn called "functional complexes" consist of adapted behaviors and the adapted structures that serve them. As he explained to a colleague,

> Evolution is the process by which more and more effective systems of behavior have appeared. Selection is always for better function, not for

> some anatomical or physiological variation for itself. Viewed in this way, the path of [human] evolution becomes a record of our emerging ability to exert agency in our social and natural environments.
>
> (Washburn to Irving Hallowell, October 20, 1952, Hallowell Papers)

Once again appealing to Darwin's authority, Washburn claimed that the kind of anatomy in which bones, muscles, ligaments, and so forth are treated in biology and medical textbooks as separate objects of adaptation "became obsolete with the publication of the *Origin of Species*" (Washburn 1951, 304).[8]

This methodological principle had real bite. Washburn noted that Weidenreich's excellent descriptions of Peking Man describe "lower jaws, teeth, and skulls in separate monographs" (Washburn 1983, 7). But writing separate monographs on each morphological item left an impression that they are separately evolving traits, making it difficult to determine how morphological patterns that involve more than one supposed trait are dynamically reshaped by changing selection pressures exerted at a number of anatomical points. Suppose, for example, someone was attempting to track the evolution of chins through a sequence of hominid species. The labor will be in vain because the chin is not a trait in its own right, but the result of a shifting movement over evolutionary time of "two relatively independent areas of the jaw," which change their orientation in different species because the teeth that the jaws hold in place are called on to perform different tasks in different species (Washburn 1983, 7–8). When Gould and Lewontin used this example to attack sociobiologists for too casually identifying traits to which they precipitously ascribed selected functions or adaptations, they were channeling, if not quite quoting, Washburn.

This was a large change from standard operating procedure in physical anthropology, but Washburn characteristically reassured his colleagues that no new methods are needed:

> The four major methods for factoring complexes out of the body are: (1) comparison and evolution; (2) development; (3) variability; and (4) experiment. All of these have been used by numerous investigators, but, to the best of my knowledge, they have not been combined into a working system. All must be used to gain an understanding of the human body.
>
> (Washburn 1951, 300)

The most striking implication of the New Physical Anthropology is Washburn's insistence that changes in the functional complexes *H. sapiens* inherits from our hominid predecessors are products of human cultural practices. "Our brains, then, are not just enlarged," Washburn argued, "but the increase in size is directly related to tool use, speech, and increased memory and planning" (Washburn 1959, 28–29). The reward for embracing this inference, he promised, would be greater unity among the various aspects of anthropology than the field had previously enjoyed. The New Physical Anthropology cannot be produced in the anatomist's laboratory alone. It also requires deep understanding of human culture, past and present. So just as cultural anthropologists would be well served by learning up-to-date biology, especially population genetics, biologically oriented anthropologists will have to learn cultural anthropology and archaeology, which studies material traces of cultural practices. Washburn told a 1951 Wenner-Gren symposium that

> Insofar as man has adapted by his way of life, the study of human evolution is inseparably bound to the study of archeology and ethnology. If we would understand the process of human evolution, we need a modern dynamic biology and a deep appreciation of the history and functioning of culture. *It is this necessity which gives all anthropology unity as a science.*
>
> (Washburn 1953, 726, our italics)

Here was a defense of four-field anthropology more compelling than any before it.

His move to the University of Chicago in 1947 catalyzed these insights. At Chicago, sociologists and anthropologists were yoked together in productive interanimation. Robert Redfield, the department's seminal anthropologist, adopted sociology's congenital stress on social change, especially the social changes that attend the process of modernization. This led him not only to support his sociologist colleague Robert Park's way of recording social changes in the city of Chicago by ethnographically and quantitatively charting shifts in the composition of its neighborhoods, but also to study ways in which Mexican villages situated some distance from urban centers, but not far enough to be uninfluenced by them, achieve social integration by adapting to change (Wilcox 2004). The Chicago anthropologist Sol Tax, who organized the 1959 Darwin Centennial Celebration at the University of Chicago, followed suit in his studies of villages in semi-rural Mexico and closer to home in Native American towns in Iowa. Stressing dynamic interaction between pre-modern, modernizing, and modern cultures more or less automatically put Chicago-style anthropologists at odds with a tendency among Boas's students to write *pathos*-laden ethnographies of presumptively doomed primitive cultures that, in being cast as frozen in time and isolated in space, were atomized, fitted with unique psychologies that contrasted with modern habits of mind, and treated in a hands-off "objective" way (Chapter 3 of this book). By contrast, the Chicago approach led to what Tax called "action anthropology," a prototype of contemporary forms of engaged and applied anthropology (Stocking 2000).

Washburn was the first physical anthropologist to be appointed to the Chicago department. Originally, he was also to have a lab in the Department of Anatomy, but that fell through.[9] So he turned from experimenting on functional structures to the other side of his "functional complexes:" observing functional behaviors. Washburn had been stimulated by Malinowski's *Argonauts of the Western Pacific* and so was aware of standing criticisms accusing Malinowski's disciples, if not Malinowski himself, of assuming rather than proving that each and every belief, practice, and institution in a culture must be performing a function in maintaining the whole (Washburn 1983, 6). Accordingly, he was pleased to see that under the influence of the British anthropologist A. R. Radcliffe-Brown, who was a member of the department in the 1930s, Chicago anthropologists scrupulously restricted functional explanations of social integration to independently identifiable practices and institutions whose relation to a culture's functioning bears a real analogy to the tight integration of the parts of organisms. The Chicago School may not have been as historicist as the Boasians, whom they, too, satirized as viewing cultures as "things of shreds and patches," but neither were they as circularly functionalist as Malinowski's epigones tended to be (Lowie 1920, 441; Chapter 3 of this book).

The structural-functionalism Washburn encountered at Chicago led him to reframe his approach to physical anthropology in a way that highlighted the connection between cultural and physical anthropology. Just as Chicago-style anthropologists were told not to identify cultural functions without identifying the structures that carried out these functions as parts of coherent social systems, so Washburn's students were tutored not to postulate biological adaptations without making sure that they are based on anatomically possible, phylogenetically accessible, and genetically well-founded complexes that perform identifiable functions in the particular environments to which they were hypothesized to be adapted. Comparative anatomical experimentation was indispensable in meeting this criterion.

So was observation of primates in the wild (Washburn 1983, 18). In 1948, Paul Fejos – who wished to establish anthropology as a force for postwar global unity because of its embrace of, not mere tolerance for cultural differences – found money from his Viking Fund (renamed The Wenner-Gren Foundation for Anthropological Research in 1951) to enable Washburn to visit East Africa. He returned in 1955, where he hit on the idea of studying baboon behavior. He went to Africa again in 1958, this time with funding from the Ford Foundation, to help his student Irwin DeVore with field observations suggesting the adaptive function of social hierarchy in baboon troops. These experiences led Washburn to remark, "Malinowski's functional theory probably works more usefully for monkeys than for human beings" because "Radcliffe-Brown's analogies [to organic bodies] are quite unnecessary when studying non-human primates." They are not analogies at all, but literal descriptions (Washburn 1983, 17).

All of Washburn's signature themes had been woven together by May 1962, when the Executive Board of the AAA, on which he served in virtue having been elected the Association's incoming President, asked him to devote his forthcoming Presidential Address, scheduled for November of that year, to defending an official resolution passed at the AAA's 1961 meeting expressing support for *Brown* v. *Board of Education* and opposing pseudo-biological defenses of segregation such as Carleton Putnam's *Race and Reason* (Washburn 1983, 19; Putnam 1961; Chapter 6 of this book). Washburn's decision to take on a topic of such great public moment went somewhat against the grain (De Vore and Washburn 1992, 422). He was comfortable catechizing his fellow anthropologists, but, as Jonathan Marks acknowledged in eulogizing him, Washburn's idea of addressing his fellow citizens went no further than an occasional article in *Scientific American* (Marks 2000, 226). As it happened, however, the Executive Board was having difficulty getting from the membership a convincing statement of precisely why, from a scientific point of view, the AAA was supporting *Brown* and opposing Putnam (De Vore and Washburn 1992, 422). According to notes of its May, 1962 meeting, when the question of how to proceed came up Washburn remarked, "Nobody on the biological side [of the profession] should draw up a statement [because of] the tremendous importance of the changing cultural environment upon man" (Minutes of the AAA Executive Board Meeting, Mead Papers). Knowing as they did that among physical anthropologists Washburn was the exception who proved the rule, his fellow members, including Margaret Mead, pleaded with him to use his Presidential Address to make the case. When it appeared in *American Anthropologist*, it was to serve as the Association's official argument for why segregation was not only ethically, but also biologically unjustified. Washburn complied because he couldn't help agreeing that his colleagues were right about his qualifications.

Between the time Washburn began working on his Address and November 1962, when he delivered it, Dobzhansky's *Mankind Evolving* and Coon's *The Origin of Races* appeared in print, the former about the time his speech was commissioned, the latter only a few weeks before the AAA was to meet. The importance of Washburn's Address was greatly amplified by the fact that Coon's book argued for the separate evolution of five morphologically distinct races of *H. erectus* into *H. sapiens.* These races persist today, he claimed, sub-Saharan Africans being the youngest and least advanced (Coon 1962; Chapter 6 of this book). Putnam's Southern supporters, including segregationist politicians like Governor George Wallace of Alabama, had a field day with the boost to their effort to have *Brown* v. *Board of Education* overturned that was handed to them in timely way by a professor of physical anthropology from an Ivy League university (Jackson 2005; Chapter 6 of this book). The matter quickly became even more pressing. Coon seemed in no hurry to distance himself from the uses to which his book was being put. Dobzhansky's *Mankind Evolving* had appeared too soon to include a rebuttal of it in advance. So the publication of *The Origin of Races* put Washburn in a position in which he had to persuade his colleagues to dissociate not just from ideologues like Putnam, but more importantly from a respected colleague.

Washburn did not mention Coon by name. As students of Hooton, they went back a long way. Washburn had even served as Coon's teaching assistant.[10] But he didn't have to. Everyone knew he who he was talking about when he argued that typological racial classifications are a poor guide to finding adaptations in functional complexes. He gave two examples. Coon claimed that the infraorbital foramen in Eskimos shows how their anatomy conforms to strict laws of climatic adaptation because "a strong flow of blood through that hole helps keep" their cheeks warm (Coon 1962, 61). This adaptationist tale, Washburn said, "reveals a lack of any kind of reasonable understanding of the structure of the human face.... In actual fact, most of the blood to our face does not go through that artery.... We [still] have people writing about human faces who are anatomically illiterate" (Washburn 1963b, 526). His other example was skin color, about which people like Coon too blithely told just-so stories based on race and climate. The fact that "chimps and gorillas live in precisely the same climatic conditions ... but the gorilla has one of the ... most deeply pigmented skins among the primates and the chimpanzee a very light skin" shows the peril of neglecting comparative anatomy viewed less from the perspective of museum collections than ways of life (Washburn 1963b, 526). When he "read the descriptions of adaptations [in Coon's book]," Washburn remarked, he was "not sure that there has been any progress since the nineteenth century" (Washburn 1963b, 525).

High dudgeon rather than his usual inclusive style was in order, Washburn intimated, because retaining even shadowy forms of typological thinking facilitates denial of or indifference to the fact that the well-integrated functional complexes of "human biology find [their] realization in [our] culturally determined way of life" rather than in piecemeal, trait-by-trait adaptation of conventionally identified races to particular climatic conditions, as Coon had it (Washburn 1963b, 531). Washburn proclaimed that, "[o]ur species only survives in culture and in a profound sense we are the product of new selection pressures that came with culture" (Washburn 1963b, 528). He backed this claim by citing Dobzhansky. His stringent criteria for identifying and explaining adaptations show that the decisive role of our shared capacity for cultural life can come into view only if "typology [is] removed from our thinking." Tracking gene frequency changes in human populations suggests that, "Man and his capacity for culture have evolved together" in ways that transcend racial divisions (Washburn 1963b, 521–523).

By putting his points so trenchantly, Washburn was challenging his audience to make a choice. If they did not recognize as a condition of membership in their community of inquiry that all human groups are in equal possession of our species-specific biological capacity for culture they would forfeit their right, and by implication the right of the entire profession of anthropology, to make any "authoritative and useful" interventions in public issues (Washburn 1963b, 531). In virtue of irreversible advances of biological knowledge made under the auspices of the Modern Evolutionary Synthesis, waffling between typology and adaptation by population-genetic natural selection could no longer serve any useful role in facilitating the acquisition of knowledge. On the contrary, those remaining even residually faithful to the old biology would be entangling themselves in forms of culpable ignorance whose dissemination could serve only the interests of racists. When he was young, Washburn said, "[t]here were naive racial interpretations based on the metrical data. Today we have naive concepts of adaptation taking the place of these earlier interpretations, and [with them] recrudescence of racial thinking" (Washburn 1963b, 526). The first complaint alluded to Hooton, the second to Coon. Either way, such ideas were no longer innocent:

> A lynching stirs the whole community to action, yet only a single life is lost. Discrimination, through denying education, medical care, and economic progress, kills at a far higher rate. A ghetto of hatred kills more surely than a concentration camp, because it kills by accepted custom, and it kills every day in the year.
>
> (Washburn 1963b, 530)

Washburn's Address was greeted with prolonged applause except by a number of physical anthropologists who sat on their hands (De Vore and Washburn 1992, 422). At first sight, it may seem that by so decisively shifting the boundaries of orthodoxy Washburn was stepping out of character in pushing people like Hooton and Coon to the margins of a community in which until that very moment they had been playing prominent roles and to whom he was personally connected, even indebted. His speech to the AAA did not sound his usual reassuring themes of amplifying one of anthropology's approaches without denigrating others. Ritual expulsion could not be evaded, however, if anthropology was to be a discipline that spoke with an authoritative voice to the American public about its most persistent and vexing problem: legalized racial inequality. Under the circumstances, the possibility of handing out praise depended on handing out blame. This was not to say that disagreement was unwelcome or that Hooton and Coon were not invited to close ranks with the New Physical Anthropology. It was to say that only by narrowing the permissible range of acceptable theories and methods, as the Modern Synthesis had done in evolutionary biology itself (Provine 1988; Smocovitis 1995, 130), could space open up for productive disagreements in anthropology. Discussion of well-framed issues within and between its four fields would make anthropology the coherent whole it needed to be while also making it socially praiseworthy in the eyes of the democratic pluralist polity that was being asked to trust it. In a letter to Hooton, Washburn made this point in his usual congenial way:

> The new physical anthropology has eliminated a lot of theories: Lamarckian ideas, orthogenesis, irreversibility, typological thinking, and over-emphasis on so-called non-adaptive characters. Positively, the new [physical anthropology] stresses the population as the unit of study; natural selection as the primary cause of evolution; and drift, migration, and mixture as less important factors. I think it is clear that traditional thinking in physical anthropology contained all the new ideas, but also encompassed the incorrect ones. Genetics and experiments have simply changed the emphasis.
>
> (Washburn to Hooton, 28 August 1951, Washburn Papers)

The shadow of eugenics: Washburn *contra* Sheldon and Hooton

The intensity of Washburn's animus against Hooton, and by extension Coon, arose at several points of contestation. One was Hooton's view that classifying hominids is the aim and point of physical anthropology. A decade before he addressed the AAA, Washburn was saying that the New Physical Anthropology's most pointed contrast with the old "really lies in [its] attitude toward classification" (Washburn 1953, 718). Demoting it from end to means was one of the first topics dealt with at the Summer Seminars in Physical Anthropology that Washburn conducted between 1945 and 1952 in collaboration with Lasker and Bernice "Bunny" Kaplan, Lasker's wife and colleague at Wayne State University (Lasker 1999; Little and Kaplan 2010). At the time, challenging the primacy of classification meant challenging Hooton's idea that species are marked off by non-adaptive traits. Hence Kaplan and Lasker's log of the first Summer Seminar stated,

> Typologies and other classificatory schemes, as they stand at present, seemed to some of the members to be based on the unproved – perhaps unprovable – assumption that the array of traits of which each grouping unit is composed tends to remain together as a recognizable complex which is reproduced in essentially the same form in successive generations.
>
> (Kaplan *et al.* 1946, 6)

By the end of the second Summer Seminar, the log reports that its mostly young participants had arrived at an even more pointedly anti-Hooton conclusion: "Classification for its own sake was held to be unproductive" for the field of physical anthropology (Kaplan 1947a, 9). In his Address to the AAA, Washburn saw classifying as the aim of physical anthropology as a root cause of Coon's inability to come up with accounts of adaptations that would satisfy the demands of the Modern Synthesis and his own New Physical Anthropology.

It is difficult to explain the depth of Washburn's opposition to Hooton without referring to a more distasteful aspect of Hooton's research. His methodological assumption that morphological structures determine behavioral and cognitive abilities was in the service of finding in physical abnormalities signs of psychological deficiencies that would advance the cause of eugenics. If Washburn built the New Physical Anthropology on the opposite assumption that activities determine morphological traits, it was in part to undercut Hooton's dogged attempt to find heritable psychological defects in heritable physical differences. After the War, Hooton's aim was buried in a shallow grave in the research program he called "constitutional anthropology," which stressed individual rather than group or racial types. Washburn's seemingly out-of-character effort to deprive Coon of the support of his colleagues is best explained by his perception that the inadequate conception of adaptation in *The Origin of Races* traded not only on Coon's acceptance of Hooton's assumption that classifying is the aim of the physical anthropology, but also on the eugenics and racial typing harbored by this assumption. In 1946, Coon assured Dobzhansky that he was an adaptationist by citing his intention "to find out how variations in physique correspond to differences in behavior" (Coon to Dobzhansky, February 18, 1946, Coon Papers). He was putting an adaptationist veneer on Hooton's agenda.

A long history lay behind this agenda. As a young man, Hooton worked in a prison. There he became convinced that certain anatomical characteristics are robust indicators of underlying criminal traits (Giles 2010, 148). In the 1930s he devoted much of his time to a massive anthropometric study of 13,000 criminals in an attempt to discover a criminal physical type. The upshot was a large academic book and another for the popular audience, both of which were well described in a co-authored review article by Montagu and the Columbia University sociologist of science Robert Merton as "vigorously tendentious" (Hooton 1939a, 1939b; Merton and Montagu 1940, 384). In its scholarly version, *The American Criminal: An Anthropological Study*, Hooton claimed that "[s]tudies of the relationship of man's bodily characters to his behavior have fallen into scientific disrepute and consequent neglect" not for any scientific reason but because "a species of moral sanction has been lent to the disavowals of psychophysical interdependence by the democratic doctrine of human equality" advanced most vigorously by Boas and his followers (Hooton 1939a, 252). Race as well as eugenics was implicated. "In the field of serious racial studies," Hooton wrote,

> any suggestion that the physical features which constitute the outward signs of a common inheritance are accompanied by, or indicative of, psychological or sociological tendencies is treated as a sin against the Holy Ghost of science – unforgivable, inexpiable, and utterly damning.
>
> (Hooton 1939a, 3)

As a result, racial studies were restricted to "mere anatomical and biometric description" and to the "supposed environmental modifications of human types due to climate, diet, functional adaptation, or what not" (Hooton 1939a, 3). Hooton said that such studies robbed physical anthropology of its authority to address the hereditary burdens on society imposed by generations of defectives. "Those who study crime" regarded "an anthropological interest in the criminal as a species of wanton trespass upon their professional preserves," which in his view was unjustifiably biased toward racial egalitarianism (Hooton 1939a, 4).

Hooton meant to change all that by studying "the physical characteristics of criminals with the purpose of discovering whether or not these are wholly irrelevant to their antisocial conduct." It is a reasonable assumption, he wrote, that physical types correlate with behavior. After all, it is a biological commonplace that "[t]he behavior of an animal arises from its general bodily organization" (Hooton 1939a, 5). If physical signifiers of inherited criminality could be found, it was all the more likely that racial determinants of behavior and cognition could also be identified. Accordingly, "[i]t would appear that racial physical differences should naturally be associated with racial psychological differences and that the behavior of distinct racial stocks should vary in accordance with their physical and psychological differences." That no one had yet discovered any "demonstrated racial differences in psychology," he said, could not "be attributed to a de facto absence of such differences" (Hooton 1939a, 6). Investigators simply hadn't gone about the job the right way.

In arguing this way, Hooton was shifting the burden of proof onto Boas. It was, he claimed, too soon to close the book on racially marked differences in behavior. In 1936, Hooton conceded that "[p]hysical anthropologists as yet are unable precisely to grade existing human races upon an evolutionary scale, upon the basis of the sum total of their anatomical deviations from apes and lower animals." But this was because "[a]nthropologists have found *as yet* no relationship between any physical criterion of race and mental capacity, whether in individuals or in groups" (Hooton 1936, 512, our italics). By 1939 he identified methodology as the source of the problem. Failure to find what he was looking for "may be ascribed with certainty to the crudity and general inadequacy not only of methods hitherto devised for the measurement of psychological differences, but also of the anthropological technique employed for isolation and determination of racial physical types" (Hooton 1939a, 7). The blame for failing to discover racial differences that Hooton thought must be there was laid at the feet of physical anthropologists whose "slipshod thinking, faulty technique, and meager achievement ... in the definition, classification, and study of racial types are responsible for the failure of the psychologist.... It should be the duty of the physical anthropologist," he concluded, "to provide an accurate classification of physical human types, both racial and individual, based upon the exhaustive analysis of abundant metric and morphological data" (Hooton 1939a, 7).

The future orientation in these passages arises from Hooton's disappointment that by the 1930s his youthful hopes of using behavioral typology of criminality as a wedge into racialized types of behavior and cognitive capacities hadn't produced much. Still, his conviction about where the burden of proof falls was strong enough to lead him to oppose the resolution "That democracy should deny and should disregard racial differences" in a radio debate with Montagu in 1939 (Hooton to Montagu, October 28, 1939, Montagu Papers) and to continue to uphold eugenicists' calls to "limit the reproduction of criminals and mental defectives":

> Let us cease to delude ourselves with the belief that education, religion, or other measures of social amelioration can transform base metal into gold. Public enemies must be destroyed, not reformed. We need a biological New Deal that will segregate and sterilize the anti-social and the mentally unfit.
>
> (Hooton 1935, 31)

When the war ended, Hooton's problems became even more problematic. The egalitarian spirit of Montagu's *Man's Most Dangerous Myth* and soon the

UNESCO Statements on Race meant that racially hierarchical anthropology now faced too many barriers to acceptance to maintain any pretense about where the burden of proof lies (Chapter 4 of this book). Hooton continued to believe that physical anthropologists of his persuasion had cards to play in their running battle against Boasian cultural anthropology. He was as convinced as ever that the "physical anthropologist alone" is "qualified" to discover real racial types, including different mental abilities, because "he derives his data from caliper measurements, indices, morphological observations, and statistical analysis" (Hooton 1937, 188). Nonetheless, in the second, postwar edition of his physical anthropology textbook, *Up From the Ape*, he backed off searching for racial hierarchy:

> Science can make no valid assertion that this or that race is either superior or inferior to another.... For that matter it is equally unable to put forward the claim that all races are equal biologically or in cultural capacity.
>
> (Hooton 1946, 452, 660)

He returned to the study of criminality in ways less overtly eugenic and racialist in what he called "constitutional anthropology."

Hooton's "constitutional anthropology" was an anthropological appropriation of the "constitutional psychology" of William H. Sheldon in an era when it no was longer politically advantageous to speak about eugenics or race. Bernice Kaplan characterized it in the log of the Summer Seminar of 1946 as "that aspect of the science which today is most insistent on typologies and classifications of mankind as immediate goals" (Kaplan 1947a, 9). Sheldon (1898–1977) acquired his Ph.D., in Psychology at the University of Chicago, taught there from 1936 to 1938, and then went to Harvard, where he met Hooton. He joined the army when America entered World War II, rising to the rank of lieutenant colonel. In 1945 he relocated to Columbia, where he was appointed Director of the Constitutional Clinic in the College of Physicians and Surgeons. Sheldon described the research he pursued at Columbia as "the study of the psychological aspects of human behavior as they are related to the morphology and physiology of the body" (Sheldon *et al.* 1940, 1). Put more straightforwardly, Sheldon claimed to be investigating in a stringently empirical way the relationship between body build, behavior, and styles of social interaction with a view to predicting criminal behavior from observations of "somatic types" (Sheldon 1949). Throughout the 1940s and into the 1950s he was a force to be reckoned with. His approach migrated into physical anthropology when Hooton championed it. With its show of measurement, statistics, and empirical evidence, constitutional psychology gave Hooton's old ambitions cover in a rhetorical situation uncongenial to them. This was not fortuitous. Sheldon had originally been inspired by Hooton's prewar program in physical anthropology.

Like Hooton, Sheldon believed that "a unifying conceptual schema" of human behavior had to "seek anchorage in the solid flesh and bone of the individual" (Sheldon *et al.* 1940, 3). Accordingly, his constitutional psychology made

> the assumption that human behavior, whether physiological, hormonal, or mental is always structure in action.... [We] elect to proceed as if it were known, a thing given, that biological and moral matters on this planet lie in a continuum and that human structure in action is human personality.
>
> (Sheldon 1951, 373)

As we have seen, Washburn's "functional complexes," too, are jointly structural and actional. For Washburn, however, activity, especially social activity viewed in the light of evolutionary dynamics, is a prior condition for identifying functional somatic units as adaptive. Assuming the opposite, Hooton and Sheldon claimed on reductionist grounds that bodily structure determines behavior because physical anthropology is prior to and more objective than psychology. It was because the issue was so clearly posed between these opposing ways of framing the relationship between behavior and bodily form that Washburn could so categorically dismiss constitutional psychology and, when Hooton appropriated it, constitutional anthropology. He saw Coon's *The Origin of Races* as bearing this legacy.

The legacy in question substitutes individual psychology for cultural activity. "The physical anthropologist strips off the cultural veil and examines each individual man in his organic nakedness, for which there is no alibi," Hooton wrote (Hooton 1937, 189). For Sheldon, who as a psychologist was indifferent to the evolutionary and cultural locus of learning, cognition, and behavior, this "nakedness" and "veiling" was not in the least metaphorical. His research was based on thousands of photographs he claimed could lead to a descriptive classification based on the morphology of the photographic subjects. He claimed to have discovered a three-fold typology of body builds, or "somatotypes," a term he borrowed from Hooton. He lifted his three types, ectomorphs, endomorphs, and mesomorphs, from the work of Ernst Kretschmer (1888–1964), whom he had met while visiting Europe in the 1930s (Rafter 2007). In various degrees and numerically graded combinations Sheldon took some of these types to correlate with criminal behavior. Sheldon's was not the only approach to "constitution" in the 1940s, but largely through Hooton it was the most widely known (see the broad surveys in Lessa 1943; Tucker and Lessa 1940a, b).

Washburn was appalled by Sheldon's behavioral stereotypes, which were even cruder than Hooton's typological approach to comparative anatomy. He was even more upset by Hooton's concurring with Sheldon that one could strip a veil off culture and see the literally naked individual. He tried to get his former mentor to back off (Washburn to Hooton, 28 August 1951, Washburn Papers). How could any self-respecting anthropologist agree with this sort of methodological individualism and anatomical reductionism? Hooton, however, held his ground, leading Washburn to harden his heart not only against his teacher's appropriation of constitutional psychology as constitutional anthropology, but against the very possibility that his typological approach to physical anthropology might be a candidate for unifying the cultural and biological sides of their discipline and beyond that the natural and social sciences generally. It can seem to historians of evolutionary biology an exercise in philosophical high-handedness that in the 1950s and 1960s Mayr, Dobzhansky, Simpson, Montagu, and Washburn all pitted the virtues of a somewhat ill defined "population thinking" against the vices of an even less well defined "typological essentialism" (Winsor 2006; Chapter 6 of this book). But an examination of the struggle to unify anthropology with the Modern Synthesis shows that real and pressing issues were fought out in terms of this duality. For good reason Hooton complained to Washburn after the 1950 Cold Spring Harbor Symposium, "Sherry, I hope I never to hear the word 'population' again" (Marks 2000, 225).

Hooton's way of responding to an invitation to speak at Cold Spring Harbor contrasts with the responses of Montagu, who viewed it as an opportunity to show how genetically literate he had become (Chapter 4 of this book); of Mayr, who was at first wary about transferring his knowledge of avian systematics to hominids (Chapter 6); and of Coon, whose contempt for genetics led him to decline the invitation until Washburn bucked him up (Chapter 6). Hooton eagerly accepted, but immediately began badgering Dobzhansky, Washburn's co-organizer, to replace established anthropologists as participants with Sheldon's and his own graduate students in constitution. Unaware of his ideological baggage until Montagu and Dunn wised him up, Dobzhansky had expressed interest soon after Sheldon arrived at Columbia in 1947 in whether his data could be made sense of genetically (Dobzhansky to Montagu, March 11, 1947, Dobzhansky Papers). Accordingly, it was to Dobzhansky rather than to his former student Washburn that Hooton complained about what he feared would be underrepresentation of constitutional theory at Cold Spring Harbor. In correspondence between the principals after the event, Hooton's motive became clear. Not without reason – after all, his psychologizing prescinded from the idea of culture on any account of it – he had come to believe that Washburn wanted to keep constitutional anthropology out of the discipline and so presumably out of the Symposium. Accordingly, Hooton "kicked like hell to Dobzhansky," as Washburn put it, because he thought the geneticist might be more pliable (Washburn to Coon, May 24, 1952, Coon Papers). He made his pitch in such a high-handed way, however, insisting on rights earned by what he cited as the impressive rise in constitutional research in recent years, that Dobzhansky did not break ranks with Washburn. A survey of anthropological works conducted in 1950 does indeed show a sudden rise in the number of constitutional studies (Simon 1950, 294). What it does not show is that these publications were made possible almost entirely by Sheldon's appointment to direct a program at Columbia dedicated exclusively to his own ideas (Washburn to Coon, May 24, 1952, Coon Papers). Still, Hooton got his way. The organizers scheduled a session on constitution in which Sheldon was to be the principal speaker, a few of his acolytes would also present, and Hooton would be the principal commentator.

Hooton's insistence on securing a prominent place for constitutionism at the 1950 Symposium illustrates the maxim that you should be careful what you wish for, since you might get it. Sheldon's warm-up act, a paper by Carl Seltzer on the correlation between mesomorphs and criminality, was greeted by an objection from the floor that all the speaker had shown was that, "Particularly aggressive delinquents are well qualified for their profession by being predominantly mesomorphic in body build:" neither too puny nor too clumsy to mug people or climb through the windows of their homes (Warren, in Seltzer 1951, 371). Sheldon's own talk provoked doubts in the audience about whether he had provided any evidence that his three body types are inherited even when they are parsed into numerically graded sub-types. Correlations between these types and behavioral tendencies were even more dubiously heritable. Would it not be better, suggested the geneticist Adriano Buzzati-Traverso, first to conduct laboratory studies of particular physical and behavioral traits on animal models to see if any correlations that showed up were also correlated with genetic changes (Buzzati-Traverso, in Sheldon 1951, 378; see also Kaplan 1951, 33–34: "Many … have strong reservations about using the morphological approach … to constitution until the genetic factors involved have been demonstrated")? Even then, to be of any value in the case of the highly polymorphic *H. sapiens* such studies would have to discount phenotypic plasticity of the sort that Boas had discovered in head shapes, which Dobzhansky had now given good grounds for thinking express the same genotypes in different environments unless proven otherwise (Chapter 4 of this book).

Sheldon's attempts to forestall objections like these had the effect of provoking his audience to press their objections further. There were few if any real questions in the discussion period. Instead, what looked like questions were contemptuously framed assertions, putting Sheldon on the defensive. To the suggestion that his project should be delayed until genetics had testified about heritability he replied not with empirical arguments, but by trying to shift the burden of proof. To wait until genetics had weighed in, he said, was to "purchase the advantage of perfect objectivity … at the price of relevancy." It was to walk on "dangerous ground" by endorsing "*not* carrying out the converse, or the complement of this kind of work" (Sheldon 1951, 378). To the objection that classification based on whole body types was of less value than investigation into particular traits, he responded with another example of rhetorical table turning. Restricting investigation to parts of the body was akin to saying, "In order to study the relationship between constitution and anything we must omit studying the constitution – for the constitution is really the phenotype *as a whole*" (Sheldon 1951, 380, emphasis in original). To the objection that his classifications put people into boxes, he replied by saying they were provisional. With further inquiry they would be replaced by a smooth, non-typological distribution (Sheldon 1951; Sheldon *et al.* 1940, 27).

Still, one might well wonder whether having started with classifying types of humans Sheldon, or for that matter Hooton or Coon, would ever cease doing it. Washburn, for one, went beyond doubting it to flatly denying it. In an exchange after Cold Spring Harbor, he complained to Hooton that, "One gets the feeling [in reading Sheldon] that all that is sought after is difference, any difference, with all critical standards thrown to the winds" (Washburn to Hooton, August 28, 1951, Washburn Papers). "If one believes that classifications alone give understanding," he wrote a few years later, "then one will make the classifications more and more complicated" (Washburn 1953, 718). He was especially convinced of this because in an imprudent moment at Cold Spring Harbor Sheldon had warned his audience that only by endorsing constitutional studies could society hope to address the "central problem of social science, which undoubtedly is that of controlled human breeding" (Sheldon 1951, 374). It was a eugenic dream, and a crude one at that, that had been motivating Sheldon all along. Eugenic dreams require classifying people into types. Their purpose would be defeated if they gave way to continuous, dynamically changing distributions.

Washburn rejected as unfair Hooton's accusation that he had a "fanatical opposition to constitution." Had not he and Dobzhansky done even better than stack the Symposium with graduate students by inviting Sheldon himself to take the podium at Cold Spring Harbor? Had not Sheldon's approach been seconded by several other invited speakers? Had not Hooton delivered the official response at the session? As to supporting younger scholars of constitution, Washburn reminded his former mentor that he had invited them "to the Wenner-Gren [Summer] Seminars, getting their expenses paid, recommending that they get research grants, and assigning their work to my classes?" "I object to being called fanatically opposed to constitution.... It is an odd kind of fanatic opposition which behaves this way" (Washburn, August 15, 1951, Washburn Papers).

Writing again to Hooton a few weeks later, Washburn was more frank. Constitutional anthropology and psychology, he said, share the same logic as classification for eugenic purposes, which in turn rests on unacknowledged racialism:

> The parallel to the history of racial studies is very close. Both the early racial and constitutional studies were pre-modern biology. Both developed a variety of taxonomic schemes. Both had the choice of trying to reduce the contradictions and increase the utility of the early schemes either by constructing more elaborate classifications or by trying to understand the processes that produced the groupings. As I read the history of both, further elaboration of classification without biological justification of the categories would be futile.
>
> (Washburn to Hooton, August 28, 1951, Washburn Papers)

He was on solid ground. In 1942 Sheldon had written for all to see:

> If constitutional studies can lead to the establishment of a rational foundation for a science of heredity and eugenics, we may then hope ... to eliminate the principal constitutional and degenerative physical scourges of the race.... But of greater importance than that, it might then also be possible by discriminate breeding to strengthen the mental and spiritual fiber of the race.
>
> (Sheldon and Stevens 1942, 437)

Hoping to pry him loose from Sheldon, Washburn ended his letter by asking Hooton to remember that "[s]cientific progress can be speeded by free and frank discussion" (Washburn to Hooton, August 28, 1951, Washburn Papers). His effort speaks well of Washburn, but it was in vain, as was his similar attempt to push Coon toward real population thinking. By 1962, Hooton's continued embrace of Sheldon and Coon's fidelity to Hooton's underlying assumptions had pushed Washburn to the edge of his commitment to communitarianism in science.

Hooton remained bonded to Sheldon in spite of Washburn's protestations because he was the anointed heir of Hooton's youthful eugenic and racialist hopes (Tucker 1940, 432). After the war, these hopes went into retreat, decline, and eclipse (Littlefield *et al.* 1982; Lieberman *et al.* 2003; Barken 1992). Hooton died in 1954, leaving constitutionalism with no advocates in physical anthropology. Sheldon received no invitations to speak to anthropologists after Cold Spring Harbor. In the proceedings of major anthropological conferences, such as the 1952 Wenner-Gren meetings that integrated physical with cultural anthropology on Washburn's terms, we find no mention of constitution (Kroeber 1953; Tax 1953).

By the mid-fifties, Mead, speaking from the side of cultural anthropology, made it clear that "reintegration of the branches of anthropology" depended on the new evolutionary biology. The new biology demonstrated more clearly than her mentor Boas "the independence of patterned cultural behavior from the racial constitution of the particular carriers" of a behavior (Mead 1958, 481). It carried none of the baggage of nineteenth century notions of race and racial inequality that led Boas to steer clear of evolutionary dogmatism. Mead's assessment shows that the postwar decline of scientific racism and eugenics so visible in the marginalization of Hooton and Sheldon, and in 1962 Coon, was more than an ideological shift. It resulted from a great deal of situated, addressed, contentious, but nonetheless scientific argumentation. If they attend only to developments in biology and theoretical population genetics historians can too quickly attribute this change solely to blowing in the political wind (Provine 1973; Provine and Russell 1986). Much of this discussion and the resulting sea change took place under Washburn's leadership in providing argumentative occasions for it to happen. As early as the first Summer Seminar in Physical Anthropology anthropologists had recognized that Sheldon's titillating photographs were masquerading as scientific objectivity and hiding "some basic assumptions … that should be the target of criticism." They already knew, too, that population "geneticists … were not hoping to contribute better racial classification to physical anthropology, but rather were looking forward to a shift in emphasis from classification toward … understanding … the processes of race formation" in the sense of "race" defined by the Modern Evolutionary Synthesis (Kaplan 1951, 34, 30).

Washburn and the participants in his Summer Seminars were as aware as Kroeber that whenever anthropologists appeal to the psychological traits of typified individuals, groups, or races they inevitably undercut the concept of culture that demarcates their discipline (Chapter 3 of this book). In the 1970s, Washburn returned to the tension between culture and psychology by helping galvanize anthropologists from all schools to oppose the Sociobiology of W. D. Hamilton,

Robert Trivers, E. O. Wilson, and, painfully, Washburn's prize student Irven DeVore (Segerstråle 2000; Chapter 7 of this book). Fairly or not, Washburn viewed Sociobiology through the lens of Arthur Jensen's 1969 statistical study of race and IQ. He judged it, too, as a recrudescence in superficial population genetic clothing of Sheldon, Hooton, and Coon's eugenics – and racially tinged typologizing. If there are "genes for altruistic acts," he told the American Psychological Association, which had invited him to share his views on Sociobiology, "there must be criminal genes … for criminal acts" (Washburn 1978b, 416). He was quick to notice a few years later that,

> The first section of [Julius] Wilson and [Jensen's disciple Richard] Herrnstein's *Crime and Human Nature* [1985] is on biological causes of crime, and in this they go back to the Sheldonian system as a reputable, defensible reference.... This is a very dangerous kind of perspective. If people think there really is a relationship between crime and a biological cause the next step in reducing crime surely suggests eliminating or controlling people who exhibit factors that someone believes are indicators of a potential for criminal behavior.
>
> (De Vore and Washburn, 1992, 422)

Characteristically, Washburn summed up his objection in terms of what is good for the communities of inquiry in whose harmonizing he played a key role. If anthropologists embrace Sociobiology, he said, "Social anthropology will regress at least fifty years" (Washburn 1978a, 36).

Varieties of scientific unity in postwar America

From the perspective of the logical empiricist view of science that became ascendant in America by about 1960, the syncretism of Washburn's effort to integrate anthropology does not look much like scientific unification. For logical empiricists, the unity of science requires that the generalizations of a science of lesser scope be subsumed under, indeed be derivable from, the laws of fields whose range is wider and presumably deeper. It is worth pointing out, accordingly, that in the period in which Washburn was exerting himself to integrate anthropology's four fields and relate them to evolutionary biology the effort to unify the sciences was oriented more to coordination than theory reduction.

This was true even of the "unity of science movement" itself. George A. Reisch has made a good case that the aim of its principle orchestrator, Otto Neurath, was to educate the public on the importance of science in guiding public policy in a world that remained as dangerous as the one from he and other scientist-philosophers had fled for their lives. When they arrived in America they made common cause with the pragmatists who helped them escape, who envisioned scientific unification as socially conscious collaborative problem solving. It was only in reaction to the anti-Communism of the fifties, Reisch argues, that the left-leaning logical empiricists retreated to universities, turning the unity of science movement into an apolitical epistemology and subverting pragmatism by empowering scientists to solve peoples' problems for them (Reisch 2005, 305).

Even in its pluralistic period, the unity of science movement met resistance in academia. Horace Kallen, who championed and even named the 'cultural pluralist' conception of democracy (Kallen 1924), saw in the émigrés traces of the very totalitarianism from which they were fleeing. Properly characterized, Kallen wrote in 1940, "'The unity of science' means … no more than the mutual guarantee of … liberty by each science to each, collective security for the scientific spirit from dogmatic aggression" (Kallen 1940, 83; Reisch 2005, 167–175). The Harvard "Red Book" of 1945, which set the stage for the growth of higher education in the postwar decades, endorsed the "modernizing and broadening" of university education by celebrating its expanding disciplinary diversity. When its authors raised "the question of unity" among these fields they were not expressing a desire to damp that diversity down. They wanted only to ensure that collegiate general education was coherent enough to promote, extend, and unify democratic society: "We are faced with a diversity of education which, if it has many virtues, nevertheless works against the good of society by helping to destroy the common ground of training and outlook on which any society depends" (Harvard University Committee on the Objectives of a General Education in a Free Society 1945, 43; Hollinger 1996, 161). Worries about excessive unity intensified in the 1950s, fueled by perceived threats to existing disciplines from logical empiricists, who took physics to be the paradigmatic science. A co-organizer of a 1955 conference commemorating the bicentennial of Columbia University by discussing the theme of "unity of knowledge" reported that many of its participants, including Kroeber, Merton, W. V. O. Quine, Julian Huxley, and even Neils Bohr greeted reductionistic notions of scientific unity warily: "We were reminded that unity may be a word of threat as well as a word of promise; that it may signify the end of searching as well as the relentless quest" (Leary 1955a, xi).

These issues took on a different hue in anthropology. Constituted as it was by four fields straddling natural, social, and humanistic forms of inquiry, it lived in daily intimacy with the problem of unity. But it also entertained ambitions as hegemonic of those of physics-oriented logical empiricists. Its movers and shakers wanted anthropology to be the integrating discourse of the emerging post-colonial global order. This ambition burned especially brightly in Fejos, who was at the controls of one of the principal sources of funding for anthropological research, the Viking Fund, renamed the Wenner-Gren Foundation in honor of the Swedish vacuum cleaner magnate whose money Fejos dispersed. The editor of a 1945 Wenner-Gren sponsored symposium advocating anthropology's key role in the new global order, Ralph Linton, wrote that those who turned to the sciences for help in rebuilding the world "find themselves in the position of a sick man shifted from specialist to specialist without obtaining any over-all picture of his illness or any one plan for its cure. The time is ripe," he said, "for a new synthesis of science, especially of those sciences which deal with human beings and their problems." Because anthropology has always "tried to understand all sorts of phenomena as they affected [human beings]," it was only natural that "by its very definition the science of anthropology makes a bid for this position." Linton identified Sherwood Washburn's New Physical Anthropology as the best expression of the synthesis he called for (Linton 1945, 3–4).

In monitoring anthropology's fortunes in the following decades, Washburn struck a characteristically personal note. In his closing remarks to a 1983 conference, he "looked out over the room and [saw] many were friends who had helped me with my career." But he also sensed that this conference "marked the end of ... a time in which anthropologists could know most other anthropologists" (Washburn 1983, 12). With its success, tensions and misunderstandings had grown among American anthropology's constituent fields. As early as 1955, Tax asked,

> What do techniques in linguistics have to do with primates or primates with style or cultural values?... A graduate student impatient to get on with his archaeological digging chafes through courses in linguistics, culture-and-personality, and folklore, less convinced of the logic of the enterprise than of the perverseness of the older generation.... Can such heterogeneity be maintained in a single discipline?
>
> (Tax 1955, 313)

Three decades later Washburn still hoped so. He unleashed a torrent of blame against Sociobiology because he was aware that its appeal to psychological adaptationism came at the expense of the culture concept and shifted the boundary between biology and culture erected by Kroeber (Chapter 3 of this book). His proclamation that "social anthropology will regress at least fifty years" if it embraced Sociobiology predicted intensification of the fissioning tendencies already in play in the discipline in that event. These tendencies would even grow worse if biological anthropologists adopted a conception of scientific unity like the scholasticized version of scientific reductionism in which Wilson tried to give his Sociobiology philosophical protection in his 1990 book *Consilience*.[11]

Notes

1 In 1957 Washburn used the term "culture" to name our species' unique ability to pass innovations across generations by symbolic communication (Washburn 1959). Working from his definition of man as a 'symboling' animal, White agreed. So did Marshall Sahlins (Washburn, 1959, 55, 74–75). But other participants in the same discussion found enough of what Washburn found in other species to say that they too had culture in his sense (Harlow 1959, 44). Some used "culture" even more widely to identify all forms of mimetic learning, even if it doesn't cumulate. After the explosion of ethological studies following Jane Goodall's observations of chimpanzee life, the wider use became common, forcing those who wish to signal our uniqueness to use phrases like "cumulative cultural adaptation" as a *differentia specifica* of our cultural way of life (Hill *et al.*, 2009, for example). Washburn responded by distinguishing between species that have culture and (the only) one that possesses articulated language (De Vore and Washburn 1992, 420–421).

2 Three of Washburn's papers, including "The Strategy of Physical Anthropology," are referenced in Dobzhansky's *Mankind Evolving* (1962) and his name occurs in six passages. The phrase "hastily made over apes" is quoted from Muller 1960, 458 on page 330. See Chapter 4 of this book.

3 Dale Sullivan has usefully enumerated five aspects of the rhetoric of scientific orthodoxy: "Education (initiating the neophyte), legitimation (establishing authority to speak), demonstration (exhibiting the truth as defined by the relevant culture), celebration (rehearsing the victories and praising the heroes of the culture), and criticism (establishing new knowledge and demarcating borders)" (Sullivan 1994, 285–286). All these aspects can be seen in different proportions in the conferences we have listed.

4 In this Washburn was implicitly following a pattern of argument identified by Chaim Perelman: retaining a point or value dear to the heart of an opponent, but reordering its importance (Perelman and Olbrrechts-Tyteca 1969).

5 We capitalize Evolutionary Psychology and Sociobiology when we refer to research programs going by these names. In the lower case we use the names to refer to subject matters and issues about which a variety of research programs and theories might contend.

6 Smocovitis 2012 is one of the few scholarly works recounting efforts to align anthropology with the Modern Synthesis. In relating interesting details about the same cast of characters and events we discuss, she agrees with us that among the makers of the Synthesis, "Dobzhansky is the one who did the most to integrate with anthropology" (Smokovitis 2012, S111–112). She takes the leading motivation for unification within evolutionary biology and between it and anthropology to be pursuit of "the Enlightenment project of the unity of knowledge with a positivistic worldview" (S213; see also Yudell 2014, 161–162). In making the problematic of race and racism in America focal, we argue in this chapter for a more engaged view of scientific unification and in the next for a greater diversity of opinion within Synthesis itself.

7 Lasker recalls it similarly:

> [Washburn and I] have agreed on most of the issues that have faced physical anthropology over the years: the problem with "race" as the unit for organizing human biological variability and the value of organic evolution for understanding temporal variation.
>
> (Lasker 1953)

In 1954, Lasker got the American Association for the Advancement of Science to stop holding meetings at segregated hotels and helped elect W. Montagu Cobb of Howard University, an African-American, as his successor in representing Anthropology in the AAAS (Bogin 2003, 196). We are grateful to Herbert Lewis for helpful private communication on Lasker's influence on Washburn.

8 W. E. Le Gros Clark shared Washburn's view of functional complexes (Washburn and Moore 1974, 159–161).

9 Washburn did some experimental work on pig anatomy with his students at the University of Chicago, but not on the scale he originally envisioned (De Vore and Washburn 1992).

10 Washburn later called Coon "a very nice guy and a good friend" (De Vore and Washburn 1992, 422).

11 *Consilience* is sometimes called non- or even anti-reductionistic. That is because it argues for theoretical reduction without entity reduction. The objects of each field remain its objects. This does not prevent Wilson from misappropriating Whewell's concept of consilience, which opposes all three kinds of reductionism, theoretical, entitative, and methodological (Whewell 1847).

References

Primary sources

Irving Hallowell Papers, American Philosophical Society, Philadelphia, PA.
Margaret Mead Papers, Manuscript Division, Library of Congress, Washington, DC.
Sherwood Washburn Papers, Bancroft Library, University of California-Berkeley, Berkeley, CA.

Secondary sources

Abir-Am, Pnina. 1999. "Introduction." *Osiris* 14 (1): 1–33.
Anthropological Society of Washington. 1959. *Evolution and Anthropology: A Centennial Appraisal*. Washington DC: Anthropological Society of Washington.
Barkan, Elazar. 1992. *The Retreat of Scientific Racism: Changing Concepts of Race in Britain and the United States between the World Wars*. Cambridge: Cambridge University Press.
Beasley, Edward. 2010. *The Victorian Reinvention of Race: New Racisms and the Problem of Grouping in the Human Sciences*. New York: Routledge.
Bogin, J. 2003. "In Memoriam: Gabriel Ward Lasker (April 19, 1912–August 27, 2002)." *American Journal of Physical Anthropology* 121: 195–197.
Bowler, Peter. 1988. *The Non-Darwinian Revolution*. Baltimore: Johns Hopkins University Press.
Brace, C. Loring. 2005. *"Race" is a Four Letter Word: The Genesis of the Concept*. Oxford: Oxford University Press.
Casper, Christian F. 2007. "In Praise of Carbon, in Praise of Science: The Epideictic Rhetoric of the 1996 Nobel Lectures in Chemistry." *Journal of Business and Technical Communication* 21 (3): 303–323.
Ceccarelli, Leah. 2001. *Shaping Science with Rhetoric: The Cases of Dobzhansky, Schrödinger, and Wilson*. Chicago: University of Chicago Press.
Condit, Celeste M. 1985. "The Functions of Epideictic: The Boston Massacre Orations as Exemplar." *Communication Quarterly* 33 (4): 284–298.
Coon, Carleton S. 1962. *The Origin of Races*. New York: Knopf.
Darwin, Charles. 1859. *On the Origin of Species by Means of Natural Selection, Or the Preservation of Favoured Races in the Struggle for Life*. London: John Murray.
Darwin, Charles. 1871. *The Descent of Man and Selection in Relation to Sex*. London: John Murray.
Demerec, Milislav, ed. 1950. *Origin and Evolution of Man. Cold Spring Harbor Symposia on Quantitative Biology*, 15. Cold Spring Harbor, NY: The Biological Laboratory.
Detwiler, Samuel R. 1929. "Anatomy as a Science." *Science* 70 (1824): 563–566.
De Vore, Irven and Sherwood Washburn. 1992. "An Interview with Sherwood Washburn." *Current Anthropology* 33 (1992): 411–423.
Dobzhansky, Theodosius. 1962. *Mankind Evolving*. New Haven: Yale University Press.
Dobzhansky, Theodosius. 1973. "Nothing in Biology Makes Sense Except in the Light of Evolution." *American Biology Teacher*. 35 (3): 125–129.
Fuentes, Augustin. 2010. "The New Biological Anthropology: Bringing Washburn's New Physical Anthropology Into 2010 and Beyond – The 2008 AAPA Luncheon Lecture." *American Journal of Physical Anthropology* 143 (S51): 2–12.
Giles, Eugene. 2000. "Coon, Carleton Stevens." Available online at www.anb.org.libraries.colorado.edu/articles/14/14-00951.html

Giles, Eugene. 2010. "Principal Figures in Physical Anthropology Before and During World War II." In *Histories of American Physical Anthropology in the Twentieth Century*, edited by Michael A. Little and Kenneth A. R. Kennedy, 141–154. Lanham, MD: Lexington Books.

Haller, John S. 1971. *Outcasts from Evolution: Scientific Attitudes of Racial Inferiority 1859–1900*. Urbana: University of Illinois Press.

Haraway, Donna J. 1988. "Remodelling the Human Way of Life: Sherwood Washburn and the New Physical Anthropology, 1950–1980." In *Bones, Bodies, Behavior: Essays on Biological Anthropology*, edited by George W. Stocking Jr., 206–259. Madison: University of Wisconsin Press.

Harlow, Harry F. 1959. "Basic Social Capacity of Primates." *Human Biology* 31 (1): 40–53

Harvard University Committee on the Objectives of a General Education in a Free Society. 1945. *General Education in a Free Society: Report of the Harvard Committee*. Cambridge: Harvard University Press.

Hollinger, David A. 1996. *Science, Jews, and Secular Culture: Studies in Mid-Twentieth-Century American Intellectual History*. Princeton: Princeton University Press.

Hooton, Earnest Albert. 1926. "Methods of Racial Analysis." *Science* 63 (1621): 75–81.

Hooton, Earnest Albert. 1935. "Homo Sapiens – Whence and Whither?" *Science* 82: 19–31.

Hooton, Earnest Albert. 1936. "Plain Statements about Race." *Science* 83 (2161): 511–513.

Hooton, Earnest Albert. 1937. *Apes, Men, and Morons*. New York: Putnam.

Hooton, Earnest Albert. 1939a. *The American Criminal: An Anthropological Study*. Cambridge: Harvard University Press.

Hooton, Earnest Albert. 1939b. *Crime and the Man*. Cambridge: Harvard University Press.

Hooton, Earnest Albert. 1946. *Up from the Ape*. Revised edition. New York: Macmillan.

Jackson Jr., John P. 2005. *Science for Segregation: Race, Law, and the Case Against Brown v. Board of Education*. New York: New York University Press.

Jackson Jr., John P. and Nadine M. Weidman. 2004. *Science, Race and Racism*. Santa Barbara: ABC-Clio Press.

Kallen, Horace M. 1924. *Culture and Democracy in the United States*. New York: Boni and Liveright.

Kallen, Horace M. 1940. "The Meanings of 'Unity' Among the Sciences." *Educational Administration and Supervision* 26: 81–97.

Kaplan, Bernice. 1947a. "Second Summer Seminar in Physical Anthropology." *Yearbook of Physical Anthropology* 2: 9–15.

Kaplan, Bernice. 1947b. "The Third Summer Seminar in Physical Anthropology." *Yearbook of Physical Anthropology* 3: 11–24.

Kaplan, Bernice. 1948. "The Fourth Summer Seminar in Physical Anthropology." *Yearbook of Physical Anthropology* 5: 22–39.

Kaplan, Bernice. 1949. "New Techniques in Physical Anthropology: A Report on the Fifth Summer Seminar in Physical Anthropology." *Yearbook of Physical Anthropology* 5: 14–33.

Kaplan, Bernice. 1951. "The Scope of Physical Anthropology: What is to Be Taught? A Report of the Sixth Annual Summer Seminar in Physical Anthropology." *Yearbook of Physical Anthropology* 6: 25–37.

Kaplan, Bernice A. 1954. "Environment and Human Plasticity." *American Anthropologist* 56 (5): 780–800.

Kaplan, Bernice, Elizabeth Richards and Gabriel Lasker. 1946. "A Seminar in Physical Anthropology." *Yearbook of Physical Anthropology* 2: 5–11.

Kroeber, Alfred L., ed. 1953. *Anthropology Today: An Encyclopedic Inventory*. Chicago: University of Chicago Press.

Lasker, Gabriel 1999. *Happenings and Hearsay: Experiences of a Biological Anthropologist*. Detroit: Savoyard.

Leary, Lewis. 1955a. "Preface." In *The Unity of Knowledge*, edited by Lewis Leary, ix–xii. Garden City, NY: Doubleday.

Leary, Lewis, ed. 1955b. *The Unity of Knowledge*. Garden City, NY: Doubleday.

Lee, Richard B. and Irwin DeVore, eds. 1968. *Man the Hunter: Proceedings of a Symposium held at the University of Chicago in 1966*. Chicago: Aldine.

Lessa, William. 1943. *An Appraisal of Constitutional Typologies*. Menasha: American Anthropological Association.

Lieberman, Leonard, R. C. Kirk and M. Corcoran. 2003. "The Decline of Race in American Physical Anthropology." *Prezeglod Anthropologiczny* [*Anthropological Review*] 66: 3–21.

Linton, Ralph. 1945. "The Scope and Aims of Anthropology." In *The Science of Man in the World Crisis*, edited by Ralph Linton, 3–18. New York: Columbia University Press.

Little, Michael A. and Bernice Kaplan. 2010. "The Immediate Postwar Years: The *Yearbook of Physical Anthropology* and the Summer Seminars." In *Histories of American Physical Anthropology in the Twentieth Century*, edited by Michael A. Little and Kenneth A. R. Kennedy, 155–172. Lanham: Lexington.

Littlefield, Alice, Leonard Lieberman and Larry T. Reynolds. 1982. "Redefining Race: The Potential Demise of a Concept in Physical Anthropology." *Current Anthropology* 23: 641–655.

Lorimer, Douglas. 1988. "Theoretical Racism in Late-Victorian Anthropology, 1870–1900." *Victorian Studies* 31: 405–430.

Lowie, Robert. 1920. *Primitive Society*. London: Boni and Liveright.

Marks, Jonathan. 2000. "Sherwood Washburn, 1911–2000." *Evolutionary Anthropology* 9: 225–226.

Mead, Margaret. 1958. "Cultural Determinants of Behavior." In *Behavior and Evolution*, edited by Anne Roe and George Gaylord Simpson, 480–503. New Haven: Yale University Press.

Merton, Robert K. and M. F. Ashley-Montagu. 1940. "Crime and the Anthropologist." *American Anthropologist* 42 (3): 384–408.

Muller, Hermann J. 1960. "The Guidance of Human Evolution." In *Evolution of Man* (vol. 2. of *Evolution after Darwin*, edited by Sol Tax and Charles Callender). Chicago: University of Chicago Press.

Perelman, Chaïm and Lucie Olbrechts-Tyteca. 1969. *The New Rhetoric: A Treatise on Argumentation*. Translated by John Wilkinson and Purcell Weaver. Notre Dame, IN: University of Notre Dame Press.

Provine, William B. 1973. "Geneticists and the Biology of Race Crossing." *Science* 182: 790–796.

Provine, William B. and Elizabeth S. Russell. 1986. "Geneticists and Race." *American Zoologist* 26: 857–887.

Putnam, Carleton. 1961. *Race and Reason: A Yankee View*. Washington, DC: Public Affairs Press.

Rafter, Nicole. 2007. "Somatyping, Antimodernism, and the Production of Criminological Knowledge." *Criminology* 45 (4): 805–833.

Rainger, Ronald. 1978. "Race, Politics, and Science: The Anthropological Society of London in the 1860s." *Victorian Studies* 22: 51–70.
Redfield, Robert. 1942. "Introduction." In *Levels of Integration in Biological and Social Systems*, edited by Robert Redfield. Lancaster: Jaques Cattell Press.
Reisch, George A. 2005. *How the Cold War Transformed Philosophy of Science: To the Icy Slopes of Logic*. Cambridge: Cambridge University Press.
Richards, Robert. 1987. *Darwin and the Emergence of Evolutionary Theories of Mind and Behavior*. Chicago: University of Chicago Press.
Roe, Anne and George Gaylord Simpson, eds. 1958. *Behavior and Evolution*. New Haven: Yale University Press.
Segerstråle, Ullica. 2000. *Defenders of the Truth: The Battle for Science in the Sociobiology Debate and Beyond*. Oxford: Oxford University Press.
Seltzer, Carl C. 1951. "Constitutional Aspects of Juvenile Delinquency." In *Origin and Evolution of Man: Cold Spring Harbor Symposia on Quantitative Biology*, edited by Milislav Demerec, 15: 361–372. Cold Spring Harbor: Long Island Biological Association.
Sheldon, William Herbert. 1949. *Varieties of Delinquent Youth: An Introduction to Constitutional Psychiatry*. New York: Harper.
Sheldon, William Herbert. 1951. "The Somatotype, the Morphophenotype, and the Morphogenotype." In *Origin and Evolution of Man: Cold Spring Harbor Symposia on Quantitative Biology*, edited by Milislav Demerec, 15, 373–382. Cold Spring Harbor: Long Island Biological Association.
Sheldon, William Herbert and S. S. Stevens. 1942. *The Varieties of Temperament: A Psychology of Constitutional Differences*. New York: Harper.
Sheldon, William Herbert, S. S. Stevens and W. B. Tucker. 1940. *The Varieties of Human Physique: An Introduction to Constitutional Psychology*. New York: Harper.
Simon, Florence. 1950. "Review of Form and Function in Physical Anthropology." *American Anthropologist* 52 (2): 291–295.
Smedley, Audrey. 1998. *Race in North America: Origin and Evolution of a Worldview*. 2nd edition. Boulder: Westview.
Smocovitis, Vassiliki Betty. 1995. *Unifying Biology: The Evolutionary Synthesis and Evolutionary Biology*. Princeton: Princeton University Press.
Smocovitis, Vassiliki Betty. 1999. "The 1959 Darwin Centennial Celebration in America." *Osiris* 14: 274–323.
Smocovitis, Vassiliki Betty. 2012. "Humanizing Evolution: Anthropology, the Evolutionary Synthesis, and the Prehistory of Biological Anthropology, 1927–1962." *Current Anthropology* 53 (S5): S108–S125.
Spuhler, J. N., ed. 1959. *The Evolution of Man's Capacity for Culture*. Detroit: Wayne State University Press.
Stocking Jr., George W. 2000. " 'Do Good, Young Man': Sol Tax and the World Mission of Liberal Democratic Anthropology." In *Excluded Ancestors: Inventable Traditions: Essays Toward a More Inclusive History of Anthropology*, edited by Richard Handler, 171–264. Madison: University of Wisconsin Press.
Sullivan, Dale L. 1991. "The Epideictic Rhetoric of Science." *Journal of Business and Technical Communication* 5 (3): 229–245.
Sullivan, Dale L. 1994. "Exclusionary Epideictic: NOVA's Narrative Excommunication of Fleischmann and Pons." *Science, Technology, & Human Values* 19 (3): 283–306.
Tax, Sol, ed. 1953. *An Appraisal of Anthropology Today*. Chicago: University of Chicago Press.

Tax, Sol. 1955. "The Integration of Anthropology." *Yearbook of Anthropology* 1: 313–328.

Tax, Sol and Charles Callendar. 1960. *Evolution after Darwin: The University of Chicago Centennial*. Chicago: University of Chicago Press.

Tucker, William B. 1940. "Is There Evidence of a Physical Basis for Criminal Behavior?" *Journal of Criminal Law and Criminology (1931–1951)* 31 (4): 427–437.

Tucker, William B. and William A. Lessa. 1940a. "Man: A Constitutional Investigation." *The Quarterly Review of Biology* 1 (3): 265–289.

Tucker, William B. and William A. Lessa. 1940b. "Man: A Constitutional Investigation (Concluded)." *The Quarterly Review of Biology* 15 (4): 411–455.

Vivian, Bradford. 2006. "Neoliberal Epideictic: Rhetorical Form and Commemorative Politics on September 11, 2002." *Quarterly Journal of Speech* 92 (1): 1–26.

Washburn, Sherwood L. 1951. "The New Physical Anthropology." *Transactions of the New York Academy of Sciences* 13: 298–304.

Washburn, Sherwood L. 1953. "The Strategy of Physical Anthropology." In *Anthropology Today: An Encyclopedic Inventory*, edited by Alfred L. Kroeber, 714–727. Chicago: University of Chicago Press.

Washburn, Sherwood L. 1959. "Speculations on the Interrelations of the History of Tools and Biological Evolution." *Human Biology* 31 (1): 21–31.

Washburn, Sherwood L., ed. 1961. *Social Life of Early Man*. Chicago: Aldine.

Washburn, Sherwood L., ed. 1963a. *Classification and Human Evolution*. Chicago: Aldine.

Washburn, Sherwood L. 1963b. "The Study of Race." *American Anthropologist* 65 (3, Part 1): 521–531.

Washburn, Sherwood L. 1978a. "Animal Behavior and Social Anthropology." *Society* 15 (6): 35–41.

Washburn, Sherwood L. 1978b. "Human Behavior and the Behavior of Other Animals." *American Psychologist* 33 (5): 405–418.

Washburn, Sherwood L. 1983. "Evolution of a Teacher." *Annual Review of Anthropology* 12 (1): 1–25.

Washburn, Sherwood L. and S. R. Detwiler. 1943. "An Experiment Bearing on the Problems of Physical Anthropology." *American Journal of Physical Anthropology* 1 (2): 171–190.

Whewell, William. 1847. *History of the Inductive Sciences from the Earliest to the Present Times*. London: J.W. Parker.

Wilcox, C. 2004. *Robert Redfield and the Development of American Anthropology*. Lanham: Lexington Books.

Winsor, M. 2006. "Creation of the Essentialism Story: An Exercise in Metahistory." *History and Philosophy of the Life Sciences* 28: 149–174.

Wolpoff, Milford and Rachel Caspari. 1997. *Race and Human Evolution: A Fatal Attraction*. New York: Simon and Schuster.

Woolridge, Clara, ed. 1959. *Genetics and Twentieth Century Darwinism. Cold Spring Harbor Symposia on Quantitative Biology*, vol. 24. Cold Spring Harbor, NY: The Biological Laboratory.

Yudell, Michael. 2014. *Race Unmasked: Biology and Race in the Twentieth Century*. New York: Columbia University Press.

6 A *kairos* moment unmet and met

The controversy over Carleton Coon's *The Origin of Races*

Coon's untimely project

Carleton Stevens Coon's (1904–1981) *The Origin of Races* (Coon 1962a) appeared a few months after Dobzhansky's *Mankind Evolving*. It brewed up a scientific and political storm that forms a watershed moment in our unfolding story. Coon, a physical anthropologist who in 1948 moved from Harvard to assume a professorship and museum curatorship at the University of Pennsylvania, had long maintained that "[a]natomical differences between living races" of *H. sapiens* are "self-evident.... They have long been recognized" (Coon 1955, 264). In his new book he argued that *H. erectus*, our immediate predecessor, split into races that subsequently evolved independently into *H. sapiens* "as each subspecies, living in its own territory, passed the critical threshold from a more brutal to a more sapient state" (Coon 1962a, 657). We are now a single species. But, Coon argued, our unification took place *after* the evolution of the great races whose traces we still bear.

Like Johann Friedrich Blumenbach in the eighteenth century, Coon identified five such races. Using the acquisition of fire as an index of intelligence, intelligence as an index of evolutionary superiority, and (estimated) IQ as an index of intelligence, he went on to propose that "Congoids" crossed the sapient threshold about 200,000 years after "Causasoids" and 300,000 after "Mongoloids," the most successfully dispersed (and so presumably the fittest) of the major races. He claimed to be relying on up-to-date fossil finds and up-to-date population-based evolutionary theory in making his case. In Chapter 5 we explained why Washburn believed that Coon was too out of sync with the Modern Evolutionary Synthesis and the New Physical Anthropology to make that claim stick. In this chapter we will show why Dobzhansky, too, judged Coon to be dead wrong about the history of our species, and why Simpson and Mayr, Dobzhansky's closest allies, greeted Coon's hypothesis more positively.

At the time, Coon was widely regarded as the most accomplished physical anthropologist to have been schooled by Hooton, who dominated the field in America in the middle decades of the century (Chapter 5 of this book). Coon's fieldwork was in the Middle East and North Africa, where he was not only an anthropologist, but a secret agent whose exploits in World War I might have been lifted from pulp fiction (Coon 1931, 1935, 1951, 1980). His dissertation focused on the cultural and racial composition of Berbers of the Rif valley. Even so, his "chief concern," as he acknowledged, was racial anatomy (Coon 1931, vii). In reviewing this early work, the Boas-trained anthropologist Melville Herskovits noted that, "One feels that Dr. Coon is more at home in [his] section" on race than on culture (Herskovits 1933, 374).

In the 1930s. Coon took on the task of updating William Ripley's then-classic work, *The Races of Europe* (1899). Ripley's book was a major influence on America's Ur-racist, Madison Grant (Spiro 2008, 92–97). Grant himself urged Coon on. "I'm glad to know," he wrote "that the continuation of his [Ripley's] work is in such good hands" (Grant to Coon, April 4, 1933, Coon Papers). In rewriting *The Races of Europe*, Coon sounded themes that would recur, albeit reworked, in his postwar work. First, there was an emphasis on climate. "It is not easy to overemphasize the importance of climate in human history, particularly in the earliest times when man was merely a numerically unimportant parasite in the total fauna" (Coon 1939, 19). Second, migration was more important than cultural innovation:

> With changes in climate, [Pleistocene man] was forced to migrate with the animals and plants on which he lived, and the hunting and gathering of which he was adept. The only alternative was to stay on and adapt his culture to a new food supply, which would need new implements and new methods. On the whole, it was easier to move.
>
> (Coon 1939, 19)

Third, the major races evolved separately from a northern prototype:

> The earliest *Homo sapiens* known was an ancestral long-headed white man of short stature and moderately great brain size.... The negro group probably evolved parallel to this white strain from a related *sapiens* ancestor. At what point the ancestors of negroes and whites diverged is not known.
>
> (Coon 1939, 50–51)

Fourth, Coon held out hope for finding laws that would explain racial classification: "Laws in biology and in its sub-division sociology when once understood are seen to be as invariable and as valid as laws in physics" even if they do not supplant historical accounts of culture and cultures (Coon 1939, 251). Finally, Coon paid little more than lip service to the role of genetics in evolution. Here was the source of much of the trouble we will be recounting.

After World War II, Coon departed from Hooton's conviction that the races of man are separated by stable, because non-adaptive, traits by putting an adaptationist spin on their evolution. Still, an unpublished abstract of a proposed book to be called *Races of the World* shows that even in the mid-1950s he persisted as much as Hooton in making classification the point of physical anthropology:

> The fundamental procedure of science is classification. We are interested in classifying the living races of man on every available basis, anatomical, physiological, and behavioral. By applying the zoological test for subspecies … on critical genetically determined variables we find that modern man is divided into five clear subspecies, the Causasoid, Mongoloid, Negroid, Capoid, and Austroloid, each of which has racial subdivisions of lesser rank.
>
> (Coon Papers)

In this respect, Coon's approach to physical anthropology contrasts with that of his fellow Hooton student Washburn. Washburn, too, took an adaptationist turn, but used Dobzhansky's approach to population genetics to bring his New Physical Anthropology into full conformity with the principles of the Modern Evolutionary Synthesis (Chapter 5 of this book). This meant treating classification (specifically of primates and hominids) not as an end, but as a tool for understanding how the acquisition of culture reshaped human morphology. By contrast, Coon used evolution, understood in terms of the adaptive effects of a set of rigid climatic laws on organic form, to get at his end of racial classification. In consequence, his postwar work retained elements of typology and so in Washburn and Dobzhansky's view failed to learn the anti-racialist lessons taught by anthropology-informed population-genetic evolutionary biology.

We will reconstruct Coon's ideas and reactions to them through the lens of the rhetorical concept of *kairos* or timeliness. "Ideas have their place in time," writes John Poulakos. "Unless they are voiced at the precise moment they are called upon they miss their chance to satisfy situationally shared voids within a particular audience" (Poulakos 1983, 39). It might be imagined that the *kairo*s concept plays little and ideally no role in scientific argumentation, which, it is often presumed, accumulates knowledge in proportion as it deflects what Martin Luther King called "the fierce urgency of the now" by sealing off inquiry from the pressures of the day.[1] As we have seen again and again in this study, however, this leisured condition seldom obtains even in science, least of all where the natural sciences bear on human affairs. Accordingly, we believe that rhetorical argumentation in science:

> is not restricted to places in which science intersects with the sphere of social action.… It occurs at any point within scientific inquiry – even the most theoretical – in which specific audiences are tasked with making judgments and performing actions that turn a situation this way rather than that way, and when those judgments are influenced by a whole range of rhetorical appeals that are brought to bear on a moment of choice.
>
> (Crick 2014)

Coon was not insensitive to timeliness. On the contrary, in an emerging postwar context in which the unity of science was becoming a hallmark of open societies (Chapter 5 of this book), he thought he saw an opportune moment for physical anthropologists to unite the social and biological sciences through general laws linking morphology to environment. In spite of his inviting narrative style, however, or perhaps because of it, his construal of the rhetorical situation in which he simultaneously addressed the public and his fellow professionals took insufficient account of scientific and political changes in the twenty years since America entered World War II. In the face of the consolidating Modern Evolutionary Synthesis, a rapidly growing hominid fossil record coming out of Africa, flows of genetic data (mostly serological) from there and elsewhere, and the global process of decolonization whose expression in America was the Civil Rights Movement, Coon's *The Origin of Races*, which he intended as a *magnum opus*, failed to persuade. Instead, Dobzhansky and Washburn's approach became the received view in both cultural and physical anthropology, thereby playing a key role in integrating the field (Chapter 5 of this book). Dobzhansky, in a negative review of Coon's book, and Washburn, in denouncing it in his Presidential Address to the AAA in November, 1962, portrayed his methods as hopelessly old-fashioned and not incidentally racist (Dobzhansky 1962, 1963a, b; Washburn 1963). We considered Washburn's critique in Chapter 5. In this chapter we will stress the role Dobzhansky's population genetics played in changing the rhetorical situation created by Coon's book.

The realignment Dobzhansky and Washburn catalyzed had a lasting effect on public as well as professional discourse. Having lost the cover previously afforded them by elite Northern intellectuals, Coon's ideas henceforth became hostage to white supremacists, where, if precariously, they remain to this day – for example, in the rhetoric of the latter-day Klansman David Duke (Duke 1998). To ensure that populist bigotry and newly revived religious opposition to evolution would not penetrate public education American evolutionary biologists and anthropologists made common cause in reforming biology instruction in secondary schools and colleges by making Mendelian genetics carry the egalitarian lessons imparted to it by Dobzhansky (Chapter 4 of this book). This still ongoing effort may not have succeeded in disseminating population genetic principles and egalitarian conclusions about race to the citizenry as widely and deeply as early advocates hoped (Lieberman *et al.* 2003; Cartmill and Brown 2003; Egan 2016). But this does not invalidate the fact that the initiative did succeed in depriving opponents of *Brown* v. *Board of Education*, the Civil Rights Act 1964, and the Voting Act of 1965 of any opportunity to lean on scientific authority to support racial inequality. The fact that creationism rather than racism has been since then the (overt) focus of populist opposition to evolutionary science testifies, if obliquely, to this fact.

In view of this success, it is striking that the Coon controversy put Dobzhansky, who denied Coon's credentials as a population thinker, at odds with Mayr and Simpson, who accepted them. This was an exceptional development, since to promote their scientific agenda the makers of the Modern Synthesis took care to forge agreements and publicly advertise them as unified, and hence secure, knowledge. We will argue that Dobzhansky's alertness to what was at stake in Coon's hypothesis reflected his longer and deeper interaction with American anthropologists. That his suspicion of Coon's motives proved correct is worth bearing in mind now that echoes of Coon's approach to human races have shown signs of resurfacing (Wade 2014; Chapter 1 of this book).

Pushes and pulls in postwar physical anthropology

The Cold Spring Harbor Symposium on Human Origins and Evolution (1950)

There are good reasons to view the years immediately following World War II as an opportune moment for rhetorical action in anthropology. Physical anthropologists felt keenly the imperative to unify the various natural sciences with one another and with the social sciences in the hope of creating a rationally guided postwar world order. Coon shared this ambition. He believed that physical anthropology:

> needed a completely fresh point of view about the causes of racial differences through their relationships between race and human behavior.… The whole thing needed to be handled from a fresh point of view which would tie it in with the natural sciences on the one hand and human relations on the other.
>
> (Coon to W. W. Howells, November 8, 1947, Coon Papers)

One spur toward unifying physical with other areas of anthropology and adjacent disciplines was the imperative to account for "specimens of fossil man … coming to light in gratifying numbers" in the 1930s and early 1940s (Howells 1942, 182). Franz Weidenreich (1873–1948) was a key figure in making and interpreting some of these discoveries. Fleeing Nazi Germany in 1934, he briefly became Mayr and Simpson's colleague at the American Museum of Natural History before traveling to China, where he worked on unearthing and describing *Sinanthropus pekinensis*: "Peking Man," an example of *H. erectus* (Wolpoff and Caspari 1997, 179–194). In tune with the dominant trend in paleontology between the late 1880s and the first stirrings of the Modern Synthesis, Weidenreich took natural selection to be purely eliminative. "In no instance," he wrote, is it "capable of producing new types by itself, but merely of singling out from different types already present" (Weidenreich 1939, 85). For the creative force in evolution, he assumed orthogenesis, according to which evolution manifests an internal, largely necessitated drive toward complexity.

It might seem odd that Weidenreich would persist in this theory. It had come under fierce attack from advocates of the Modern Synthesis. In 1944, Simpson persuasively argued that if natural selection is allowed to work at various rates at different times the adaptation that prevails in microevolution will also be seen in macroevolution: evolution at and above the species level (Simpson 1944). Moreover, orthogenesis carried a lot of baggage. Henry Fairfield Osborn, Edward Cope's student and Hooton's teacher, used its built-in concept of parallel evolution to distance our kind from the African genesis postulated by Darwin and Thomas Huxley. Evidence for their view was, however, accumulating rapidly. Weidenreich responded by turning orthogenesis against its former self. His multi-regional hypothesis, as it has come to be called, affirmed the unity of our species by treating *H. erectus* populations as an interbreeding array of regionally dispersed races whose contemporary descendants evolved by a continuous process into *H. sapiens* by a combination of inner drive and gene flow (Weidenreich 1946).[2] His thesis was, and was intended to be, the antithesis of Nazi racial science, which came close to viewing human races as separate species, much as American polygenists had in the period leading up to the Civil War and as the Canadian plant geneticist R. Reginald Gates was still doing (Gates 1944, 1947; Stanton 1960; Horsman 1987; Desmond and Moore 2009).[3] Between 1941 and 1950 about 13 percent of research articles in evolutionary biology still invoked orthogenesis (Brush 2009, 99–10, 132–133). Weidenreich's influence was a factor. But much of it had to do with method. Orthogenesis "made suitable the subjection of the data to a series of measurements, for under such a theory study by means of measurements is enough" (Kaplan 1951, 25). Inner tendencies not much affected by environmental contingencies or luck-of-the-draw variation meant that physical anthropologists could hold onto their calipers and not worry about learning genetics.[4]

Weidenreich's influence on the topic of the *erectus-sapiens* transition was felt at the symposium on Human Origins and Evolution Washburn and Dobzhansky mounted at Cold Spring Harbor in 1950. Since the aim of the Symposium was to align physical anthropology with the Modern Synthesis, ultimately with a view to refuting scientific racism, a recurrent issue was how and how much Weidenreich's picture would have to be changed to conform to the meanings the Synthesis assigned to key terms such as "genus," "species," and "race." Participants offered a range of answers, but for the most part the conversation was kept too pleasant to allow conceptual differences to surface sharply. Washburn later said,

> Although one of the main purposes of the conference was to stress the importance of thinking in terms of populations, typology continued to be used by a number of [unnamed] participants. Population vs. type was probably too fundamental an issue to be discussed usefully in a public meeting.
> (Washburn 1983, 15–16)

As a community builder, Washburn thought it more important to keep anthropologists and geneticists talking (Chapter 5 of this book). It can be argued that what didn't get resolved at Cold Spring Harbor mutated into the return of the repressed a decade later when Coon published *The Origin of Races* in a rhetorical situation far more laden with social tension.

Dobzhansky's talk at Cold Spring Harbor illustrates Weidenreich's ghostly presence. In first foray into anthropology, in 1944, Dobzhansky had praised him, and by implication his commitment to the unity of the species, for undermining the pride physical anthropologists took in multiplying hominid species by naming new fossils after themselves. This is just vanity, Dobzhansky argued, since it is simply an artifact of typology-infected taxonomic practices (Dobzhansky 1944). In his article in *American Anthropologist* the geneticist also expressed his belief that Weidenreich's account of hominid evolution was empirically more consonant with treating gene flow and natural selection, not orthogenesis, as its drivers. Recognizing Weidenreich's underestimation of the innovative power of natural selection, he pointed out that his "uncreative" view of selection brought with it "misunderstanding of … race and species definitions" in the senses required by the New Systematics and the Modern Synthesis (Dobzhansky 1944, 254). Dobzhansky followed up on this claim in his Address at Cold Spring Harbor by insisting that "race" and "species" – he said little about "genus" – must be defined in terms of comparative frequencies of genotypes in populations. Races are reproductively open and species reproductively closed Mendelian populations (Chapter 4 of this book).

At Cold Spring Harbor, Dobzhansky stressed four relevant implications. First, conceptions of biological categories that betray the presence, however spectral, of the idea of an original type underestimate genetic diversity and wrongly portray it as falling into discreet racial packages (Dobzhansky 1950, 387). For this reason, "One may safely reject the view that inter-populational variability in man arose through breakdown of uniform 'primary' races" (389). Third, as he and Dunn had shown in *Heredity, Race and Society*, "The difficulty of arriving at a wholly satisfactory classification of human races has … increased by taking into consideration many independent traits," since "the geographical distribution of human traits often shows quite striking independence" (391). The discovery of a plethora of hominid fossils served to underscore the point. Finally, although "conjectures," as Dobzhansky called them, about race-specific adaptations were "timely … [as a way of] stimulat[ing] interest in causal analysis of human variability … the most important trend in the human species is toward genetically determined educability" (399, 400), as he and Montagu argued earlier (Dobzhansky and Montagu 1947, Chapter 4 of this book).

These points, especially the last, were aimed at a book Coon had just co-authored with fellow physical anthropologists Stanley Garn and Joseph Birdsell (Coon *et al.* 1950). Their goal was to identify human races in terms of adaptation to different environments. To Dobzhansky, it was good that the authors were willing to multiply human races. They counted thirty of them. Less encouraging, however, was the quasi-essentialist, insufficiently biogeographical, and only superficially statistical assumption that each race is "a composite, a type specimen, a mean of the group" – the assumption Montagu mocked as the "omelette" conception of race in his own speech to the Symposium (Coon *et al.* 1950, 111; Montagu 1950, 318; Chapter 4 of this book). In *Races* the fault lines between typological and population genetic thinking are in fact hard to miss: "A race therefore is *in a sense a population.* A population is composed of individuals who normally interbreed, possess a common genetic pool, and *look alike within broad limits*" (Coon *et al.* 1950, 111, our italics). As "in a sense" and "look alike" show, pride of place is given to morphology. Genetics (mostly in the form of frequency of blood types in various populations) was nice to have, but could only supplement phenotypic differences: "*A race is a population which differs phenotypically from all others with which it has been compared.* These differences are of varying degree. If we know of genotypic differences as well, we can add these to the list" (Coon *et al.* 1950, 112, italics in original). This hybrid was Birdsell's brainchild. He worried that the morphological list Coon originally proposed lacked any foundation in or reference to population genetics. "Nowhere in our original racial scheme," he told Coon, "did we indicate the closeness of genetic ties in the major racial population pools" (Birdsell to Coon, undated, Coon Papers).

Dobzhansky read the incoherence in *Races* the same way as his congenitally *politique* friend Dunn. It betrayed "a transition stage in anthropological thinking about race" between the "old or classical view of race as a fixed type of man" and the "new or genetic view of race as a stage in the evolution of an interbreeding population" (Dunn 1951, 105). Clearly, a learning curve was inevitable as anthropologists absorbed the new genetic thinking. At Cold Spring Harbor Dobzhansky hoped to nudge the process along by warning anthropologists that there is a large gap between thinking that a trait is an adaptation and proving it, let alone identifying a particular adaptation or suite of adaptations as defining a race or a species. It was a point on which he also instructed Montagu (Chapter 4 of this book).

Dobzhansky, not being a systematist, was not expected to directly tackle hominid classification from the perspective of the Modern Synthesis, but the author of *Systematics and the Origin of Species*, Ernst Mayr, was. When Washburn and Dobzhansky first asked him to give a paper at the gathering, Mayr did not jump at the invitation. He was, he pointed out, undereducated in anthropology. Still, he accepted, admitting to Washburn, "It is a bit difficult for an outsider like myself" to navigate the relevant literature and soliciting the anthropologist's help (Mayr to Washburn, December 30, 1949, Mayr Papers). His studiousness paid off. His strategy was to apply what he had already done with "the magnificent geographic variation in South Sea Island birds" to "the magnificent body of new data" from recent fossil hominid finds (Mayr 1951, 109; compare Mayr 1980, 420). At the Symposium Mayr recast Weidenreich's account of the *erectus-sapiens* transition as a largely non-overlapping succession of non-interbreeding species belonging to a single genus.[5] He did so by taking the typological connotations out of "genus." "Genus," he argued, becomes more than a classificatory concept when it picks out "an ecological situation different from that occupied by the species of another genus [because these species] occupy a different adaptive plateau" (Mayr 1950, 110). Adaptive plateaus reflect evolutionary grades and trends that as they go up confer greater "independence from the environment" (Mayr 1950, 116). Huxley and Simpson also used this idea, thereby retaining something of the old scale of nature that ranked kinds as lower or higher depending on the depth of their psychological capacity for flexible and anticipative agency.

Mayr anchored his claim that *Homo* is a succession of species in the notion of competitive niche exclusion pioneered by the Russian ecologist-biologist Georgy

Gause (Gause 1935). Only one species at a time can occupy the same niche. Hence the evolution of *Homo* has been dominated by gradual, directional, "phyletic" evolution under the control of natural selection, which, as Washburn had been arguing, reshaped bodies into different species as increasingly more powerful communicative and cooperative forms of activity evolved (Chapter 5 of this book). Mayr did acknowledge the kernel of truth in Weidenreich's multiregionalism. There exists "a centrifugal force" of "geographical and other local variation which tries to break up the human species" and early on led to the formation of races or subspecies of *H. sapiens*, as evidenced by the existence of "pronounced racial groups, such as the Whites, Negroes, and Mongoloids" (Mayr 1950, 114). But, he argued, complete geographic and *a fortiori* genetic isolation was never reached and became increasingly unlikely as we became "less dependent on local adaptation" and more dependent on, or better liberated by, "generalized adaptive improvements such as are described by the social anthropologist" (Mayr 1950, 116). These considerations now make it close to impossible for our species to break up into races that might become separate species: "Man has, so to speak, specialized in despecialization.... If the single species man occupies successfully all the niches that are open for *Homo*-like creatures it is obvious that he cannot speciate" (Mayr 1950, 116). The growing agency of hominid species, not their passive accommodation to local environmental conditions, marks the ascent of man.

Mayr's address went a long way toward finding the common ground between biologists and anthropologists for which Washburn and Dobzhansky were looking. Among those impressed was Coon – so much so that after the Symposium he began painting himself as a population thinker in Mayr's image. The only difference, he said, was that his own focus was on the origin of modern human races rather than species. "The use of Mayr's system [of classifying humans]," he remarked, "makes my task much less revolutionary, and simpler," since he no longer had to sort out the vexed issue of different species and could keep his eye fixed clearly on his object of interest: different races (Coon 1953, 260). We may now be one species with powerful general adaptations of the sort that Dobzhansky, Montagu, and Mayr highlighted. But Coon insisted that we are also divisible into races that are (or were) adapted to different environments, not all of which are able (yet) to use those species-defining abilities as effectively as others.

Initially, Coon declined Washburn's invitation to participate in the Symposium. "I would like to give the conference a miss," he told him, because "American geneticists have become totalitarian. They have worked out a dogma and anyone who doesn't fall in with their way of thinking is unthinkable.... I want no part in an American opposite number of Lysenkoism" (Coon to Washburn, March 16, 1950, Coon Papers). Washburn talked him into participating by reminding him that, "Dobzhansky ... is especially interested in the question of adaptive characters" (Washburn to Coon, March 30, 1950, Coon Papers; see Dobzhansky to Coon, December 25, 1947, Coon Papers). Four years earlier Coon had blushingly confessed to Dobzhansky that, although he was indeed interested in adaptation, he knew no genetics, unlike the slightly younger Washburn. He told the geneticist that his plan to study the subject with J. B. S. Haldane in the UK had been scuttled by the outbreak of World War II and that his efforts to learn it on his own were "stymied by fatigue and the fruit fly. I wish some more fascinating animal had been selected" (Coon to Dobzhansky, February 18, 1946, Coon Papers). Since then, however, Coon's ignorance had turned into contempt. After the Symposium he was still complaining to Washburn that, "Geneticists have already been exposed to us for a long time, yet they reject the work we have already done. Can we merge their much more populous world to ours without being swallowed and losing sight of our objectives?" (Coon to Washburn, November 14, 1951, Coon Papers). What Coon took away from Cold Spring Harbor was a conviction that he could ignore Dobzhansky and Montagu's warning about positing adaptations without genetic proof because what his skeletons, skulls, and morphological measurements were saying counted by *Mayr's* standards as "population thinking."

Coon's confidence in this claim was strengthened by various pronouncements of Julian Huxley in the 1950s. He interpreted Huxley's statement that geneticists had provided the groundwork that allowed scientists to "pursue other problems" to mean that genetics could safely be consigned to the care of geneticists without giving it a further thought (Huxley 1954, 4). "The biologist who studies race among birds [like Mayr] and mammals [like Simpson]," Coon wrote, "is less concerned with laboratory genetics, which he can seldom arrange, than with observable variations in size, shape, and color, many if not all of which can be attributed to environmental adaptation" (Coon 1954, 188). In an apparent rejoinder to Montagu, who insisted at Cold Spring Harbor on waiting for gene frequencies to confirm adaptationist hypotheses, he remarked,

> Were we to await the day when the genetics of skin color, eye color, hair form, and hair quantity, to cite but a few variable human features, should be as well known as the inheritance of blood groups and hemophilia, we would be unable to speak of race for decades to come.
>
> (Coon 1954, 188; Montagu 1950; Chapter 4 of this book)

Coon also enlisted Huxley's help in discounting the notion that culture disrupts or at least complicates claims about racial traits. "Social anthropologists," he wrote, "should accept Julian Huxley's plea for freedom to pursue adaptation wherever it may lead us" without bowing either to genetics or culture (Coon 1955, 257, referring to Huxley 1954, 4). He was probably pleased when Huxley critiqued Dobzhansky's *Mankind Evolving* as not really evolutionary anthropology at all, but only a geneticist's "prolegomena" to it (Huxley 1962, 144–145).

What is surprising is how much support Mayr gave Coon. His alliance with Dobzhansky had long since immunized him against thinking that physical anthropologists could ignore genetics any more than paleontologists and other biologists could. His influential conception of speciation, for example, involves reorganization of the genome brought about by the stress of living at the isolated edge of a population's biogeographical range (Mayr 1954). Mayr began corresponding with Coon simply because he saw in him a potential ambassador of the Modern Synthesis to the tribe of physical anthropologists; with encouragement and direction he would presumably enlighten himself. Soon, however, their relationship turned into a mutual assistance pact and mutual admiration society.[6] When Coon published an article in *Atlantic* adumbrating the theory of raciation that would appear in *The Origin of Races*, Mayr checked his manuscript for accuracy (Coon 1957; Mayr to Coon, August 29, 1957, Coon Papers). In turn, Mayr asked Coon to vet the draft chapter on human evolution in his forthcoming *Animal Species and Evolution* and began working on a failed effort to bring him back to Harvard (Mayr to Coon, August 29, 1957; Coon to Mayr, August 31, 1957; Coon to Mayr, June 24, 1959; Mayr to Coon, May 12, 1960; Coon to Mayr, March 17, 1960, Coon Papers).[7] In none of these dealings did Mayr register an objection to Coon's idea about the antiquity and persistence of human races. On the contrary, Coon got the strong impression that Mayr agreed with him about

> how ancient *H. Sap.* must be, as well as the primary races.... Everybody else is saying we are very recent ... Washburn and Ashley Montagu and all the others chime in on that tune. Within the profession I stand practically alone on this issue.
>
> (Coon to Mayr, August 31, 1957, Coon Papers)[8]

In *The Origin of Races*, accordingly, we find Coon leaning with confidence on Mayr's (and Simpson's) authority in asserting that the continental races preceded speciation and survived it. "Realizing the enormity of my discovery in terms of its divergence from accepted dogma," he wrote in the introduction to his book,

> I knew that I must provide a theoretical foundation for the facts I had unearthed. The possibility that races can be older than a species had to be explored. I soon found, *by reading and through conversations with Mayr and Simpson* and other biologists, that what I had thought a revolutionary concept was so common an event in nature that others rarely bothered to mention it: to wit, that a species which is divided into geographical races can evolve into a daughter species while retaining the same geographical races.
>
> (Coon 1962a, viii-ix, our italics; note the silence on Dobzhansky)

To explain Mayr's resonance with Coon we turn to their shared interest in connecting typology-based ecological rules with evolutionary grades. Soon after the Symposium, Marshall T. Newman, an anthropologist who classified American Indian races at the Smithsonian Institution, reviewed Coon, Garn, and Birdsell's *Races* (Newman 1950b, 189, 1948, 1950a; Willey and Newman 1947). Newman appreciated how they explained that "the outstanding bodily characteristics of the present-day races are adaptations to differing environments, especially the climatic extremes" and wondered why such "neo-Darwinian principles … have been so little used in racial anthropology" (Newman 1950b, 189, 190). So entranced was Newman by Coon's book that he found things in it that weren't there. He pointed, for example, to its implicit appeal to nineteenth century ecological rules such as Allen's Rule (the proportional reduction of bodily protuberances and appendages as climates grow colder) as applied to the Mongoloid face and body. "The adaptational correlation between heavy surface pigmentation and excessive light and heat (Gloger's Rule)," he claimed, "is obviously a good one in man" (Newman 1950b, 191). So is Bergmann's rule that warm-blooded animals living in a cold climate will be larger than individuals of the same species living in warm ones.

Coon was floored by Newman's review. "At the time he wrote it I … had never heard of Allen, Gloger, or Bergmann" (Coon 1953, 15). Soon Coon was incorporating their rules into his account of human raciation, thinking of them as the key to the law-governed and progress-oriented approach to a unified theory of evolution that, should he succeed in validating it, would confer great authority on him. Coon argued, for example, that

> once a species or subspecies which possesses a wide and climatically varied geographical area has become established at an optimum body size range … then the total mass of the organism, all else being equal, follows the ecological rule of Bergmann, postulated for non-migratory and non-hibernating warm blooded animals.
>
> (Coon 1953, 262)

If ecological rules seem *prima facie* at odds with the Modern Synthesis it is partly because they were either products of pre-Darwinian thinking (Bergmann 1847, 1848; Gloger 1833) or, in the case of Allen's rule, brain children of someone who embraced orthogenesis even after Darwinian selectionism became influential (Allen 1877, 1905; Glaubrecht and Haffer 2010; Watt *et al.* 2010). One might well imagine that turning these rules into laws led Coon to regress toward Hooton's typologizing instead of pushing him further toward population thinking. Alert to this threat, Washburn wrote, "If a new physical anthropology is to differ effectively from the old it must change its ways of doing things to conform with the implications of modern evolutionary theory" and not drift back toward nineteenth century preconceptions (Washburn 1951, 299). Mayr agreed with this sentiment, but he was also impressed by how well his fellow German ornithologist Bernard Rensch (1900–1990), even after his embrace of natural selection and renunciation of orthogenesis, was able to use ecological rules as warrants for adaptationist explanations of macro-evolutionary trends. (Rensch even discovered a rule of his own; "Rensch's rule" correlates increased sexual dimorphism when males are larger and decreased dimorphism when females are bulkier.) So impressed was Mayr that he encouraged Columbia University Press to publish an English translation of Rensch's *Neuere Probleme der Abstammung* in the same series as Dobzhansky's, Simpson's, and his own canonical book of the Modern Synthesis. *Evolution Above the Species Level*, as Rensch's book was tellingly retitled, was published just in time to affect a 1947 Princeton conference that Mayr later cast as the Modern Synthesis's catalytic moment, and just in time, too, to welcome contrite postwar German evolutionists to the sort of Darwinism they should have embraced long before. When he took to writing the history of the formation of the Synthesis, Mayr even cast Rensch as a founder of the Synthesis (Rensch 1947a, 1959b; Mayr 1980). Coon, too, soon began praising Rensch, but not for the same reason. Mayr's aim was to connect ecological rules about morphology with grades of agency over the environment, not with racially correlated adaptations to particular environments (Coon 1954, 200). In giving Coon cover, Mayr failed to deter him from adopting notions of racial inequality to which Dobzhansky was preternaturally alert, tutored as he was by Montagu and Washburn, and to insulate himself from them.

Richard Delisle has argued that even after his conversion Rensch conceived "the concept of 'natural selection' … as a law in its full, deterministic sense; a law among many others in the biological and physico-chemical realms which bind together all cosmic entities in a single tight causal nexus" (Delisle 2009, 126). Accordingly, his turn to natural selection remained close enough to its orthogenetic predecessor to qualify as a target for Washburn's complaint that some biologists and anthropologists were merely dressing up old ideas in new terminology and declaring themselves population thinkers in good standing. Some recent commentators say just this:

> Rensch retains the terminology he used before his selectionist turn.… He imports all the empirical generalizations such as Bergmann's, Allen's, and Gloger's rules mentioned in his earlier work, and repeats his argument that geographically gradual variation can take place with only minor influences of natural selection, and partly uses examples from his "pre-Darwinian" work.
>
> (Levit *et al.* 2008, 311)

Rensch confirmed the complaint in advance at the 1959 Darwin Centennial Celebration at Chicago:

> Bergmann's, Allen's, Glazer's and other climatic rules … show that evolutionary progress was not accidental, but forced by the interaction of the law and of steady mutation and selection.… Thus animals with more rational structures and functions arose.… Hence I see the development of higher types of mammals and to some extent of a being like man as necessitated.
>
> (Rensch, in Tax and Callendar 1960, III, 151)

This was not Mayr's view. Ecological rules conform to population thinking, he argued, because they "have only a statistical validity" and respect the inherent contingency of Darwinian evolutionary dynamics (Mayr 1956, 10; Beatty 1995).[9] Nonetheless, Mayr argued that traits conforming to well-founded ecological rules can be presumed to be adaptive because they run along contours already carved by natural selection over macro-evolutionary time scales and mark taxa above the species level. In defending this view against the physiologist P. F. Scholander, who maintained that detailed physiological and morphological information about each case is necessary before any adaptive inferences can be made, Mayr insisted that the probative obligation runs the other way:

> The hypothesis … that Bergmann's rule is the result of natural selection in favor of an optimal surface to volume ratio is a legitimate one. It is axiomatic in scientific methodology that a hypothesis is considered valid until it has either been disproven or until a better one has been proposed.
>
> (Mayr 1956, 106)

Once validated, Mayr was confident that ecological rules could safely be applied as *explanantia* to particular cases.

An important motive for recasting ecological rules this way was that, as the prominence of the topic of evolutionary progress at the University of Chicago Darwin Centennial Celebration in 1959 shows, it was widely believed in the 1950s and 1960s that if the Modern Synthesis was to present itself as a complete and unified theory of evolution at all scales it would have to treat gradual evolutionary advance not only in particular lineages, such as the equines that Simpson used as a case study (Simpson 1944), but as a phenomenon general enough to need explaining in its own right (Ruse 1996). Evolutionary progress was not to be dismissed as an artifact of orthogenetic dogma. This turn toward evolutionary trends, grades, and progress met an enthusiastic welcome from many physical anthropologists, whose work invited ordering fossils into progressive series. Like Coon and Rensch, some of them did not fully abjure morphological definitions of races and other taxonomic categories when they signed on to the Modern Synthesis (Gladwin 1947; Roberts 1952a, 1952b, 1953; Newman 1953, 1954, 1956, 1963; Baker 1962). Coon acknowledged that ecological rules "cannot be called laws in the sense of Newton's laws or the second law of thermodynamics" (Coon 1953, 16). But this did not prevent him any more than it prevented Rensch from casting rule-covered adaptive rationales in the language of law-like necessity and temporal invariance on which typological thinking depends. For example, he wrote that racial differentiation "was once necessary" because "it made possible the opening up of all areas of the earth not covered by the ice [and] the domestication of many kinds of plants and animals." From this environmentally deterministic, remarkably teleological, crypto-orthogenetic idea, he concluded that, "The races of man have failed to change since the beginning of present history because no further changes were needed" (Coon 1954, 183). This claim had been made by racial typologists in the previous century and still undergirded Hooton's stress on the persistence of non-adaptive traits (Chapter 5 of this book).

Dobzhansky *Agonistes*: confronting Carleton Coon and Carleton Putnam

Dobzhansky brought to the rhetorical situation precipitated by Coon's *The Origin of Races* a professional and personal relationship with its author even more complicated than Mayr's, in part because he had been dealing with anthropologists, including Coon, for a longer time. As early as 1939, Dobzhansky told Montagu that he was disappointed with Coon's retention in his earlier book, *The Races of Europe* (Coon 1939), of the presumption of old-style systematicists, including Coon's mentor Hooton, that taxonomic differences are non-adaptive. A shift to population genetic conceptions of classificatory concepts would change their minds, Dobzhansky told Montagu. He proposed to "have these matters discussed and if possible convince Coon that a methodological reform is here well overdue" (Dobzhansky to Montagu, December 7, 1944, Montagu Papers). We do not know if such a discussion ever took place, but the point is moot because, as we noted above, by 1950 Coon seemed to Dobzhansky to be moving in the right direction on his own steam. Having signed onto the Modern Synthesis after Cold Spring Harbor, he explicitly stated in his popular book *The Story of Man* (1954), "The biological forces of mutation and selection … are the only proven mechanisms of evolutionary change" (Coon 1954, 183; his omission of gene flow and genetic drift speaks to his innocence of population genetic theory). In 1959, we find Coon and Dobzhansky collaborating on a conference to celebrate the Darwin Centennial (Woolridge 1959). In *Mankind Evolving*, which appeared in May, 1962, six months before *The Origin of Races*, Dobzhansky publicly commended Coon, Birdsell, and Garn for being population-oriented enough to list thirty human races, taxing them only with failing to mention that races can be enumerated in any number of ways depending on what classification is needed to frame and answer this or that question (Dobzhansky 1962a, 265–267; Chapter 4 of this book). By adding two races Dobzhansky seemed to give their list his blessing (Dobzhansky 1962a, 263).

For his part, Coon was as eager to secure Dobzhansky's approval as he was to secure Mayr's and Simpson's, but, given his touchiness on the subject of genetics, even more eager to neutralize his opposition. In opening a correspondence with the geneticist in 1946, Coon was effusive in his praise. He told Dobzhansky that his 1944 article in *American Anthropologist*, in which he had recommended reframing Weidenreich in adaptationist terms, "should be read by all students of physical anthropology.… I shall assign it to my students" (Coon to Dobzhansky, February 13, 1946, Coon Papers). A week later Coon informed Dobzhansky that he, too, was an adaptationist (Coon to Dobzhansky, February 18, 1946, Coon Papers; Chapter 5 of this book). A year later he shared with Dobzhansky an early draft of what would become the co-authored *Race*s (Coon to Howells, November 8, 1947, Coon Papers). Dobzhansky, who at the time was more eager to get a foothold with anthropologists than to set them straight, encouraged Coon in an anodyne way to keep moving away toward adaptation:

> The problem of adaptive value of human traits will be the central problem of physical anthropology in the future.... This is indeed the common ground on which anthropologists and geneticists will eventually meet, and I personally feel that this field of study is so interesting and important that if I could I would sacrifice *Drosophila* genetics at least in part for this real human genetics. I can only welcome most enthusiastically your work.
>
> (Dobzhansky to Coon, December 25, 1947, Coon Papers)

When *Mankind Evolving* appeared, Coon was delighted to see himself praised for recognizing a larger number of human races than typological thinkers, in whose ranks he was presumably glad not to see himself listed, generally did (Dobzhansky 1962a, 263). In a letter thanking Dobzhansky for sending him a copy of the book inscribed "to C. S. Coon with warmest regards," Coon wrote that, except for identifying two fossils as *sapiens* that Coon placed on the *erectus* side, "[w]hat you say is almost identical with what I am saying in my [still in press] book *The Origin of Races*" (Coon to Dobzhansky, May 26, 1962, Coon Papers). Coon had every reason to think that his campaign had succeeded and that he and Dobzhansky were on the same page. Accordingly, it is easy to understand why he was surprised, hurt, bewildered, and lastingly angry when Dobzhansky sent him a copy of his forthcoming review of *The Origin of Races*, which sought to discredit his book with the *bien-pensant* audience of the *Saturday Review of Literature.* In it he accused Coon of regressing to the orthogenesis he claimed to have repudiated and of giving aid and comfort, wittingly or not, to the enemies of the civil rights of African-Americans. He advised him to repudiate appropriations of his work by racists in the follow-up book that Coon informed readers of *The Origin of Races* he was writing (Review of *The Origin of Races*. Unpublished. Dobzhansky Papers). Although the editors of *Saturday Review* scotched the review for reasons that are still unclear, the typescript was soon circulating among Coon's and Dobzhansky's respective allies.[10] In slightly variant versions, it appeared the following year in *Scientific American* and, together with statements by Coon and Montagu, in the anthropologist Sol Tax's controversy-promoting journal *Current Anthropology* (Dobzhansky 1963a, 1963b).

Only when we recognize the pressing public issues of the day and the accumulating pattern of distorted communication between Dobzhansky and Coon – they were constantly nudging each other toward their own positions, Coon flattering Dobzhansky, Dobzhansky patronizing Coon – can we understand how and why this seemingly sudden rupture occurred. Admittedly, as Coon's friend, editor, and fellow anthropologist W. W. Howells told Montagu, he "had a very short fuse and a gift for taking things the wrong way" and could become "violently indignant" (Howells to Montagu, December 11, 1995, Montagu Papers). Prior to 1962, however, Dobzhansky's only criticism of Coon in print or correspondence was about the draft of an article the anthropologist had shared with him in 1947. Still, his objection on that occasion foreshadowed what would divide them later.

Dobzhansky fretted at the time that Coon's call to study adaptation in humans "loses a part of its effectiveness if the depth of our ignorance [about adaptations] is a bit hidden" by "the form rather than the substance of your paper" (Dobzhansky to Coon, December 25, 1947, Coon Papers). He meant that Coon's narrative style, with its penchant for giving adaptive rationales that advanced whatever evolutionary story he was telling, led him to neglect documenting his claims and referring to the relevant literature, thereby burying the tentative character of most of that literature and recommending his scenarios to readers simply because they seemed to spin a good yarn. This was an especially grievous offense to Dobzhansky because for him adaptation is hard to prove (Chapter 4 of this book). Dobzhansky told Coon that as a result, "[i]t does not appear from the article where the known facts end and the hypothesis begins." In responding, Coon told him that his narrative style came from having been "trained as a novelist" (Coon to Dobzhansky, January 6, 1948, Coon Papers).[11] This admission didn't help.

Dobzhansky was not alone in objecting to Coon's style. Howells warned Coon in 1948 that readers of his manuscripts concurred that "[y]ou are sticking your neck out" by not making it clear that you are presenting not a finished case, but only "opening up the question of selective adaptation in a big way with no holds barred." Howells told Coon just what Dobzhansky did: It was difficult for the reader to tell where documented facts ended and hypothesizing and interpreting began (Howells to Coon, January 20, 1948, Coon Papers).

The same problem manifested itself two years later in *Races*. Readers were told with great confidence why Yahgan Indians are able to endure great cold, why measles killed off lots of "Indians" but was only a nuisance to "white men," why Arab noses would freeze in Siberia, and why "the flat-faced Tungus, whose ancestors have lived in this [Siberian] environment for many years, is able to stand this extreme of temperature" (Coon *et al.* 1950, 4). It was always climate exerting selection pressure on anatomy, leading to morphological adaptation or migration equipped with the adaptations a group already possessed. Nor was Coon's story-telling tendency corrected in *The Origin of Races*. The book freely attributed motives to hominid actors, especially male actors, and these attributions played a substantive role in driving his argument. He portrayed smartened up *erectus* males, for example, as agents of improvement as they wandered around impressing not only females, but more importantly their fathers, who, in a masculinist fantasy, he pictured as happy to "give their daughters" to superior males on the expectation that they "might bring in the most meat to feed the most people" (Coon 1962a, 86). Dobzhansky made this sort of narrativizing sleight of hand the epistemological premise of his critique of *The Origin of Races*. Coon's claim that five (and only five) geographically isolated *H. erectus* races arrived independently at the same higher, sapient evolutionary grade, with sub-Saharan Africans bringing up the rear only 30,000 years ago, struck Dobzhansky as not only repudiating Coon's recognition of at least thirty *H. sapiens* races, but, worse, as regressing beyond Weidenreich's version of orthogenesis almost to Henry Fairfield Osborn's cruder sort, according to which typologically defined races, species, and higher taxa pull themselves up parallel evolutionary ladders by their own bootstraps and push their hapless predecessors off the rungs below. Coon's skill in telling his readers about the belated evolution of "Congoids," Dobzhansky charged, left them with a strong impression that "Negroes are … evolutionarily backward and primitive" and "socially and culturally inferior" (Review of *The Origin of Races*. Unpublished. Dobzhansky Papers).

The heart of Dobzhansky's argument is his assertion that Coon "gets himself tangled in semantic mischief" in describing evolutionary processes (Review of *The Origin of Races*. Unpublished. Dobzhansky Papers). When Coon first read the draft review, he brushed the point aside. For Dobzhansky, however, semantics was not "mere semantics," as Coon assumed. It is mischief of the highest order because it bears on the definition and proper use of evolutionary biology's core theoretical concepts (Chapter 4 of this book). When fully spelled out, as Dobzhansky admittedly did not do in a review aimed at a general audience, the objection meant that Coon's claims as well as his narrative way of making them undercut his pretense to be working under terms laid down by the Modern Synthesis. In any version of the Synthesis, but especially one that highlights the temporal aspects of biogeography as much as Dobzhansky's, speciation is *by definition* an outcome of raciation. So one cannot coherently say that the continental races of *H. sapiens* are the same as races of *H. erectus* that happened to dwell in China, for example, or sub-Saharan Africa. It hurts rather than helps to reply that before and after speciation these races were morphologically similar or even identical. This claim subverts the population-genetic definition of a race and suggests that there were, and to some degree still are, five pure types of *H. sapiens*. Nothing could be less consistent with the Modern Synthesis or more suggestive of less than full devotion to the unity of our species.

Alluding to Coon's identification only a few years before of thirty races of *H. sapiens*, Dobzhansky noted in the review that, "Professor Coon knows as well as anyone else that … every anthropologist can give names to any number of racial subdivisions he chooses" (Review of *The Origin of Races*. Unpublished. Dobzhansky Papers). He might more accurately have written that Coon *should* have recognized that "classifying and systematizing are devises used to make diversity intelligible and manageable." As things stood, Coon's old-fashioned insistence that classifying races is the point of physical anthropology prevented him from recognizing classification as a process-oriented, problem-relative, pragmatic activity (Dobzhansky 1962a, 178–179). In *Mankind Evolving*, Dobzhansky made this point gently (Dobzhansky 1962a, 266–267). In reviewing *The Races of Mankind*, he made it with a certain ferocity: Whatever Coon's investigatory purpose was in lighting on five morphologically and typologically defined races he left readers with the impression that, "*Homo erectus* … woke up one fine morning and found himself transmuted into *Homo sapiens*" – or rather woke up to this discovery five times in five places (Review of *The Origin of Races*. Unpublished. Dobzhansky Papers).

Coon recognized a certain amount of gene flow between these races. But in restricting it to a few intrepid and hence intelligent male hunters who helped less advanced *erectus* races cross the line to sapience, Dobzhansky argued, he had to assume that *erectus* groups opened up to welcome quick-witted and sexually intrepid representatives of a more advanced race and then, to prevent regression by contact with less advanced *erectus* groups, just as suddenly snapped shut (Review of *The Origin of Races*. Unpublished. Dobzhansky Papers; Dobzhansky 1963a, 172; 1963b, 366). The scenario was intelligible only on the assumption that there weren't that many *erectus* populations around and that the few that existed were geographically isolated. Having recognized thirty races as long ago as 1950, why except to tell this tale would Coon now be talking about five typologically pure races arising on their own from five and only five separate *erectus* populations? The actual number of *erectus* races is in any case unknown and in virtue of the observable unity of our species must be presumed to have interbred in so many crisscrossing ways that *erectus* became *sapiens* in a continuous, gradual, protracted process that Dobzhansky compared to braiding strands of a cable or a rope (Dobzhansky 1963a, b). On the principles of the Modern Synthesis it is impermissible, too, to imply that, "Every evolutionary line sooner or later reaches a higher grade by a parallel but independent development" (Review of *The Origin of Races*. Unpublished. Dobzhansky Papers). Accordingly, in spite of Coon's advertisement of himself as embracing adaptive natural selection, his scenario was so orthogenetic that it came close to "relaps[ing] back to the old idea that there were many independent human species, which Weidenreich and Coon have done so much to invalidate" against neo-polygenists like Gates (Review of *The Origin of Races*. Unpublished. Dobzhansky Papers).

Why, then, was Coon backtracking from his own best work (Dobzhansky 1962c, 6)? Dobzhansky became suspicious about the answer as soon as he recognized an affinity between Coon's ideas and those that, even before the appearance of *The Origin of Races*, were circulating among opponents of the Supreme Court's 1954 *Brown* v. *Board of Education* through the efforts of a publicity operation owned and operated by Coon's cousin Carleton Putnam. Like Washburn and Montagu, Dobzhansky had been warning fellow professionals about the pseudo-scientific fallacies of Putnam's tract *Race and Reason* and the antics of the "National Putnam Letters Committee" for several years (Dobzhansky 1961; Jackson 2005). He was convinced on reading *The Origin of Races* that Putnam's racist themes echoed Coon's and that Coon's book expressed these themes in ways designed to lend them scholarly support. In his review, he harped on the themes of orthogenesis and polygenism to what might seem (and to Mayr and Simpson did seem) an excessive degree because he feared that the commonplaces of these old biological notions were the tacit causal-explanatory glue of the narrative of *The Origin of Races*. However much their author might deny it, Coon's book invited its readers to think of race in the nineteenth century terms that many of them still latently presumed to be the meaning and lesson of evolution. The wide circulation of *Race and Reason* pre-adapted Coon's readers to pick up these signals and to believe that Professor Coon, the eminent University of Pennsylvania anthropologist, was offering scientific support for basing public education on racial inequality. Since Dobzhansky feared that any biologist or anthropologist who gave Coon's book a pass would be implicating him or herself in this enterprise, he decided to write a harsh review.

His reading of the situation was not unjustified. An editorial in the *Charleston News and Courier* on the eve of the publication of Coon's book stated that, "Science, law, and religion are by no means on the side of those who demand integration of the races. Indeed, they seem to uphold the American tradition of racial segregation" ("A Plea for Moderation," *Charleston News and Courier*, October 5, 1962). Favorable reviews of *The Origin of Races* began to appear not only in the racist-eugenicist *Mankind Quarterly*, as might be expected, but also in William F. Buckley's high-profile *National Review*, where Nathaniel Weyl, an editor of *Mankind Quarterly*, praised Coon for refuting Montagu's claim that races in the customary (and in Weyl's view the only) sense are mythical, not real biological, entities (Weyl 1963).[12] As soon as it appeared, Putnam used Coon's book to make this very argument (Putnam 1964, 13).

Dobzhansky ended his review by hoping that in his promised companion volume Coon would "clear up the ambiguities and inconsistences of the present volume, which unfortunately lend themselves to such grievous misuses for the purpose of racist propaganda" (Review of *The Origin of Races*. Unpublished. Dobzhansky Papers). Out of professional courtesy, and because he thought it honorable to issue his request for clarification personally, Dobzhansky sent Coon a draft of his review. Offended at what he took as an attack on his professional credentials, Coon replied that he had no intention of commenting on the uses to which his results might be put by others (Coon to Dobzhansky, October 20, 29, 1962, Coon Papers). To accept Dobzhansky's suggestion would be to confess that he had no scientific integrity. For Dobzhansky, to ask him to do so constituted a challenge to his reputation that a court would surely protect. He threatened to sue for libel.

Coon's reaction changed the rhetorical situation of Dobzhansky's review. His professional request for clarification in a future publication turned into a perceived demand that Coon should immediately and publicly repudiate the uses to which his research was being put by segregationists. The accent no longer fell on the ending of the review, in which he collegially asked Coon to distance himself in a future publication from misuses of his research, but on the opening sentence, in which he proclaimed that in world whose fate hung by a nuclear thread scientists could no longer enjoy the luxury of "living in ivory towers," but must take responsibility for the consequences of what they write (Review of *The Origin of Races*. Unpublished. Dobzhansky Papers). The fierce urgency of the now spread from the threat of atomic annihilation to anxieties about desegregating schools.

A bit shaken by what he had wrought, Dobzhansky wrote Mayr, who had already published a review of Coon's book in *Science* (Mayr 1962), asking for information, advice, and support. Coon's huffy response, he told his friend, was

> more appropriate to Carleton Putnam than to Carleton Coon.... Do you think I am being unfair to Coon? Or have you deliberately avoided in your review ... in *Science* ... mentioning the uses to which Coon's book will be

> put *and for which, I am sad to say, it is clearly intended*? Please show this to George [Simpson], whose opinion I would naturally value like yours.
> (Dobzhansky to Mayr, October 23, 1962, *Dobzhansky Papers*)

Dobzhansky wanted to know why Mayr raised no concerns about segregationist appropriations of Coon's hypothesis in his review. Mayr did acknowledge that *The Origin of Races* "will stir up more than one controversy," but the context shows he meant professional controversies (Mayr 1962, 422).[13] Dobzhansky was probably wondering, too, why Mayr's review praised Coon as "at all times a superb *storyteller*" when he himself found Coon's narrative approach so offensive (Mayr 1962, 420, our italics). Calling *The Origin of Races* "bold" and "imaginative," Mayr wrote that Coon's defense of the "thesis of mankind's ancient unity and ... the corresponding antiquity of the racial diversity of mankind" was at least as good as any other, and probably better, and so deserved a thorough professional airing (Mayr 1962, 420). Nor did Mayr complain about Coon's use of simple statistical averages of multiple traits in describing races and species. On the contrary, he cast Coon as a full-fledged population thinker by writing, "The typological approach had reached the end of usefulness, and numerous new approaches were still in the data-gathering stage. Coon's great new synthesis is one of a number of recent publications that signal the arrival of a new period" (Mayr 1962, 422).

In replying, Mayr offered Dobzhansky no help. "I saw none of these implications when I read the volume which you sent," he wrote (Mayr to Dobzhansky, November 1, 1962, Dobzhansky Papers). He advised him to list offending passages; if he published his review legal consequences in which the onus would fall on him were indeed likely to follow.[14] He didn't tell him that privately he had informed "Carl" that he had "done an absolutely remarkable job" in showing "how consistent a *story* emerges when all the evidence is related to all the other evidence" (Mayr to Coon, October 11, 1962, Coon Papers, our italics).

If anything, Simpson, to whom Mayr duly forwarded Dobzhansky's letter, was even less supportive. "Since you and I usually agree quite closely on such matters," the paleontologist wrote, "I was surprised that I could have read [Coon's] book and formed an impression so unlike yours.... Yes, I do think your review is unfair to Coon" (Simpson to Dobzhansky, November 1, 1962, Dobzhansky Papers). The evolution of races from less to more advanced grades, he told him, is not inconsistent with the modes and tempos of evolutionary change Simpson himself advocated. Ascent through grades is sprinkled throughout with speciation events. Whether and when race formation occurs in the process seemed to Simpson an empirical, not a definitional or ideological, issue, thereby reinforcing Mayr's call for professional discussion of Coon's hypothesis. Not only that. When Simpson's review of Coon appeared in *Perspectives in Science and Medicine* in late 1963 it was in effect a rebuttal of Dobzhansky's. As a vertebrate paleontologist, he was professionally better qualified than either Mayr or Dobzhansky to back up or challenge Coon's claim to be an objective, disinterested scientist. Excusing Coon's wavering between morphological and biological species concepts, he declared *The Origin of Races* to be "an honest and substantial contribution to the scientific study of races and nearly free from aprioristic bias" (Simpson 1963, 269). He implies that not orthogenesis, but the concentrated directional selection he called "orthoselection" in his 1944 *Tempo and Mode in Evolution* drove the races toward more or less equal sapience (271). If so, Simpson endorsed the coherence, testability, and even plausibility of Coon's hypothesis by recasting it in his own explanatory framework. Simpson's "evolutionary species concept" differed from Mayr's and Dobzhansky's "biological species concept"[15] in his 1944 *Tempo and Mode in Evolution*. By implication so did his definition of "race."

Rebuffed by his friends, but convinced more than ever of the validity of his objections, Dobzhansky told Mayr and Simpson, "There are no other … people in the world whose opinions on evolutionary matters I value more highly. To disagree with you is worse than almost anything." In view of their reactions he had re-read Coon's book and done some "soul searching." But because "the harm [Coon has] done by how he chooses to say" what he does "must be corrected" he informed them that, "I feel obliged to stick to my guns" (Dobzhansky to Simpson, Mayr, and Wm. Strauss, November 9, 1962, Dobzhansky Papers). Not to protest the effects of Coon's book would make him complicit in them in ways from which he was hoping in vain to deflect Simpson and Mayr.

Accordingly, Dobzhansky allowed his still unpublished review to circulate widely and redrafted it for publication elsewhere with his opening remark about scientists no longer living in ivory towers placed imperatively at the end (Dobzhansky 1963a, 172; 1963b, 366).[16] This time he didn't make the mistake "of sending a copy to Coon. His behavior," he told Washburn, "was not that of either a gentleman or a scientist" (Dozhansky to Washburn, December 20, 1962, Washburn Papers). Still, word of what had happened quickly came Coon's way. Coon wrote Putnam to tell him (perhaps incorrectly) that "dopey Dobbie" had been forced to withdraw his review from *The Saturday Review* because he was "privately flattened by both Mayr and Simpson" (Coon to Harold Strauss, November 4, 1962, Coon Papers; Coon to Putnam, January 22, 1963, George Papers). He told his editor,

> I have felt for some time that Dobzhansky has passed his peak. I was also dimly aware that, simpleton that he is, he was well under the hairy thumb of Ashley Montagu, and now I am pretty sure of it.

Coon even toyed with the idea that Dobzhansky was a tool of the Soviets. "Does [he] have kinfolk as hostages behind that old fence?" (Coon to Strauss, February 4, 1963, Coon Papers). He wrote to the President of Dobzhansky's new home institution, Rockefeller University, asking him to bring Dobzhansky to heel for having acted in an unprofessional manner. In doing so he exaggerated "the acceptance [of his own ideas] by George Simpson, Ernst Mayr, and Sir Julian Huxley" (Coon to Detlev Bronk, February 25, 1963, Coon Papers; for more of this mudslinging, Jackson 2005, 167–170; Collopy 2015).

There, amid hurt feelings on all sides, the matter might have rested were it not for the fact that in Spring, 1962, the Executive Committee of the AAA commissioned Washburn to use his Inaugural Address as incoming President to explain and justify an official Statement rejecting scientific racism that the AAA had issued at its annual meeting in November, 1961 (Chapter 5 of this book). Aimed at Putnam's growing influence, the Statement read:

> In view of the statements now appearing in the United States that Negroes are biologically and in innate mental ability inferior to whites, the Executive Board of the American Anthropological Association takes the position that: There are no scientifically established facts which justify the exclusion of any race from the rights guaranteed by the Constitution of the United States. The basic principles of equality of opportunity and equality before the law are compatible with all that is known about human biology. We are certain that the members of any race can fully participate in the democratic way of life and in modern technological civilization.
>
> (Quoted in Jackson 2005, 158)

We observed in Chapter 5 that between May, when Washburn was commissioned to give his Address, and its delivery in November 1962, the rhetorical situation was affected by the publication of Coon's book.[17] Washburn, we saw, was aware that he now had to do more than destroy Putnam's pseudo-scientific racism. He also had to convince his fellow anthropologists that even if it was not being exploited by segregationists Coon's *Origin of Races* should not get the professional airing that Mayr, Simpson, and others thought it deserved. Washburn made his case by forcing a stark choice on members of the AAA – and by extension members of the American Association of Physical Anthropologists (AAPA), from which Coon had recently resigned his presidency, huffily calling his fellow members "a craven lot" when they followed the lead of the AAA by issuing a Statement of their own in support of African-American equality (Jackson 2005, 159–160). Anthropologists must give up every trace of typological and racialist thinking, he argued, including the not inconsiderable traces of it in Coon's book, or else become complicit with racism without advancing scientific knowledge a whit (Washburn 1963, 521). Coon was still stuck in nineteenth century ways of thinking that are at odds what our morphology and population thinking actually reveal: "Human biology finds its realization in a culturally determined way of life." This claim, he said, is well established in Dobzhansky's *Mankind Evolving*, which he called "a great book" (Washburn 1963 521–523, 531). He didn't overtly say so, but Washburn also relied on Dobzhansky's unpublished review of Coon's book to make the point that we became a single species by co-evolving the species-wide modifications of hominid anatomy that support our species-wide cultural life (Washburn 1963, 521–523, 531; Chapter 5 of this book).

The beleaguered Dobzhansky was so pleased by Washburn's speech that he suggested it be "paste[d] into each copy of Coon's books" (Dobzhansky to

Washburn, January 4, 1963, Washburn Papers). In his *Reminiscences*, which were tape recorded not long after these events, he expressed gratitude for his support. He was also pleased that Coon's former collaborators Birdsell and Garn now opposed him and that Herskovits, the leading authority on African-American anthropology, took his side. So did the social psychologist Otto Klineberg, who had demonstrated the invalidity of IQ tests administered to African Americans (Dobzhansky 1962–1963, 618–619). So, of course, did Montagu (Dobzhansky 1962–1963, 620).[18] The resulting consensus has held to this day.

Dobzhansky's persistence at the potential cost of old friendships, congenial collaborations, possible professional humiliation, and the threat of having to mount a legal defense is not hard to explain. He felt that his old prediction that apologists for racism would make a comeback as soon after World War II as an opportunity presented itself was coming to pass before his eyes in the scientific cover Coon was giving Putnam and other opponents of desegregation (Dobzhansky to Montagu, May 22, 1944, May 1, 1948, Chapter 4 of this book). His expressions of gratitude suggest that he didn't know for sure that Coon was fronting for Putnam. He knew only that Coon's way of talking about race had gone backward and that Putnam always managed to sound a bit too scientific for an airline executive – he was a founder of Delta Airlines – who was spending his money trying to prove that the Supreme Court had wrongly decided *Brown vs. Board of Education.* (In the course of relating these points to an interviewer, Dobzhansky paused to exclaim, "Let's not fly Delta!" (Dobzhansky 1962–1963, 451).) He died without knowing that Putnam's show of scientific knowledge came directly from Coon and had its roots in their shared hatred for Boas, Montagu, cultural anthropology, the AAA, and a deep well of anti-Semitism (Jackson 2005). Nor did he ever know that it was Mayr who advised Coon to make "major race groups" the focus of *The Origin of Races* rather than get himself tangled up in the statistical problems that dogged *Races* (Mayr *et al.* 1953, 100, 146; Coon to Mayr, 27 January 27, 1957; Mayr to Coon, February 9, 1958; Coon to Mayr, February 12, 1958, Coon Papers). Ironically, Mayr's advice led Coon further away from the population-genetic approach he was promoting.

The disunity of the Modern Synthesis: Mayr and Dobzhansky on population thinking

Why, we might ask, did Dobzhansky and Mayr disagree so fundamentally about Coon's book, and why until recently has their disagreement not been discussed by historians of the Modern Synthesis (Jackson 2005; Collopy 2015)? Answering the first question simply by citing differences on social issues is not enough. It implicitly answers the second question by turning these differences into personal opinions that are irrelevant to the history of evolutionary science as such – not least to historians as intent as Mayr on proving the theoretical unity and empirical adequacy of the Modern Synthesis. Having repeatedly argued in this book that scientific, philosophical, and ideological issues are often tightly intertwined, we think it likely that the split among close allies on the Coon affair reflected unacknowledged differences in what Simpson, Mayr, and Dobzhansky meant by population versus typological thinking, the touchstone they all used to judge whether a biologist was or was not in line with the New Systematics and the Modern Synthesis. If true, it is likely that Dobzhansky's, Mayr's, and Simpson's differences about population thinking informed their responses to the Civil Rights Movement as much as the other way around.

The hypothesis that the founders of the Modern Synthesis differed in what they meant by population thinking has been articulated by Joeri Witteven (Witteven 2013). It was Simpson, he argues, who seems to have been the first to draw the typological-population contrast, although not in so many words. As one who made his living classifying specimens in a museum, he was aware that as a signatory to the New Systematics he had no reason to regard as typical what in the trade were wrongly called type specimens, many of which were treated by their discoverers as new species to which their own names would hopefully be attached by the museum-based specialists to whom they consigned their finds. Simpson proposed to treat such specimens as mere name-bearers that serve as placeholders for natural-historical investigation of ranges of variation within interbreeding populations. This range is always underrepresented in museum samples and supposed exemplars (Simpson 1940, 1944). This is a population-type dichotomy *avant la lettre* (Witteveen 2013).

Mayr, his colleague at the American Museum of Natural History, was just the kind of naturalist Simpson was calling for. His fieldwork on South Pacific birds explored the biogeographical distribution of variation within populations. The budding philosopher of biology in Mayr was interested in using the population-type distinction not just to reform classificatory methods, however, but to recast the very idea of what a species is. As outlined in his *Systematics and the Origin of Species*, Mayr's biological species concept – so called because it is rooted in biological causes of biogeographical phenomena, such as natural selection, migration, and genetic drift – defines a species as a geographically dispersed population chock full of phenotypic and genotypic variation but closed to interbreeding with neighboring, successive, and even sympatric populations (Mayr 1942). Only when statistical evidence shows a continuous range of variation within such populations and gaps between them are we entitled to speak of a species. From this ontological rather than Simpson's methodological point of departure, Mayr insisted that there is no such thing as a type specimen. Each member of a Mendelian population is different. This became the obsessive focus of Mayr's subsequent effort to extirpate as typological anything that stood in the way of the Modern Synthesis. Finding metaphysical essentialism lurking in all forms of typological thinking, especially the mutationism of De Vries and Goldschmidt, and running together the many kinds of essentialism that litter the history of philosophy, he claimed that before Darwin virtually everyone in biology was a Platonist (Mayr 1980, 1982). In his 1944 *Tempo and Mode in Evolution* this semantic inflation gave him philosophical cover to use the type-population equivocally in different contexts, as Witteveen suggests he did.

Unlike Simpson's, Dobzhansky's species definition was close enough to Mayr's to enable them to form a united front. The diachronicity of Dobzhansky's "race" concept, however, spilled over onto his formulation of the biological species concept. In *Genetics and the Origin of Species*, he viewed species, too, as a "stage" in the evolutionary process (Dobzhansky 1937, 312). He subsequently knuckled under to Mayr's objection that a species is not a process but the result of one and agreed to think of his races as synonymous with Mayr's sub-species (Mayr 1942, 119; Chapter 4 of this book). Still, Dobzhansky's processive way of thinking about evolutionary concepts probably didn't change much. For him races and species are inherently historical entities.

This difference bears on Mayr's and Dobzhansky's diverging reactions to Coon. Witteveen points out that as a process-oriented thinker types are historical prototypes for Dobzhansky (Witteveen 2013). Outside of laboratories, however, where pure lines are constructed by subtracting natural diversity, historical prototypes have never existed and so cannot be invoked to speak invidiously about half-breeds, quatroons, or other ways of classifying people that depart from racial purity in ways dear to the heart of scientific racists (as Boas also held, Chapter 2 of this book).[19] If there ever were pure types in nature they would not have had an evolutionary future in any case, since by definition where there is no variation for it to work on there can be no natural selection, evolution's principal motor.

We have shown that Dobzhansky saw traces of typology in the sense of prototypes in the use of "mean values of [the varying traits of] groups of individuals." Montagu was taking his cues from Dobzhansky in satirizing this practice as "the omelette conception," even if Dobzhansky (and we) found traces of it in Montagu himself (Chapter 4 of this book). If he wanted to he could also have found traces of it in Mayr. They are what led Mayr to treat Coon as a population thinker and dues-paid-up adherent of the Modern Synthesis. In a revealing letter written on November 1, 1962, one of the few occasions on which he and Dobzhansky aired their differences about Coon, Mayr claimed that identifying races and species by averaging over multiple traits "is a logical inference from the non-identity of individuals within populations," that is, from population thinking as *Mayr* construed it. He even implicated himself in the conception of typological thinking Dobzhansky disdained by remarking that "[a]dvanced … means further removed from the ancestral condition" (Mayr to Dobzhansky, November 1, 1962, Dobzhansky Papers). To Dobzhansky's historically tuned ears, there never was any such condition. By his lights, Mayr failed to make Coon a population thinker and implicated himself in his shortcomings – an uncomfortable discovery for both if they had pursued the matter, which they did not.

It is possible that Mayr and Dobzhansky's close relationship was affected by the Coon controversy. For the next decade their correspondence dwindled to holiday greetings. True, Mayr was busy. He became Director of Harvard's Museum of Comparative Anatomy in 1961. In 1966, Dobzhansky plaintively asked him whether he had become such a big shot that he was having difficulty finding time to write his old comrade (Dobzhansky to Mayr, October 3, 1966, December 14, 1967, Dobzhansky Papers). To the apparent relief of both, the flow of letters resumed in the early 1970s, when Mayr dedicated a book to Dobzhansky and engaged him in helping Provine and himself prepare a major conference on the history of the Modern Synthesis (Mayr 1970; Mayr and Provine 1980). The issue of polysemy in the notion of population thinking, however, was never raised in these discussions. On the contrary, Mayr's overdrawn and underdetermined contrast between typological essentialism and population thinking enforced on the public face of the Modern Synthesis a superficial and decidedly apolitical unity that only recently has been subjected to concerted criticism (Winsor 2006). It may be tempting to dismiss "population thinking" as too polysemic even to submit to rational reconstruction. Rhetorically, however, the phrase does real work precisely in its semantic diversity.

When Mayr learned in 1975 that Dobzhansky would soon succumb to cancer, he made a point of writing, "Our friendship has lasted forty years and there have never been any jealousies" to mar it. They sometimes disagreed, he acknowledged, but always found ways to work out their differences (Mayr to Dobzhansky, February 3, 1975, Dobzhansky Papers). An entrenched habit of translating what they were saying into each other's preferred terminology and disciplinary concerns was at the heart of this effort. "Your founder effect," Dobzhansky wrote, "is a special case of genetic drift" (Dobzhansky to Mayr, November 16, 1961, Dobzhansky Papers). Dobzhansky's diachronic races were Mayr's synchronic sub-species.

Although Mayr and Dobzhanksy collaborated on physical anthropology at the 1950 Cold Spring Harbor Symposium, they worked together not at all on issues of cultural anthropology or public policy. By contrast, Dobzhansky's work with anthropologists of both stripes stimulated his search for evolutionary reasons to support his passion for democracy (Beatty 1994; Chapter 4 of this book). He developed an interpretive framework that put genetic legs under the Boasian theorems we identified in Chapter 2. This is what he meant when he told his fellow geneticists that he was "undeservedly honored by being [wrongly] listed [by Putnam] among the students of Boas" (Dobzhansky 1968, 103). We have argued that, paradoxically, it was Dobzhansky's strict genetic definitions of evolutionary terms that enabled him to sniff out deviations from commitment to human equality in his fellow scientists. He found definitional deficiencies not only in overt enemies of egalitarian anthropology like Darlington and Gates and covert ones like Coon, but more subtly in friends and allies like Montagu and Muller (Chapter 4 of this book). Regrettably, we must add Mayr to this list even if Dobzhansky didn't.

In articles on human evolution addressed to the public in the early two thousands, Mayr used the concept of humans as a polytypic species to endorse egalitarian commonplaces in politically correct ways (Mayr 2000, 2002). But his road to the new orthodoxy was rocky. In responding to the combustible rhetorical situation of 1963, he used his undemanding contrast between essentialism and population thinking to charge those who think humans are equal in the sense of identical with committing the same "typological fallacy" as the Nazis. It is "particularly pernicious," he wrote, to call for "the same education for everybody." This, he said, actually "denied equal opportunity because differently endowed pupils would undoubtedly obtain different kinds, rates, and degrees of education if truly given 'equal opportunities.' Educational identicism is antidemocratic" (Mayr 1963, 649).

Mayr did not believe that individual differences line up with races, but his remarks about education show that in 1963 he was tone deaf to the fact that under the circumstances he would be read that way. Tailoring appropriate educations to different groups presumes that inborn capacities and incapacities can be known in advance. So it is not surprising that Coon brought Mayr's statement to the attention of the segregationist Weyl along with Julian Huxley's reiterated call for positive eugenic measures (Huxley 1960, 1963). Coon assured Weyl, "You are on firm ground with both Huxley and Mayr agreeing with you" (Coon to Weyl, April 6, 1963, Weyl Papers). Mayr's remarks occupied a prominent place in Weyl's next book, which dubiously had him supporting Weyl's claim that "[m]uch rapid progress would presumably be made if the sperm-bank proposal of Muller were adopted" (Mayr 1963, 662; Weyl and Possony 1963, xiii; on Muller, Chapter 5 of this book). Mayr ceased forthwith making his argument about equal education in public. Armed with the indiscriminate weaponry of his philosophical distinction between typological essentialism and population thinking, however, he made it to a correspondent in 1979: "'All men are created equal' … was quite literally believed in [because it was] reinforced by the philosophy of essentialism. Natural selection … implicitly claims the opposite: 'No two individuals are created equal'" (Mayr to Loren Graham, August, 1979). A year earlier, we find him privately expressing agreement with Darlington's opposition to the immigration of former colonial subjects into Britain on the grounds that dysgenic miscegenation might result. "I am delighted you have said all these things," Mayr confided to the cytologist, "which are so true but which are simply suppressed in the 'egalitarian' mass media" (Mayr to Darlington, November 28, 1978, responding to Darlington 1978; quoted in Harman 2004, 260). If Provine was right that some biologists privately retained older views even while marching in the postwar antiracist parade we might have to count Mayr in their number.[20]

Notes

1 The phrase comes from a speech King delivered at Riverside Church in New York, April 4, 1967.

2 Multi-regionalism has made something of a comeback (Stringer 2012). DNA data suggesting Neanderthals interbred with modern humans is what it predicts.

3 Gates was asked to leave a visiting appointment at Howard University about this time because of his racist beliefs (Schaffer 2007).

4 In1948, S. J. Holmes pointed out that the "luxuriant crop of rival theories" that flourished in Darwin's wake included many falling under "the general heading of orthogenesis" (Holmes 1948, 324). Orthogenesis has especially deep roots in American

paleontology, which for a long time was under the influence of Neo-Lamarckians like Alpheus Hyatt (1838–1902) and Edward Drinker Cope (1840–1897). As late as 1935, A. Franklin Shull declared in his Presidential Address to the American Society of Naturalists, "What the world most needs ... is not good five-cent cigar, but a workable and correct theory of orthogenesis" (Shull 1935, 449).

5 Jeffrey Schwartz acknowledges Mayr's influence, but thinks that its compression of hominid diversity had a pernicious effect on the Modern Synthesis (Schwartz 1999; see Delisle 2001; Stringer 2012; Templeton 2013 on hominid diversity). The failing was not his alone. It was only in the 1950s that Dobzhansky came around to the view that the *Australopithecenes* are a separate, albeit bipedal, sub-family (Dobzhansky 1962, 174).

6 Mayr acknowledged being "a good friend of Carleton Coon" (Witteveen 2013, 138–140).

7 Mayr and Coon's correspondence suggests that they were mystified about why Harvard failed to extend an offer.

8 Huxley charged that as a result of his focus on genotypes Dobzhansky "misunderstood some of the basic ideas of modern evolutionary theory," including defining fitness in terms of comparative reproductive success and thereby excluding the physical robustness that allows individuals to win "in the struggle for existence" (Huxley 1962, 144–145; Huxley 1955). Astonished, Dobzhansky replied that he was far from alone in defining fitness as comparative reproductive success. "Population genetics has 'foisted' this idea quite firmly" on everyone who signs on to the Modern Synthesis (Dobzhansky 1962b, 275; see Simpson 1959; Chapter 4 of this book).

9 Mayr turned philosopher of biology partly to justify reading ecological rules as interpretive concepts, not laws (Mayr 1988).

10 It is possible that *Saturday Review* didn't publish Dobzhansky's review because he unwittingly violated its protocols when he shared his draft with Coon (Jackson 2005, 243–244, n. 51).

11 Coon wrote two fictional books about people in North Africa (Collopy 2015).

12 Weyl's forte was retailing supposedly scientific arguments against segregation (Jackson 2005, 181–182). In *The Negro in American Civilization* he passed on Coon's just-so stories of racial adaptations and his broadsides against Boasian anthropology (Weyl 1960, 10, 160–161). He regularly corresponded with Coon. When Weyl informed him that he had sent a copy of the draft of his review of *The Origin of Races* to Dobzhansky, Coon warned him, "Sending a copy to Dobzhansky is like telling the enemy where you will be shootable at any given moment" (Coon to Weyl, April 6, 1963, Weyl Papers).

13 Most reviewers – Mead, Birdsell, Howells, and Simpson, among others – at least mentioned the uses segregationists were making of Coon's book. Alluding to the nuclear terrors of the day, John Maddocks remarked that Coon "may become kind of a Hermann Kahn of anthropology, remembered for a great thick book distinguished mostly by its tactlessness" (quoted in Jackson 2005, 164; Kahn's 1960 *On Thermonuclear War* argued for the feasibility of fighting nuclear wars. He was satirized in Stanley Kubrick's classic Cold War film *Dr. Strangelove*.)

14 Nothing came of this threat.

15 An evolutionary species is an ancestral-descendant sequence of populations (a lineage) evolving separately from others and with its own unitary evolutionary role and tendencies" (Simpson 1961, 153).

16 In addition to Mayr and Simpson, Dobzhansky sent the draft review to eighteen other colleagues. "Of these, fourteen thought that the review was fair, three considered it unfair, and one did not reply" (Dobzhansky to "Dear Colleague," December 17, 1962, Washburn Papers).

17 The assembled anthropologists were also feeling the lifted weight of the hair-raising Cuban Missile Crisis, which was resolved only a few weeks before the AAA meetings.

18 "I am sorry that you feel like this about Coon," Huxley wrote Montagu. "I agree that Coon has overplayed ... temperamental differences between 'races,' but a considerable period of evolutionary isolation ... was bound to have had some correlated effects ... such as intelligence and temperament" (Huxley to Montagu, January 17, 1963, Montagu Papers). In a handwritten marginal note in the same letter Huxley refers Montagu to his negative review of Dobzhansky's *Mankind Evolving* (Chapter 4 of this book and present chapter).

19 Dobzhansky has been taken to task for describing people he encountered in his travels in South America as combining features commonly associated with several races. These descriptions occur in unpublished, but privately circulated journals loosely modeled on Darwin's *Voyage of the Beagle*. This is crude talk, but Dobzhansky's point was not to ratify stereotypes, but to testify about how well-adapted people are who from the perspective of pure racial (proto)types would be classed as presumptively unfit half-breeds. Boas made the same point (Chapter 2 of this book).

References

Primary sources

Carleton Coon Papers, National Anthropology Archives, Smithsonian Institution, Washington, DC.

Theodosius Dobzhansky Papers, American Philosophical Society, Philadelphia, PA.

Wesley Critz George Papers, Southern Historical Collection, University of North Carolina, Chapel Hill, NC.

Ashley Montagu Papers, American Philosophical Society, Philadelphia, PA.

Sherwood Washburn Papers, Bancroft Library, University of California-Berkeley, Berkeley, CA.

Nathaniel Weyl Papers, Hoover Institution Archives, Stanford University, Stanford, CA.

Secondary sources

Allen, Joel Asaph. 1877. "The Influence of Physical Conditions in the Genesis of Species." *Radical Review* 1: 108–140.

Allen, Joel Asaph. 1905. "The Influence of Physical Conditions on the Genesis of Species." In *Annual Report of the Smithsonian Institution for 1905*, 375–402. Washington, DC: Government Printing Office.

Baker, Paul T. 1962. "The Application of Ecological Theory to Anthropology." *American Anthropologist* 64 (1): 15–22.

Beatty, John. 1994. "Dobzhansky and the Biology of Democracy: The Moral and Political Significance of Genetic Variation." In *The Evolution of Theodosius Dobzhansky: Essays on His Life and Thought in Russia and America*, edited by Mark B. Adams, 195–218. Princeton: Princeton University Press.

Beatty, John. 1995. "The Evolutionary Contingency Thesis." In *Concepts, Theories, and Rationality in the Biological Sciences: The Second Pittsburgh-Konstanz Colloquium in the Philosophy of Science*, edited by Gereon Wolters and James G. Lennox, 45–81 Pittsburgh: University of Pittsburgh Press.

Bergmann, Carl. 1847. "Über Die Verhältnisse Der Wärmeökonomie Der Thiere Zu Ihrer Größe." *Göttinger Studien* 3: 595–708.

Bergmann, Carl. 1848. *Über Die Verhältnisse Der Wärmeökonomie Der Thiere Zu Ihrer Größe*. Göttingen: Vanderhoeck und Ruprecht.

Brush, Stephen G. 2009. *Choosing Selection: The Revival of Natural Selection in Anglo-American Evolutionary Biology, 1930–1970*. Philadelphia: American Philosophical Society

Cartmill, Matt and Kaye Brown. 2003. "Surveying the Race Concept: A Reply to Lieberman, Kirk, and Littlefield." *American Anthropologist* 105 (March): 114–115.

Charleston News and Courier, Editorial, "A Plea for Moderation, October 5, 1962, A8.

Collopy, Peter Sachs. 2015. "Race Relationships: Collegiality and Demarcation in Physical Anthropology." *Journal of the History of the Behavioral Sciences* 51 (3): 237–260.

Coon, Carleton S. 1931. *Tribes of the Rif*. Harvard African Studies, vol. 9. Cambridge, MA: Peabody Museum of Harvard University.

Coon, Carleton S. 1935. *Measuring Ethiopia and Flight into Arabia*. Boston: Little Brown.

Coon, Carleton S. 1939. *The Races of Europe*. New York: Macmillan.

Coon, Carleton S. 1951. *Caravan: The Story of the Middle East*. New York: Holt.

Coon, Carleton S. 1953. "Climate and Race." In *Climatic Change: Evidence, Causes, and Effects*, edited by Harlow Shapley, vol. 1, 13–34. Cambridge: Harvard University Press.

Coon, Carleton S. 1954. *The Story of Man: From the First Human to Primitive Culture and Beyond*. New York: Knopf.

Coon, Carleton S. 1955. "Some Problems of Human Variability and Natural Selection in Climate and Culture." *American Naturalist* 89 (848): 257–279.

Coon, Carleton S. 1957. "What is Race?" *Atlantic* 200: 103–108.

Coon, Carleton S. 1962a. *The Origin of Races*. New York: Knopf.

Coon, Carleton S. 1962b. *The Story of Man: From the First Human to Primitive Culture and Beyond*. 2nd revised edition. New York: Knopf.

Coon, Carleton S. 1980. *A North Africa Story: The Anthropologist as OSS Agent, 1941–43*. Ipswich: Gambit.

Coon, Carleton S., Stanley M. Garn and J. B. Birdsell. 1950. *Races: A Study of the Problems of Race Formation in Man*. Springfield: Thomas.

Crick, Nathan. 2014. "When We Can't Wait on Truth: The Nature of Rhetoric in The Rhetoric of Science." *POROI* 10, 2. Available online at http://ir.uiowa.edu/poroi/vol10/iss2/8.

Delisle, Richard G. 2001. "Adaptationism versus Cladism in Human Evolution Studies." In *Studying Human Origins: Disciplinary History and Epistemology*, edited by R. Corbey and W. Roebroeks, 107–121. Amsterdam: Amsterdam University Press.

Delisle, Richard G. 2009. "The Uncertain Foundation of Neo-Darwinism: Metaphysical and Epistemological Pluralism in the Evolutionary Synthesis." *Studies in History and Philosophy of Science Part C: Studies in History and Philosophy of Biological and Biomedical Sciences* 40 (2): 119–132.

Desmond, Adrian and James Moore. 2009. *Darwin's Sacred Cause*. Chicago: University of Chicago Press.

Dobzhansky, Theodosius. 1937. *Genetics and the Origin of Species*. 1st edition. New York: Columbia University Press.

Dobzhansky, Theodosius. 1944. "On Species and Races of Living and Fossil Man." *American Journal of Physical Anthropology* 2: 251–265.

Dobzhansky, Theodosius. 1950. "Human Diversity and Adaptation." In *Origin and Evolution of Man: Cold Spring Harbor Symposia on Quantitative Biology*, edited by Milislav Demerec, 15, 385–400. Cold Spring Harbor: Long Island Biological Association.

Dobzhansky, Theodosius. 1962a. *Mankind Evolving*. New Haven: Yale University Press.

Dobzhansky, Theodosius. 1962b. "*Mankind Evolving* – A Rejoinder." *Perspectives in Biology and Medicine* 6: 274–275.

Dobzhansky, Theodosius. 1962–1963. "The Reminiscences of Theodosius Dobzhansky, conducted by B. Land for the Oral History Research Office of Columbia University." Dobzhansky papers.

Dobzhansky, Theodosius. 1963a. "Possibility that *Homo Sapiens* Evolved Independently Five Times is Vanishingly Small." *Current Anthropology* 4: 360.

Dobzhansky, Theodosius. 1963b. "A Debatable Account of the Origin of Species." *Scientific American* 208: 169–172.

Dobzhansky, Theodosius. 1968. "More Bogus 'Science' of Race Prejudice." *Journal of Heredity* 59: 102–104.

Dobzhansky, Theodosius and Ashley Montagu. 1947. "Natural Selection and the Mental Capacities of Mankind." *Science* 105: 587–590.

Duke, David. 1998. *My Awakening: A Path to Racial Understanding.* Covington: Free Speech Press.

Dunn, L. C. 1951. "Review of *Races, A Study of the Problems of Race Formation in Man* by Carleton S. Coon, Stanley M. Garn and Joseph B. Birdsell." *American Anthropologist* 53 (1): 105–106.

Egan, Deirdre. 2016. "Haunted by the Bell Curve: Race, Genes, and Gender in American Higher Education." Unpublished Ph.D. dissertation, University of Iowa.

Gates, R. Ruggles. 1944. "Phylogeny and Classification of Hominids and Anthropoids." *American Journal of Physical Anthropology* 2 (3): 279–292.

Gates, R. Ruggles. 1947. "Specific and Racial Characters in Human Evolution." *American Journal of Physical Anthropology* 5 (2): 221–224.

Gause, G. F. 1935. "Experimental Studies on the Struggle for existence." *Journal of Experimental Biology* 9: 389–402.

Gladwin, Thomas. 1947. "Climate and Anthropology." *American Anthropologist* 49 (4): 601–611.

Glaubrecht, Matthias and Jürgen Haffer. 2010. "Classifying Nature: Constantin W. L. Gloger's (1803–1863) "Tapestry of a 'Natural System of the Animal Kingdom'." *Zoosystematics and Evolution* 86 (1): 81–115.

Gloger, Constantin W. L. 1833. *Das Abändern Der Vögel Durch Einfluss Des Klima's.* Breslau: In Commission bei August Schulz.

Harman, Oren Solomon. 2004. *The Man Who Invented the Chromosome: The Life of Cyril Darlington.* Cambridge: Harvard University Press.

Herskovits, Melville J. 1933. "Review of Carleton Coon's *Tribes of the Rif.*" *American Anthropologist* 35 (2): 373–377.

Holmes, S. J. 1948. "The Principle of Stability as a Cause of Evolution: A Review of Some Theories." *Quarterly Review of Biology* 23 (4): 324–332.

Horsman, Reginald. 1987. *Josiah Nott of Mobile; Southerner, Physician, and Racial Theorist.* Baton Rouge: University of Louisiana Press.

Howells, W. W. 1942. "Fossil Man and the Origin of Races." *American Anthropologist* 44 (2): 182–193.

Huxley, Julian. 1954. "The Evolutionary Process." In *Evolution as a Process*, edited by Julian Huxley, A. C. Hardy and E. B. Ford, 9–33. London: Allen & Unwin.

Huxley, Julian. 1955. "Evolution and Genetics." In *What is Science?* edited by James Newman, 256–293. New York: Simon and Schuster.

Huxley, Julian. 1960. "The Evolutionary Vision." In *Issues in Evolution* (vol. 3 of *Evolution after Darwin*, edited by Sol Tax and Charles Callendar), 249–261. Chicago: University of Chicago Press.

Huxley, Julian. 1962. "Review of Dobzhansky's *Mankind Evolving. Perspectives in Biology and Medicine* 6: 144–148.

Huxley, Julian. 1963. "Eugenics in Eolutionary Perspective." *Perspectives in Biology and Medicine* 6: 155–187.

Jackson Jr., John P. 2005. *Science for Segregation: Race, Law, and the Case Against Brown v. Board of Education*. New York: New York University Press.

Kaplan, Bernice. 1951. "The Scope of Physical Anthropology: What is to Be Taught? A Report of the Sixth Annual Summer Seminar in Physical Anthropology." *Yearbook of Physical Anthropology* 6: 25–37.

Levit, Georgy S., Michal Simunek and Uwe Hossfeld. 2008. "Psychoontogeny and Psychophylogeny: Bernhard Rensch's (1900–1990) Selectionist Turn through the Prism of Panpsychistic Identism." *Theory in Biosciences* 127 (4): 297–322.

Lieberman, Leonard, R. C. Kirk and M. Corcoran. 2003. "The Decline of Race in American Physical Anthropology." *Anthropological Review* 66: 3–21.

Mayr, Ernst. 1942. *Systematics and the Origin of Species*. New York: Columbia University Press.

Mayr, Ernst. 1950. "Taxonomic Categories in Fossil Hominids." In *Origin and Evolution of Man: Cold Spring Harbor Symposia on Quantitative Biology*, edited by Milislav Demerec, 15, 109–118. Cold Spring Harbor: Long Island Biological Association.

Mayr, Ernst. 1954. "Changes in genetic environment and evolution." In *Evolution as a Process*, edited by Julian Huxley, A. C. Hardy and E. B. Ford, 157–180. London: Allen and Unwin.

Mayr, Ernst. 1956. "Geographical Character Gradients and Climatic Adaptation." *Evolution* 10 (1): 105–108.

Mayr, Ernst. 1962. "Review: *Origin of the Human Races*." *Science* 138: 420–422.

Mayr, Ernst. 1963. *Animal Species and Evolution*. Cambridge: Belknap Press of Harvard University Press.

Mayr, Ernst. 1970. *Populations, Species, and Evolution: An Abridgment of Animal Species and Evolution*. Cambridge: Belknap Press of Harvard University Press.

Mayr, Ernst. 1980. "How I Became a Darwinian." In *The Evolutionary Synthesis: Perspectives on the Unification of Biology*, edited by Ernst Mayr and William B Provine, 413–429. Cambridge: Harvard University Press.

Mayr, Ernst. 1982. *The Growth of Biological Thought: Diversity, Evolution, and Inheritance*. Cambridge: Harvard University Press.

Mayr, Ernst. 1988. *Toward a New Philosophy of Biology*. Cambridge: Harvard University Press.

Mayr, Ernst. 2000. "Biology in the Twenty-first Century." *BioScience* 50: 895–897.

Mayr, Ernst. 2002. "The Biology of Race and the Concept of Equality." *Daedalus* 131 (1) 89–94.

Mayr, Ernst, E. Gorton Linsley, and Robert L. Usinger. 1953. *Methods and Principles of Systematic Zoology*. New York: McGraw-Hill.

Montagu, Ashley. 1950. "A Consideration of the Concept of Race." In *Origin and Evolution of Man: Cold Spring Harbor Symposia on Quantitative Biology*, edited by Milislav Demerec, 15, 315–334. Cold Spring Harbor: Long Island Biological Association.

Newman, Marshall T. 1948. "A Summary of the Racial History of the Peruvian Area." *Memoirs of the Society for American Archaeology* (4): 16–19.

Newman, Marshall T. 1950a. "The Blond Mandan: A Critical Review of an Old Problem." *Southwestern Journal of Anthropology* 6 (3): 255–272.

Newman, Marshall T. 1950b. "Coon, C. S., S. M. Garn and J. B. Birdsell: *Races: A Study of the Problems of Race Formation in Man* (Book Review)." *Boletín Bibliográfico De Antropología Americana* 13 (2): 188–192.

Newman, Marshall T. 1953. "The Application of Ecological Rules to the Racial Anthropology of the Aboriginal New World." *American Anthropologist* 55 (3): 311–327.

Newman, Marshall T. 1954. "Comments on the Application of Ecological Rules to the Racial Anthropology of the Aboriginal New World: Reply." *American Anthropologist* 56 (3): 458–459.

Newman, Marshall T. 1956. "Adaptation of Man to Cold Climates." *Evolution* 10 (1): 101–105.

Newman, Marshall T. 1962. "Evolutionary Changes in Body Size and Head Form in American Indians." *American Anthropologist* 64 (2): 237–257.

Newman, Marshall T. 1963. "Geographic and Microgeographic Races." *Current Anthropology* 4 (2): 189–207.

Newman, Marshall T. and Gordon R. Willey. 1947. *Indian Skeletal Material from the Central Coast of Peru.* Papers of the Peabody Museum of American Archaeology and Ethnology, Harvard University, vol. 27, no. 4. Cambridge, MA: Peabody Museum.

Poulakos, John. 1983. "Toward a Sophistic Definition of Rhetoric." *Philosophy & Rhetoric* 16 (1): 35–48.

Putnam, Carleton. 1964. *Framework for Love: A Study of Racial Realities.* Washington: National Putnam Letters Committee.

Rensch, Bernhard. 1947. *Neuere Probleme Der Abstammungslehre: die Transspezifische Evolution.* Stuttgart: F. Enke.

Rensch, Bernhard. 1959. *Evolution above the Species Level.* New York: Columbia University Press.

Ripley, William Zebina. 1899. *The Races of Europe: A Sociological Study.* New York: Appleton.

Roberts, D. F. 1952a. "Basal Metabolism, Race and Climate." *The Journal of the Royal Anthropological Institute of Great Britain and Ireland* 82 (2): 169–183.

Roberts, D. F. 1952b. "Basal Metabolism, Race and Climate." *Man* 52: 169–170.

Roberts, D. F. 1953. "Body Weight, Race and Climate." *American Journal of Physical Anthropology* 11 (4): 533–558.

Ruse, Michael. 1996. *Monad to Man: The Concept of Progress in Evolutionary Biology.* Cambridge: Harvard University Press.

Schaffer, Gavin. 2007. "'Scientific' Racism again?: Reginald Gates, the *Mankind Quarterly* and the Question of 'Race' in Science After the Second World War." *Journal of American Studies* 41 (2): 253–278.

Schwartz, Jeffrey H. 2006. "Race and the Odd History of Human Paleontology." *The Anatomical Record Part B: The New Anatomist* 289 (6): 225–240.

Shull, A. Franklin. 1935. "Weismann and Haeckel: One Hundred Years." *Science* 81 (2106): 443–452.

Simpson, George Gaylord. 1940. "Types in Modern Taxonomy." *American Journal of Science* 238: 413–426.

Simpson, George Gaylord. 1944. *Tempo and Mode in Evolution.* New York: Columbia University Press.

Simpson, George Gaylord. 1959. "Forward" to *The Life and Letters of Charles Darwin.* New York: Basic Books, pp. v-xvi.

Simpson, George Gaylord. 1961. *Principles of Animal Taxonomy.* New York: Columbia University Press.

Simpson, George Gaylord. 1963. "Review of *Origin of Races*." *Perspectives in Biology and Medicine* 6: 268–272.
Spiro, Jonathan P. 2008. *Defending the Master Race: Conservation, Eugenics, and the Legacy of Madison Grant*. Burlington: University of Vermont Press.
Stanton, William. 1960. *Leopard's Spots: Scientific Attitudes Toward Race in America: 1815–59*. Chicago: University of Chicago Press.
Stringer, Chris. 2012. *Lone Survivors: How We Came to Be the Only Humans on Earth*. London: Times Books.
Templeton, Alan R. 2013. "Biological Races in Humans." *Studies in History and Philosophy of Science Part C: Studies in History and Philosophy of Biological and Biomedical Sciences* 44 (3): 262–271.
Tax, Sol and Charles Callendar. 1960. "Panel Three: Man as an Organism." In *Issues in Evolution* (vol. 3 of *Evolution after Darwin*, edited by Sol Tax and Charles Callendar), 145–174. Chicago: University of Chicago Press.
Wade, Nicholas. 2014. *A Troublesome Inheritance: Genes, Race and Human History*. New York: Penguin.
Washburn, Sherwood. 1951. "The New Physical Anthropology." *Transactions of the New York Academy of Sciences* 13: 298–304.
Washburn, Sherwood. 1963. "The Study of Race." *American Anthropologist* 65 (3, Part 1): 521–531.
Washburn, Sherwood. 1983. "Evolution of a Teacher." *Annual Review of Anthropology* 12 (1): 1–25.
Watt, Cortney, Sean Mitchell and Voker Salewski. 2010. "Bergmann's Rule: A Concept Cluster?" *Oikos* 119 (1): 89–100.
Weidenreich, Franz. 1939. "The Phylogenetic Development of Man and the General Theories on Evolution." *Bulletin of the Geological Society of China* 19 (1): 76–92.
Weidenreich, Franz. 1946. "Generic, Specific and Subspecific Characters in Human Evolution." *American Journal of Physical Anthropology* 4 (4): 413–432.
Weyl, Nathaniel. 1963. "The Reality of Race." *National Review* 14: 33–35.
Weyl, Nathaniel and Stefan Thomas Possony. 1963. *The Geography of Intellect*. Chicago: Regnery.
Winsor, Mary. 2006. "The Creation of the Essentialism Story: An Exercise in Metahistory." *History and Philosophy of the Life Sciences* 28: 149–174.
Witteveen 2013. "Rethinking 'Typological' vs. 'Population' Thinking: An Historical and Philosophical Reassessment of a Troubled Dichotomy." Ph.D. dissertation, Cambridge University.
Wolpoff, Milford and Rachel Caspari. 1997. *Race and Human Evolution: A Fatal Attraction*. New York: Simon and Schuster.
Woolridge, Clara, ed. 1959. *Genetics and Twentieth Century Darwinism. Cold Spring Harbor Symposia on Quantitative Biology*, vol. 24. Cold Spring Harbor: The Biological Laboratory.

7 Epilogue

The roots of the Sociobiology controversy, the infirmities of Evolutionary Psychology, and the unity of anthropology

In the 1960s and 1970s, heirs of the Modern Synthesis's founding generation were intent on bringing under its sway areas of inquiry that were not part of its original dispensation, but were (and are) essential to vindicating it as a general theory of biological evolution. Ecology – Darwin's "tangled bank" of many mutually entwined species in a given area – was one focus. Another was the evolution of cooperation, which Darwin himself regarded as *prima facie* anomalous for his competition-based theory (Darwin 1868). Having made a seminal contribution to "population ecology" in his "theory of island biogeography" (MacArthur and Wilson 1967), Edward O. Wilson, a Harvard entomologist, went on to "extend population biology and evolutionary theory to social organization" in *Sociobiology: The New Synthesis* (Wilson 1975, x). He endorsed William Hamilton's ideas of "inclusive fitness," in which reproductive success is proportioned to genetic relatedness rather than restricted to immediate offspring, and "kin selection," in which adaptations evolve whereby relatives help raise offspring, thereby enhancing the chances of a hive or other collectivity. As a leading authority on the natural history of ants, Wilson lent empirical support to Hamilton's mathematical calculation that social insects are more likely to evolve cooperative role-division because they belong to the class *Hymenoptera*, whose haplodiploid chromosomal structure makes it easier for kin selection to evolve sterile and hence "altruistic" castes (Hamilton 1968a, b).[1] Humans are not *Hymenoptera*. But at the end of *Sociobiology* Wilson invoked reciprocal altruism, which depends on anticipated rewards in ways modeled by Cold War game theorists, to help kin selection evolve heritable cooperation among genetically related humans and explain conflict with more distant outsiders (Wilson 1975; Trivers 1971).

In its rhetorical situation, Sociobiology was irenic. It aligned Darwinism against Konrad Lorenz's theory of instincts as hair-trigger "cathartic discharges" and of humans as instinctively aggressive (Lorenz 1963; Segerstråle 2000, 28, 95). Widely disseminated in pop evolutionary tracts in the 1960s, Lorenz's ethology was simultaneously celebrated and satirized in Stanley Kubrick's films *Dr. Strangelove* and *2001: A Space Odyssey* (Ardrey 1961, 1966; Morris 1967). Progressives saw "the naked ape hypothesis" as making nuclear war more likely by giving it seemingly ineluctable biological causes. In view of its ideological drift and its roots in the Modern Synthesis, one might imagine that the left-leaning geneticist Richard Lewontin would have endorsed Sociobiology, all the more so because he and Wilson had personal ties. Wilson had enticed Lewontin to move to Harvard from the University of Chicago to help him develop population ecology and, as a cracker-jack molecular experimentalist, to beat back the growing influence of James Watson's molecular reductionism in the department (Wilson 1994, 219; Lewontin 1989, 40). Accordingly, no one was more surprised than Wilson to discover that Lewontin rejected *Sociobiology* lock, stock, and barrel (Lewontin 1976). "The book has a lot of science in it," he informed his colleague, "but it is not of science. It not only not of science, but [it is] a religion. [It is] a piece of scientific public relations" (Lewontin to Wilson, October 28, 1976, Lewontin Papers).

In critiquing Wilson, Lewontin recycled lines of argument he had developed in sparring about genetic determinism with his mentor Dobzhansky (Chapter 4 of this book). He and Harvard paleontologist Stephen Jay Gould deployed the same arguments against the psychologists Arthur Jensen and Richard Herrnstein, who claimed that, as a group, African-Americans possess less of the (supposedly) objectively measurable and strongly heritable trait of analytical intelligence (IQ) than whites. From this statistical finding, Jensen inferred that Great Society programs like Head Start are a waste of time and money. Herrnstein, for his part, inferred that in sorting people out by IQ meritocratic liberalism also sorts them by differential racial capabilities (Jensen 1969; Herrnstein 1971; Herrnstein and Murray 1994). It is not hard to link "Jensenism" to eugenic and racialist strains of thought.[2] Associating Wilson with these strains, however, was at odds with his aims and with what he took to be Sociobiology's implications.

As a population biologist, Wilson knew as well as Lewontin and Gould that correlations between traits depend on identifying the same environments and that IQ, if it is a single trait at all, is especially sensitive to environmental variation; that even when correlations between traits are found they might result not from adaptive natural selection, but from genetic drift or the way linkage-filled genomes carry along traits that are not directly under selection (pleiotropy); that to have biological significance populations have to be drawn more accurately than between "blacks" and "whites"; that except under stringently defined conditions it is impossible to move from population-level genotype-phenotype correlations to individual differences without importing non-scientific theorems from determinist philosophy; and, importantly, that heritability does not imply fixity.[3] It is significant that when Gould attacked Sociobiology's adaptationism – the hypothesis that most or at least most important traits are adaptations in the genetics-grounded sense – he tended to pick targets other than Wilson (Gould and Lewontin 1979). It was otherwise with Lewontin. Having endured personal slights and attacks by "Science for the People" and other New Left groups inspired and to some extent directed by him, Wilson had cause to attribute his colleague's critique not to his scientific expertise, but to the neo-Marxist politics that, unlike Gould's harmless leftish gestures, Lewontin used as an interpretive lens and brandished as a weapon (Wilson to Lewontin, January 5, 1975, Lewontin Papers).

Wilson's assessment of Lewontin's attack continues to be widely accepted by biologists and others. Even Lewontin's admirers admit there may be something to it (Lewontin *et al.* 2001, 5). Ullica Segerstråle's comprehensive recounting of the Sociobiology Controversy seeks a balanced account, but tilts toward Wilson, if only because her reconstructions of Lewontin's arguments are not always clear (Segerstråle 2000). They are not always clear because, although anthropologists figure in her story, she does insufficient justice to the protracted interaction between anthropologists and population genetic biologists that has been our subject. In this Epilogue we will revisit these interactions in order to put Lewontin's arguments in a more favorable light and to suggest that they are even more telling against Sociobiology's successor, Evolutionary Psychology. Since Sociobiology and Evolutionary Psychology have helped drive a wedge between biological and cultural anthropologists, the tendency of our argument is to hope they will close ranks.

Anthropologists were among Sociobiology's earliest and severest critics. The fact that similar objections came not just from scions of Boas's school, but from many quarters of American anthropology testifies to the strength of the consensus that formed in the wake of Washburn's denunciation of Coon (Chapters 5 and 6 of this book). In one way or another all argued that even if key behavioral traits are cooperative adaptationist explanations of them come at the expense of the culture concept that has demarcated anthropology since Kroeber (Chapter 3 of this book). Marshall Sahlins made this point in his 1976 polemic *The Use and Abuse of Biology: An Anthropological Critique of Sociobiology*:

> Consider the relation between warfare and human aggression – which Wilson at one point calls "the true biological joy of warfare." It is evident that people engaged in fighting wars … are by no means necessarily aggressive. Many are plainly terrified.… Men may be moved to fight out of love (as of country) or humaneness (in light of the brutality attributed to the enemy), for honor or some sort of self-esteem, from feelings of guilt, or to save the world for democracy.… The reason is that the biology of mankind has been shaped by culture.
>
> (Sahlins 1976, 8–9)

The shaping affects not only our morphology, as we saw Washburn arguing in Chapter 5, but also our reproductive practices. Sahlins reminded his readers that "[t]he actual systems of kinship and heredity in human society [seldom] conform to the biological coefficients of relationship" built into inclusive fitness and kin selection (Sahlins 1976, 27, 25). These notions may relieve the Darwinian worldview of a troubling anomaly, but, since kin selection proportions cooperation to genetic relatedness, they also give renewed visibility to the eugenic hypothesis that uncooperative, anti-social behavior, too, may have genetic roots, and may be linked to intelligence and race. We noted earlier how unsurprised Washburn was to find Herrnstein using the prestige of Sociobiology to press his case for biological causes of criminal behavior and how ready he was to correlate it to race (De Vore and Washburn, 1992, 422; Chapter 5 of this book). The webs of culture within which we live, move, and have our being include both selfish and selfless tendencies, but neither can be decomposed into separately evolved adaptations. Even Irven De Vore, the anthropologist who told Wilson about Robert Trivers's notion of reciprocal altruism with an enthusiasm that dismayed Washburn, his former mentor, worried about this. De Vore informed Segerstråle that he was often in a quandary about Wilson's use of anthropological information. She reports that she herself

> had a glimpse of this during my interview with [DeVore] in his office. [He] took a phone call and briefly answered some question. "That was Wilson," he told me. "He asked me, 'Isn't it true that cousin incest is a taboo in almost all societies?' He always does that." It was typical of Wilson to call DeVore up to check some fact in just this way. "The problem was, you never knew what he was going to do with the information you provided," DeVore complained. "And then he always thanked you in his acknowledgements!"
>
> (Segerstråle 2000, 81)

In "The Spandrels of San Marco and the Panglossian Paradigm," their cheeky misperformance of the genre of scientific review article in the sober *Proceedings of the Royal Society of London*, Gould and Lewontin generalized to all of evolutionary biology the fallacy of atomizing traits and presuming that there are "genes for" them. They treated this picture as continuous with eighteenth century natural theology, according to which everything happens for the best in the best of all possible worlds (Gould and Lewontin 1979). Lewontin's retrospective insistence that "Spandrels" was almost entirely Gould's doing reflects his continued hostility to the very idea of adaptation (Wilson 2015). Gould stressed how difficult it is to give solid adaptationist explanations, opening him to Mayr's lecturing about "How to Carry Out the Adaptationist Program" (Mayr 1983). Lewontin's objection cut deeper. To identify adaptations at all populations must be portrayed as solving problems posed to them by their environments, turning organisms into passive aggregates of traits rather than ontogenetically dynamic makers of the species-specific niches from which they draw the resources that enable them to exert agency (Lewontin 1982, 2001; Levins and Lewontin 1985).

The "lock and key" model, as Lewontin calls it, disrupts the thoroughgoing interaction between environment, development, and genotypes on which Dobzhansky insisted, but without enough conviction to keep him from privileging genotypes, even if they are phenotypically flexible, over other developmental resources when he took to justifying liberal democracy as the ideal economic, social, and political system for matching individual talents to social circumstances (Chapter 4 of this book).[4] The fault, Lewontin and the Harvard population ecologist Richard Levins argue, lies in liberal democracy itself. It cannot fulfill its promises of freedom, equality, and plenty and so socially constructs a discourse that explains the gap between promise and performance by putting the blame for tuberculosis, for example, on a bacillus rather than on amendable social conditions and the onus for inequality on the genetic limitations of individuals and groups (Levins and Lewontin 1985; Lewontin 1993). Selective perception of this sort is inevitable whenever enculturation, a powerful locus of gene-organism-environment interaction, is reduced to a social contract between presumably autonomous individuals, as it has been in political liberalism since the time of Locke (Levins and Lewontin 1985, 270; Lewontin 1993, 45). Behavior Genetics, Jensenism, Sociobiology, and Evolutionary Psychology – all are expectable results for Lewontin and Levins of a conceptual scheme that disseminates itself more effectively through well-intentioned but ideologically unconscious biologists like Wilson than through overtly anti-democratic racists and eugenicists.

Lewontin's antidote was to conduct rigorous genetic research without an adaptationist filter in order to support the anti-racist and anti-eugenic theorems of Boasian anthropology and ground them in a version of Dobzhansky's evolutionary biology purified of his effort to redefine "race" genetically (Chapter 4 of this book).[5] Lewontin made his reputation by experimentally verifying Boas's theorem that variation is always greater within conventionally construed but biologically meaningless races than between them. The theorem holds at a level as basic as protein evolution:

> Although there is variation between loci in their relative contributions, the average values show that 85 percent of human genetic diversity is within national populations and only 7.5 percent between nations within races and 7.5 between major races.... Since most of the world's population is made of Chinese, Indians, Europeans, and the recently hybridized populations of South America, who vary less from each other than do small isolated groups, the correct proportion of human genetic variation that is within nations or tribes is [actually] closer to 95 than to 85 percent.
>
> (Lewontin 1974, 155–156, 1972)[6]

Lewontin countered Motoo Kimura's claim that genetic variation in proteins is mostly selectively neutral by turning the argument on its head: Neutral mutationism, he argued, which spread among molecular geneticists in the 1970s, tacitly promotes Muller's view of adaptation, according to which there is only one truly fit genotype for each chromosomal locus (Lewontin 1974; Chapter 4 of this book). This optimizing view of adaptive natural selection figures not only in Muller's eugenics, but also in the notion of "gene for" appealed to by adaptationists, including Sociobiologists (Chapter 4 of this book). Lewontin challenged it by citing instances of group selection that do not reduce to individual or kin selection and by arguing that Dobzhansky's view of the genome as an integrated whole can be defended without his problematic idea of the adaptive superiority of heterozygotes or their quantitative pre-eminence in the genome (Lewontin 1970, 1974; Chapter 4 of this book). He put this point in a striking way at the end of *The Genetic Basis of Evolutionary Change*, the last of the canonical texts of the Modern Synthesis published by Columbia University Press:

> Fitness at a single locus ripped from its interactive context is about as relevant to the real problems of evolutionary genetics as the study of the psychology of individuals isolated from their social context is to understanding man's sociopolitical evolution.
>
> (Lewontin 1974, 318)

Like Dobzhansky, Lewontin is attuned to anthropology. The dictum just quoted is as hostile to stadial cultural evolutionism as it is to fracturing the genome into adapted traits. Lewontin knows that Boasites, led by Kroeber, purged cultural evolutionism from anthropology, but that it made a comeback after the 1959 Darwin Centennial Celebration (Fracchia and Lewontin 1999, 52; Chapters 3 and 5 of this book). Sociobiology, he argues, is in part an effort to free the idea of cultural evolution from "transformational" theories of social change that contaminate biological evolution, such as Leslie White's, by outfitting it with a respectably Darwinian "variational" theory in which selection evolves adaptations by discriminating between heritable units (Fracchia and Lewontin 1999, 53; Chapter 3 of this book). The trouble is that culture does not come in heritable units ("culturgens" and such items):

> Acculturation occurs though a process of constant immersion of each person in a sea of cultural phenomena smells, tastes, postures, the appearance of a building, the rise and fall of spoken utterances. [It] requires a complex mode of acquisition from family, social class, institutions, communications media, the work place, the street.
>
> (Fracchia and Lewontin 1999, 73)

Lewontin proposes what amounts to a corollary to Kroeber's axiom that the contingent historicity of cultural change will inevitably be compromised by evolutionary progressivism unless the study of organic evolution rests exclusively on genetic inheritance (Chapter 3 of this book). Until the concept of adaptation, he adds, and *a fortiori* adaptationism, has been purged from the genetic theory of natural selection the study of biological evolution and cultural history will both be undermined by traces of the stadial progressivism on which racism and eugenics were first predicated and in which they still lurk.

The slippery slope that leads to clearing conceptual space for these distortions to resurface in the way they do in Jensen, Herrnstein, and, unwittingly, Wilson begins with the idea that there are two interacting forms of inheritance, genetic and cultural. Even when he was proposing cultural evolution as a stage beyond biological evolution, Dobzhansky, perhaps following Kroeber, rejected "dual inheritance theory" (Chapter 4 of this book). Wilson endorses it (Lumsden and Wilson 1983). So do other gene-culture co-evolutionists (Boyd and Richerson 1985, Richerson and Boyd 2005). By contrast, anthropologists who have learned the lessons taught by the controversies we have recounted reject dual inheritance, cultural evolutionism, and behavioral adaptationism as categorically as Kroeber and Lewontin (Chapter 3 of this book):

> Walk away from the genetic basis of specific behaviors. Just turn around and walk away. Forget the cockamamie twin-studies with the twins who marry women with the same name and give their dogs the same name (Wright 1998, 1999); forget the cockamamie gene-mapping studies that are never confirmed or are even retracted a month after they're reported (Kennedy *et al.* 1988; Holden 1991; Rice *et al.* 1999). Some of it is ideologically driven, some is financially conflicted (Tucker 2002); some of it is just incompetent; but no scientific thoughtful critical Darwinian should be citing this stuff. It is entirely alien to Darwinism.
>
> (Marks 2004, 190)

The term "sociobiology" does not appear in Evolutionary Psychology's founding text (Tooby and Cosmides 1992). Still, it became Sociobiology's heir because it obviated telling objections to it. Sociobiologists had trouble rebutting accusations that we moderns are less well adapted than our ancestors or present-day tribal societies and fending off the inference that Sociobiology isn't much less deterministic than Lorenz's ethology. Instead of allaying the objection, Wilson's reply underscored it. Saying that we can and should resist the "whisperings of nature," he argued that it is a good thing that "genes have culture on a leash," since left to its own devises culture favors run-away behaviors like ritual cannibalism or the hyperbolic competitive displays that resulted in the extinction of the population of Easter Island (Wilson 1978, 167, 207). Evolutionary Psychology, like Sociobiology, assumes that sapient humans acquired their species-specific adaptations in Pleistocene environments; not enough time, they say, has passed since the rise of agriculture for gradual directional natural selection to work its wonders. But Evolutionary Psychologists take the notion that we moderns are less well adapted than our ancestors off the table by claiming that what evolved in the Stone Age were not hard-wired behaviors, but hard-wired cognitive capacities, each encoded in a well-adapted topic-specific module located in the brain. These enable us to respond appropriately and effectively to situations that arise as often in our world as in Neolithic times (Barkow *et al.* 1992).

Our "cheater-detection module," for example, may have arisen to alert us to fellows who were not pulling their weight in the hunt or to deter wives from passing off other men's children on their husbands. Cheaters and free riders are always with us. So the cheater detection module is as useful to us as it was to our forebears (Tooby and Cosmides 1992). Accordingly, where Sociobiology sees "organisms as overtly working toward the maximization of inclusive fitness.... Evolutionary Psychology uses inclusive fitness and kin selection to validate our assumption that our thought and behavior" are oriented to consciously entertained goals which, if achieved, add up to, but are not aimed at, reproductive success (Hampton 2009, 23). Our means-end psychology is not an illusory representation of our true motives, as in Sociobiology. In this respect Evolutionary Psychology is less reductionistic, deterministic, and anachronistic than Sociobiology, and more in tune with the rise of cognitive science's embrace of a modular brain to replace the all-purpose information processor hitherto assumed by evolutionary biologists and physical anthropologists.

This advance has not kept philosophers of biology from arguing that evolutionary psychologists misuse the basic concepts of population genetic theory even more egregiously than sociobiologists. In the Modern Synthesis, a trait counts as an adaptation, and even as a trait, only if gene frequencies have been shown to shift in a particular environment in order to perform a function (Chapter 4 of this book). Hence the inference that cognitive modules are adaptations must rest on more than current and currently observed functions, effects, or uses (Buller 2005; Richardson 2007). These gene frequency changes have not been found. Nor can the supposed universality of Evolutionary Psychology's mental modules be persuasively commended on the ground that they have low heritability. Lack of differences implies lack of natural selection's genetic fuel. So to say this is actually to work against the idea that natural selection is the author of these modules (Richardson 2007, 100–103).

Only *a priori* commitment to adaptationism, which says that there *must b*e genes for what we observe about behavior because mathematical game theory says so, allows arguments like these to be persuasive. But exceeding empirical data in this way invites ideology to make up the difference. As it happens, the vast majority of cognitive modules postulated by evolutionary psychologists have focused on gender and sexuality (Freeze 2008). Psychologists like Robert Wright and David Buss generalized Trivers's game-theoretical argument that men and women evolved different mating and parenting strategies because their costs of reproduction – so much sperm, so few eggs – are different. Since then evolutionary psychologists have supposed that natural selection shaped male and female minds differently. Men, it is said, are naturally attracted to young women with high waist-to-hip ratios, large breasts, big eyes, and small noses because these are signals of health, fecundity, and nurturing skills. Women are attracted to men with high social status and manly physiques because these indicate an ability to commandeer resources (Wade 2000; Buss 2003). Men want to have a lot of sex. Women want a mate who will provide for them and their children. Men are naturally aggressive. Women are maternal and choosey. The list goes on, scandalously culminating in biologist Randy Thornhill and anthropologist Craig Palmer's claim that rape increases male fitness and accordingly must have evolved as a male reproductive strategy (Thornhill and Palmer 2000).

All these claims illustrate ideology's core argumentative strategy of inscribing culturally contingent stereotypes into a supposedly invariant "human nature." The alleged naturalness of monarchical and aristocratic politics was the target of ideological criticism in the eighteenth century. The nineteenth century upended the alleged naturalness of class differences. In our own time the naturalness of gender roles and sexual preferences are under contestation. Can it be "just a coincidence," Jonathan Marks asks, "that Evolutionary Psychology emerged just as the conservative backlash against the Equal Rights Amendment peaked in the early eighties?" (Marks 2009, 253). Feminist biologists, social scientists, and philosophers have argued that Evolutionary Psychology is nothing more than a catalogue of "just-so stories" resting on speculation about the environments in which the human mind evolved, about which not much is actually known, and projection into this void of preferences for sexual assignments and gender roles that in our era are rapidly being denaturalized (Fausto-Sterling 1992; Fausto-Sterling *et al.* 1997; McKinnon 2005). The idea that male aggression keeps fitness high regresses to Victorian commonplaces that the Modern Synthesis discarded when it redefined fitness in terms of expected differential reproduction in populations. The language in which sociobiologists and evolutionary psychologists calculate the interests of the sexes and their genes assumes the universal applicability of the individualistic competitive practices of our own economic order. The assumption that what contemporary American college students say about what men and women want, which comprises a surprising amount of the survey data on which Evolutionary Psychology relies, betrays racist, classist, and imperialist fantasies that an elementary appreciation of cultural diversity refutes.[7] Kroeber's axiom that psychology, no matter how worthwhile in its own domain, blocks access to anthropology's distinctive range of phenomena has never looked so prescient (Chapter 3 of this book).

Sociobiology and Evolutionary Psychology have disturbed the balance between cultural and biological anthropology that was put in place through the controversies whose links we have traced in this book. The New Zealand anthropologist Derek Freeman's 1983 attack on Margaret Mead's ethnographic work is a plausible point of departure for surveying these tensions, if only sketchily. Freeman claimed that Mead arrived in Samoa untrained in field research; relied exclusively on a small circle of native informants, one of whom was putting her on; implored Boas to interpret her field notes from half a world away; and argued tendentiously for a liberated view of female sexuality that she wished upon America more than she actually found in Polynesia (Freeman 1983). Many anthropologists have greeted Freeman's claims skeptically. His own fieldwork was done decades after Mead's in a different part of Samoa. The contrast he draws between a mendacious Mead and his own methodologically scrupulous empiricism conceals an imperfect understanding of the Popperian falsificationism he preaches (Patience and Smith 1986). Nor is his empiricism always a good thing. His findings may well differ from those of other workers in Samoa because his "naive positivism" leads him to rip isolated statements from the "psychological and social worlds of the involved actors" (Levy 1983, 831; Holmes 1983, 1987).

Above all, Freeman's effort to destroy Mead's credibility is not a piece of scholarship at all. It is a polemic "against Franz Boas and American anthropology in general" (Holmes 1987, 143, Baker 1984). Freeman says so himself.

> We have … reached a point at which the discipline of anthropology … must abandon the paradigm fashioned by Kroeber and other of Boas's students, and must give full cognizance to biology, as well as to culture, in the explanation of human behavior and institutions.
>
> (Freeman 1983, 297)

Showing up ethnography as prone to subjective distortions is a stalking horse for promoting a "new 'synthesis' of biology and anthropology" that emphasizes the former (Freeman 1983, 302). By suggesting that hostility to scientific method leads cultural anthropologists like Mead, on his rendering of her, to discount any role for biology in the evolution of human behavior, Freeman opened the door for sociobiologists and evolutionary psychologists to appeal to the authority of scientific method to push their genocentrism (Coe and Palmer 2011, 560). Freeman and those who follow his lead do not discuss, and perhaps do not even know, that Kroeber opposed the doctrine of *omnis cultura ex cultura* and that biology and culture have been intertwined in four-field anthropology since Washburn's New Physical Anthropology became its working paradigm (Chapters 3 and 5 of this book).

Napoleon Chagnon's 1988 sociobiological interpretation of his own widely disseminated 1968 ethnography *Yanomami: The Fierce People* was a turning point in the controversy Freeman initiated (Chagnon 1968, 1988). The fitness of this tribe, Chagnon argued, is maintained by the practice of allowing its most violent males to command a disproportionate number of females. Their superior genes are thereby over-represented in the next generation. Murderous raids aimed at stealing women from out-groups keep gene pools diverse and contribute to fitness. For Chagnon, as for his mentor and collaborator, the Muller-influenced geneticist James Neel, the Yanomami tell us how we used to be before civilization compromised our fitness (Neel 1970, 1994, 279).[8]

Chagnon's Sociobiological turn intensified the negative reaction of activist anthropologists to his ethnography. In 1989, a former president of the Brazilian Anthropological Association forwarded an official communication to *American Anthropologist* stating that "labeling the Yanomami 'the fierce people' … contributes to reinforcing the negative prejudices that usually weigh on indigenous populations," especially a population whose lands and lives were under assault by one of most violent gold rushes in history (Carniero de Cunha l989). Genetically reifying "fierce" made the problem worse. Anthropologists who take themselves to have a duty to speak on behalf of the Yanomami and other marginal populations have been willing to back off from the pose of scientific objectivity in order to make this point. Unlike Freeman, however, they believe in the capacity of ethnography to speak truth to power if it is reflexive and reflective enough to disclose and discount the writer's own subject position, interests, and manipulation of his subjects, as Chagnon failed to do.

Conflict on this point reached a boiling point when a muckraking journalist accused Chagnon and Neel in a best-selling book (and a prepublication article that somehow slipped the noose of the *New Yorker's* famed fact checkers) of intentionally allowing measles to spread among the Yanomami in order to test their thesis about the tribe's fitness (Tierney 2000a, b). The charge was so false that the AAA had no choice but to exonerate Chagnon, leaving his objectivistic ethnography and sociobiological explanation of it intact (AAA 2001). Sahlins resigned from the National Academy of Sciences when Chagnon was elected to it. He explained his decision by saying, "Chagnon's 'scientific' claims about human evolution and genetic selection for male violence … have proven to be shallow and baseless, much to the discredit of the anthropological discipline" (Sahlins 2013).

The rift between critically engaged ethnographers and biological anthropologists influenced by Evolutionary Psychology led to deleting the word "science" from the AAA's 2010 Long-Range Plan. The word remained in the Association's official definition of anthropology and the Plan was soon amended (AAA 2011), but mistrust persists. In the wake of the "science wars" of the 1990s, more biological anthropologists fear that their cultural colleagues will impute to them sociobiological, evolutionary psychological, and Chagnonesque beliefs than actually have such colleagues. Similarly, more cultural anthropologists worry about having colleagues who reduce culture to individual psychology and who think that even mentioning social constructions is anti-scientific than they can find in offices down the hall. Biological anthropologists who track the gene frequencies of human diseases are at least as upset as their critical cultural colleagues by big claims that rest on postulated rather than actual genetic data. The center looked for by Kroeber and found by Washburn still holds in most anthropology departments. Still, the fissioning of American anthropologists into non-overlapping professional organizations, splinter groups, and separate departments has taken a toll.[9]

As appreciative outsiders, we hope anthropologists will reaffirm their unity by recalling how effective they have been in beating back racism, disseminating cultural pluralism, and promoting democracy in everyday life and political practice. We have tried to help by recounting key episodes in that story. But we also take heart from the fact that evolutionary biology itself is currently in the throes of a transformation in which the very coherence of the idea of "gene for," on which most evolutionary defenses of racially correlated trait distributions are predicated, is being challenged.

Ironically, the Human Genome Project and other gene mapping and sequencing programs have undercut the assumptions of the molecular geneticists who originally promoted them by showing that humans have far fewer genes than expected; that the genes we do have are shared across a wide swath of *taxa* ranging from simple to complex organisms; that genes regulating the developmental process are more plastic, and more causally important in generating and fixing variation, than mutations in structural genes that code for proteins; and that regulatory genes are open to environmental influences in the developmental process in a way that protein-coding genes are not (Carroll 2005; Gilbert and Epel 2009). The last of these perceptions lends support to Lewontin's insistence that organisms co-evolve with their ecological niches and for this reason are agents who make their own life worlds. Instead of simply inhabiting environments to which they are fitted as key to lock, as Dobzhansky had it, they reconstruct their species-specific niches from generation to generation (Lewontin 1982, 2000; Odling-Smee *et al.* 2003). The trans-generational continuity of niche constructive activity has combined with new awareness of how protean the genome actually is to breathe new life into a conjecture known as "the Baldwin effect." So named after one of its early twentieth century advocates, the phenomenon proposes that genetic changes typically follow rather than precede adaptive paths carved out and maintained across generations by the niche-building activities of organisms (West Eberhard 2003; on the Baldwin effect, Baldwin 1896; Simpson 1953; Weber and Depew 2005; on the dynamism of the genome, Burian and Kampourakis 2013). Far from being determined by, or *a fortiori* coded in, genes these activities are made possible by phenotypically plastic genotypes whose wide norms of reactions shift over time as a result of the adaptive behaviors they enable (Schlichting and Pigliucci 1993, 1998). On these and other grounds, evolutionary biologists have begun to ask whether the Modern Evolutionary Synthesis, especially versions of it that focus on random gene-first mutations, needs a "rethink" (Laland *et al.* 2015, 2016).

The circle of conjectures and hypotheses collectively called evolutionary developmental biology ("evo-devo") is at present only beginning to be articulated in rigorously testable ways. Much revision can be expected. Still, the turn to genetic flexibility, developmental plasticity, and niche construction has perceptible implications for human evolution. Concepts as problematic as social heritability, adaptation, and cultural evolution may at last find acceptable meanings. But even if they don't, the ecological-developmental turn in evolutionary studies has already put powerful work done in an older idiom, such as Lewontin's, in a fresh, supportive light.[10]

These sources of insight have not been lost on anthropologists who recoil from genetic reductionist notions about psychological traits and who, in view of the renascent tendency to distribute such traits into conventionally defined races, are keenly aware of a pressing need to reinforce the theorems bequeathed to them by Boas, Kroeber, and Washburn. Cultural life is not an evolutionary grade whose heights have yet to be scaled by some people, but the specifically human niche shared equally by all. Cultural life keeps itself open to innovation not by some special form of heritability, but rather by changing complexes of meaning-laden symbolic activity (Deacon 1998; Fuentes 2013, 2015).[11] Hence it will not do to say with evolutionary psychologists that "[h]umans are the product of both biological and cultural evolution, in which culture evolves in interaction with human nature, innovations, and external events" (Schaik and Michel 2016). This sentence blithely defines human nature in terms of a collection of genetically fixed psychological proclivities and frames culture as something that evolves from less to more advanced. In doing so it runs roughshod over a century of anthropology and evolutionary biology.

Admittedly, diffusion of "gene talk" in the last four decades has left a huge welt on the public. The dust up caused by Wade's *An Inconvenient Truth* (2014) shows how easily this discourse can degenerate into naturalizing racial, sexual, and classist social hierarchies that were left standing even among some scientists after the formal collapse of nineteenth century stadial evolutionism (Chapters 1, 2 and 5 of this book). Nonetheless, the openness of regulatory sectors of the genome to environmental influences is making genetic determinism so passé that new versions of Lamarckian determinism, including cultural determinism, may pose a greater threat (Meloni 2016). A better inference will come from challenging determinism of both sorts by transcending the binary between nature and nurture (Keller 2010). By stressing with Lewontin that biologically evolved beings are beings that develop in conditions tailored by their progenitors to enable them to flourish we will come to appreciate that the interplay between dynamic genomes, ontogenetic differentiation, and ecological niches rules out thinking of human beings as victims either of their genes or of impinging environmental "forces." In the process, critiques of the race concept that we have recounted acquire new persuasiveness.

In Chapter 1 we criticized Degler's and Pinker's histories of how evolutionary biology and the social sciences are related. Freeman influenced both (Pinker 1997, 43). In spite of early warnings by historians of anthropology that "Freeman misrepresents the character of Boasian anthropology and the nature of scientific inquiry" (Kuklick 1984, 559), Degler accepted the story he told (Degler 1991, 347–348). So did the founders of Evolutionary Psychology (Tooby and Cosmides 1992). A few years later, Freeman's case against Mead underwrote Pinker's claim that Boasian anthropologists, like other social scientists, treat the mind as a Lockian blank slate. "Derek Freeman," Pinker writes, "showed that [Mead] got the facts spectacularly wrong" because she brought this assumption to the scene of inquiry (Pinker 1997, 43). Among evolutionary psychologists the claim has by now become textbook orthodoxy.

Freeman's portrayal of Kroeber is an important link in this chain of arguments. He claims that Kroeber "conceptually dissociated cultural anthropology from biology" by "propounding of a doctrine of absolute cultural determinism that totally excluded biological variables" (Freeman 1983, 6). Degler repeated the charge (Degler 1991, 96–101). So did Tooby and Cosmides (1992, 22). In repeating it again, Pinker says, "Boas had created a monster":

> His students came to dominate American social science, and each generation outdid the previous one in its sweeping pronouncements.... Kroeber did not just deny that social behavior could be explained by innate properties of minds. He denied that that it could be explained by *any* properties of minds. A culture, he wrote is *superorganic*, it floats in its own universe, free of the flesh and blood of actual men and women.
>
> (Pinker 2002, 23, author's italics)

In Chapter 3 we explained why this was not Kroeber's idea of the superorganic and argued that his dispute with White shows that he opposed the doctrine that culture alone causes and explains the etiology of human traits. His effort was aimed solely at demarcating anthropology's objects of inquiry. Using psychology to project heredity beyond the boundaries of biology, he insisted, renders invisible anthropology's proper object, culture. Bringing that object into view does not imply that genes play no role in human development. Pinker's misunderstanding is not hard to understand. What Kroeber actually says is not something he wants to hear, but beyond that he gives no indication of ever having read him. Every quotation of Kroeber in *The Blank Slate* is taken from Degler and none of Kroeber's 600 or so publications appear in his bibliography. Pinker even gets his name wrong, referring to him as "Albert" (Pinker 2002, 23).

The claim that anthropology and other social sciences have severed themselves from biology now lives a life of its own in Tooby and Cosmides's "Standard Social Science Model" (SSSM):

> For almost a century, adherence to the Standard Social Science Model has been strongly moralized within the scholarly world, immunizing key aspects from criticism and reform (Pinker 2002; Tooby and Cosmides 1992). As a result, in the international scholarly community, criteria for belief fixation have often strayed disturbingly far from the scientific merits of the issues involved, whenever research trajectories produce results that threaten to undermine the credibility of the Standard Social Science Model.
>
> (Tooby and Cosmides 2005, 7)

This dogma is echoed in evolutionary psychology textbooks. Conceding that Boas used the SSSM to fight racism and classism, Workman and Reader write, "From these honorable beginnings Tooby and Cosmides argue that the SSSM … tended to stifle alternative approaches" (Workman and Reader 2004). Lee Ellis imputes to social scientists an almost pathological fear of biological explanations of human behavior. He calls this disposition biophobia, a term that contrasts with Wilson's biophilia (Ellis 1996, 14; Wilson 1984). "Even those who have never taken a formal course in anthropology, psychology, or sociology," write Gandolfi, Barash and Gandolfi, "use some version of [the SSSM] in their thinking. It dominates all the social sciences (except for economics [*sic*]), including history and political science" (Gandolfi *et al.* 2002, 10). Bruce Bridgman writes:

> What [the social sciences] have in common is the SSSM, the idea that the critical variables for understanding human behavior, experience, and social structure are primarily environmental and cultural rather than biological. Human nature in this view is reduced to not much more than a capacity for culture.
>
> (2003, 6)

So "hegemonic" is the SSSM that Evolutionary Psychology requires "a near 180-degree shift in orientation" (Bridgman 2003, 4).

Assessing these claims calls for history better than their advocates have provided. Any such history will be less confident than evolutionary psychologists that biological determinants of human behavior and cognition can now come to the fore because we no longer have to worry about scientific racism or, apparently, racism itself. Such a history will inform us that in disenabling racism an evolved capacity for culture is no trivial thing to have. It opens up conceptual space for inquiring into questions whose answers sociobiologists and evolutionary psychologists have tried to preempt, but, relying as they do on inadequate evidence and outdated evolutionary theory, they have not succeeded in finding and are increasingly unlikely to find in the future.

In bad histories, race is touted as natural and inevitable, as something that can be and has been discovered. By contrast, good histories show the power-laced contingency of the race concept and so serve as excellent weapons to make it clear that race is something invented. As Jean Finot concluded in one of the first works to make this point: "The science of inequality is emphatically a science of White people. It is they who have invented it and set it going, who have maintained, cherished, and propagated it" (Finot 1907, 310–311). Throughout the twentieth century, the history of ideas was enrolled to combat what by the 1940s was called racialism. Every decade of the twentieth century has produced a work of this genre:

- Jean Finot, *Race Prejudice* (1907)
- Franz Boas, *The Mind of Primitive Man* (1911)
- Frank H. Hankins, *The Racial Basis of Civilization* (1926)
- Jacques Barzun, *Race: A Study in Modern Superstition* (1937)
- Ruth Benedict, *Race: Science and Politics* (1940)
- Oscar Handlin, *Race and Nationality in American Life* (1957)
- Thomas Gossett, *Race: The History of an Idea in America* (1963)
- Leon Kamin, *The Science and Politics of I.Q.* (1974)
- Stephen Jay Gould, *The Mismeasure of Man* (1981)
- Audrey Smedley, *Race in North America* (1992)

Undoubtedly the star of this genre is Ashley Montagu's *Man's Most Dangerous Myth: The Fallacy of Race*, which went through six editions between 1942 and 1999 (Chapter 4 of this book). The recurrent appearance of such books testifies to the persistence of racist ideologies in the United States; opposing racism is a task whose end we are far from reaching, if indeed we ever will. Our book reinforces the conclusion that we cannot begin with the idea that racist ideas are a thing of the past, as at least one recent book does (Sussman 2014). Uncritical conceptions of race have recently been making a comeback in the sciences themselves (Chapter 1 of this book; see also Bliss 2012; Morning 2011; Yudell 2014). Accordingly, close attention to the history of how biology and anthropology have dealt with this topic, especially in combination, remains essential.

As this book was going to press, political events seem to have belied a claim we made at in its opening chapter: that racism, especially racism as a social and political theory with scientific backing, no longer dares speak its name. At the end of 2016, Madison Grant's name once again graced the pages of the *New York Times*, held up as example of the heritage of "intellectual (*sic*) racism" that with Donald Trump's nomination and election has resurfaced in public discourse (Baker 2016). The label "alt-Right" is the latest in a long series of names for the persisting racist rightwing of American politics. Its partisans have seen in Trump's rise a chance to reinvigorate their political dream of a new apartheid for the United States. At a gathering in which the racist right assembled to support his ascent to the presidency many spoke of their hope that Trump would lead them closer to their dream of establishing an all-white "ethnostate" (Dartagnan 2016). There has been some press coverage of the confluence of mainstream "Tea Party," White Nationalist, anti-Semitic, and libertarian thought (Posner and Neiwert, 2016; Sheffield 2016). The last advertisement Trump sponsored before he won the 2016 Presidential election was as anti-Semitic as anything from the 1930s (Marshall 2016). We hope our book makes it clear that whenever racism rises to the level of public speech the scientific racism that evolutionary biology and anthropology combined to discredit is sure to reassert itself in one guise or another. The history of science continues to be battleground that both sides of this controversy exploit. We hope we have made it clear that we all need to continue the fight against the poor science, poor history, and even poorer politics that play key roles in denying citizens the respect and rights they deserve.

Notes

1 In 2010, Wilson withdrew kin selection as an explanation of caste in ants in favor of a form of group selection (Nowak *et al.* 2010).

2 Jensen spent a postdoctoral year with the eugenics-minded British psychologist, Hans Eysenck, and consorted at Stanford with the engineer-physicist William Shockley, who after helping invent the transistor embarked on a second career advocating eugenics and the genetic inferiority of blacks in intelligence (Tucker 1994, 195–198). He encouraged Jensen to publish the attack on Head Start that appeared in *Harvard Educational Review*. Jensen's response to criticism of his article was to cozy up to conservative public intellectuals who had no expertise in either psychology or genetics: Shockley, whose efforts were bankrolled by Wickliffe Draper's Pioneer Fund, which had been promoting strict racial segregation and "repatriation" of African Americans "back" to Africa since the 1930s (Tucker 2002); Charles Murray of the American Enterprise Institute, who with Herrnstein published in 1994 a best-selling "Jensenist" tract, *The Bell Curve* (Herrnstein and Murray 1994); and the libertarian-conservative African-American economist Thomas Sowel. As late as 2005, writing with the Canadian eugenicist J. Philippe Rushton, Jensen was claiming that the expanded opportunities afforded African-Americans by liberal policies confirm his original findings (Rushton and Jensen 2005).

3 Rather than generally falling, as Herrnstein predicted, IQs have been rising steadily since the invention of the IQ test, which is constantly being renormalized (Flynn 1987).

4 For Lewontin and Levins, Dobzhansky's dialectics of competing *ideas*, which we explicated in Chapter 4, falls short of the materialist dialectic of *things* themselves, especially organic things (Levins and Lewontin 1985). This is why they dedicated *The Dialectical Biologist* to the memory of Frederick Engels, author of *The Dialectics of Nature*. On this view, culture is not biology, but contrasting them as non-natural vs. natural makes their deep interpenetration impossible to see. Descartes's dualism between mind and matter is a frequent whipping boy.

5 "My own problematic is the problematic of my professor," Lewontin has written. "I recognize that everything I do in science I get in one way or another from the program he initiated" (Lewontin *et al.* 2000, 29; also Lewontin 1989, 44, 1994, 2). Lewontin showed filial piety by co-editing Dobzhansky's scientific papers (Dobzhansky 1981). He did so, too, when he contrasted Wilson's *Sociobiology*

> with Doby's [1937] book, which almost single handedly created a school of experimental and natural historical studies and gave strong impetus to a number of theoretical developments as the result of its re-orientation of [views] about race formation and genetic variation in populations.
>
> (Lewontin to Wilson, October 28,1976; Chapter 4 of this book)

What these testimonials don't say is that Lewontin lost confidence in Dobzhansky's ability to fulfill the promise of "the program he initiated." His flirtation with Behavior Genetics seems to have been a turning point (Chapter 4 of this book). Not without reason, Lewontin worried that the anti-racism of Dobzhansky's research program would be eroded, or lost altogether, in the new era of molecular genetics unless it was countered with a display of scientific authority equal to or greater than James Watson's and likeminded molecular biologists. Accordingly, he began building a hard-nosed persona for himself at Dobzhansky's expense. He claimed at a gathering commemorating his teacher on the occasion of the twentieth anniversary of his death that Dobzhansky was a "theorist without tools" who lacked the mathematical and experimental skills required to manage the anti-racist, anti-eugenicist cause in a context in which biology was fast becoming a techno-science (Lewontin 1995). By 2000, Lewontin was offhandedly saying that Dobzhansky could "barely add 2+2" (Lewontin *et al.* 2000, 29).

Dobzhansky and Wilson were not alone on the receiving end of Lewontin's attacks on the scientific authority of fellow scientists, that is, on what Aristotle called their *ethos* (Keränen 2010a, b; Hartelius 2011). Such attacks abound in Lewontin's correspondence and publications. For example, in a letter to an attorney Herrnstein retained with a view to suing Lewontin for slandering him by calling him a racist Lewontin wrote that [Herrnstein] "is threatened by my expertise.... I have a high status in the field and have not hesitated to trade on it to discredit the pseudo-science being peddled by Herrnstein, Jensen *et al.*" (Lewontin to Sidney Schreiberg, n. d., Lewontin Papers; the case did not go to trial). It is a compliment to the rhetorical effectiveness of Lewontin's appeal to the "argument from *ethos*" that Wilson himself expended considerable effort constructing a contrasting genial, avuncular public *persona* for himself.

6 W. F. Edwards has dubbed Lewontin's proof of the higher variability of genetic variation within groups than between them, especially when applied to human races, "Lewontin's fallacy" (Edwards 2003). Lewontin's proof is about the large number of alleles that can be substituted at each genetic locus without affecting function (Lewontin 1974). He used experimentally acquired molecular data to support Dobzhansky's contention that natural populations have sufficient variation to respond to environmental change. Edwards refutes a different claim: Clustering at two or more chromosomal loci gives a pretty good fit with racially distributed traits in the ordinary sense. Their methods of analysis are equally valid, but Edwards moves too equivocally between molecular gene loci, biological phenotypes, and social constructions of race to make his case for the biological significance of continental races (Marks 2010b, 270; Kaplan and Winther 2013; Winther 2014). The controversy is still live. We do not purport to end it, but only to show how closely Lewontin's experimental work is linked to Boasian propositions.

7 Critical-cultural scholars of race, gender, and sexuality differ from Lewontin's way of unmasking "biology as ideology" (Lewontin 1993). In meeting the challenges posed by adaptationist justifications of traditional gender roles and binary sexual preferences they have reshaped ideological criticism in ways that do not depend on Lewontin's Marxian assumption that the ideational superstructure is determined by the economic base, an assumption about social dynamics that is in any case unseemly in one who is so dismissive of determinism in biology.

8 Clifford Geertz blamed Neel for deflecting Chagnon from writing a responsible ethnography (Geertz 2000). Neel fashioned himself "Physician to the Gene Pool"

(Neel 1994). Since his medical angle on genetics assumes the distinction between normal (healthy) and abnormal (sick), it was natural for him to adopt Muller's view that selection tends to eliminate all but fit genotypes from a population (Neel 1994, 224; Chapter 4 of this book). From there it was easy to construe the physical robustness of the Yanomami as tantamount to genetic fitness (Neel 1994, 150). Neel's approach supports Lewontin's argument that neutral mutationism covertly assumes Muller's view of fitness, but takes this as a point in its favor (Neel 1994, 279).

9 Evidence exists that biological anthropologists are more comfortable explaining the role of genes in human evolution than biology teachers (Egan 2016). Fear of creationism inclines biologists more than anthropologists to steer clear of the complexity required to address the topic.

10 Developmental, niche constructionist, and ecological themes have also had a leavening effect on more traditional psychological-adaptationist approaches to social evolution (Sterelny 2012).

11 This is not to deny that epigenetic inheritance, a biological phenomenon, may be a factor in the evolutionary dynamics of humans as well as other species (Jablonka and Lamb 2014).

References

Archival collections

Richard C. Lewontin Papers, American Philosophical Society, Philadelphia, PA.

Secondary sources

AAA. 2001. "Preliminary Report of the American Anthropological Association El Dorado Task Force." November 12, 2001.

AAA. 2011. AAA Long-Range Plan: As Amended by the AAA Executive Board. May 4, 2011.

Ardrey, Robert. 1961. *African Genesis: A Personal Investigation into the Animal Origins and Nature of Man.* New York: Atheneum.

Ardrey, Robert. 1966. *The Territorial Imperative: A Personal Inquiry into the Animal Origins of Property.* New York: Atheneum.

Baker, Kelly J. 2016. "White-Collar Supremacy." *New York Times*, November 25. Available online at www.nytimes.com/2016/11/25/opinion/white-collar-supremacy.html

Baker, Thelma S. 1984. "Review." *Human Biology* 56 (2): 402–404.

Baldwin, J. Mark. 1896. "A New Factor in Evolution." *The American Naturalist* 30 (354): 441–451.

Barkow, Jerome H., Leda Cosmides and John Tooby, eds. 1992. *The Adapted Mind: Evolutionary Psychology and the Generation of Culture.* New York: Oxford University Press.

Barzun, Jacques. 1937. *Race: A Study in Modern Superstition.* New York: Harcourt, Brace.

Benedict, Ruth. 1940. *Race: Science and Politics.* New York: Modern Age.

Bliss, Catherine. 2012. *Race Decoded: The Genomic Fight for Social Justice.* Stanford: Stanford University Press.

Boas, Franz. 1911. *The Mind of Primitive Man.* New York: Macmillan.

Boyd, Robert. and Peter J. Richerson. 1985. *Culture and the Evolutionary Process.* Chicago: University of Chicago Press.

Bridgeman, Bruce. 2003. *Psychology & Evolution: The Origins of Mind.* Thousand Oaks: SAGE.

Buller, David J. 2005. *Adapting Minds: Evolutionary Psychology and the Persistent Quest for Human Nature*. Cambridge: MIT Press.

Burian, Richard and Kostas Kampourakis. 2013. "Against 'Genes For': Could an Inclusive Concept of Genetic Material Effectively Replace Gene Concepts?" In *The Philosophy of Biology: A Companion for Educators*, edited by Kostas Kampourakis, 597–628. Dordrecht: Springer.

Buss, David M. 2003. *The Evolution of Desire: Strategies of Human Mating*. New York: Basic.

Carniero de Cunha, M. 1989. "Untitled Letter to the Editor." *Anthropology Newsletter*, 30 (1), 3. Available online at www.ucsb.edu/discus/html/messages/62/112.html?98969736

Carroll, Sean B. 2005. *Endless Forms Most Beautiful: The New Science of Evo Devo and the Making of the Animal Kingdom.* New York: Norton.

Chagnon, Napoleon A. 1968. *Yanomamö: The Fierce People*. New York: Holt, Rinehart and Winston.

Chagnon, Napoleon A. 1988. "Life Histories, Blood Revenge, and Warfare in a Tribal Population." *Science*, 239 (4843): 985–992.

Coe, Kathryn and Craig Palmer. 2011. " 'Low Sex' Cultures, Religious Moral Traditions, and Evolutionary Theory: Cultural Mechanisms for Influencing Male Sexual Behavior." *Journal of Anthropological Research* 67 (4): 557–572.

Dagg, Anne Innis. 2005. *"Love of Shopping" Is Not a Gene: Problems with Darwinian Psychology*. New York: Black Rose Books.

Dartagnan. 2016. "White Supremacists Hold D.C. Press Conference To Discuss Their Plans And Their Love For Donald Trump." *Daily Kos*. September 10. Available online at www.dailykos.com/story/2016/9/10/1563919/-White-Supremacists-Hold-D-C-Press-Conference-To-Discuss-Their-Plans-And-Their-Love-For-Donald-Trump

Darwin, Charles. 1868. *The Variation of Animals and Plants under Domestication.* London: John Murray.

Deacon, Terrence. 1998. *The Symbolic Species: Coevolution of Language and the Brain.* New York: W. W. Norton.

Degler, Carl N. 1991, *In Search of Human Nature: The Decline and Revival of Darwinism in American Social Thought.* Oxford: Oxford University Press.

Depew, David J. 2012. "The Rhetoric of Evolutionary Theory." *Biological Theory* 7 (4): 380–389.

De Vore, Irven. 1992. "An Interview with Sherwood Washburn." *Current Anthropology* 33: 411–423.

Dobzhansky, Theodosius and Richard C. Lewontin. 1981. *Dobzhansky's Genetics of Natural Populations* I–XLIII. New York: Columbia University Press.

Edwards, A. W. F. 2003. "Human Genetic Diversity: Lewontin's Fallacy." *BioEssays* 25 (8): 798–780.

Egan, Deirdre. 2016. "Haunted by the Bell Curve: Race, Genes, and Gender in American Higher Education." Unpublished Ph.D. dissertation, University of Iowa.

Ellis, Lee. 1996. "A Discipline in Peril: Sociology's Future Hinges on Curing Its Biophobia." *American Sociologist* 27 (2): 21–41.

Erickson, Paul. 2015. *The World the Game Theorists Made*. Chicago: University of Chicago Press.

Fausto-Sterling, Anne. 1992. *Myths of Gender: Biological Theories about Women and Men*. New York: Basic Books.

Fausto-Sterling, Anne, Patricia Adair Gowaty and Marlene Zuk. 1997. "Evolutionary Psychology and Darwinian Feminism." *Feminist Studies* 23 (2): 403–417.

Ferguson, R. Brian. 1995. *Yanomami Warfare: A Political History*. Sante Fe: School of American Research Press.

Flynn, James R. 1987. "Massive IQ Gains in 14 Nations: What IQ Tests Really Measure." *Psychological Bulletin* 101 (2): 171–191.

Fracchia, Joseph and Richard C. Lewontin. 1999. "Does Culture Evolve?" *History and Theory* 38(4): 52–78.

Freeman, Derek. 1983. *Margaret Mead and Samoa: The Making and Unmaking of an Anthropological Myth*. Cambridge: Harvard University Press.

Fuentes, Agustín. 2013. "Blurring the Biological and Social in Human Becomings." In T. Ingold and G. Paalson, *Biosocial Becomings: Integrating Social and Biological Anthropology*, 42–58 Cambridge: Cambridge University Press.

Fuentes, Agustín. 2015. "Niche Construction and Religious Evolution." *Religion and Science* doi: 10.1093/acrefore/9780199340378.013.30.

Finot, Jean. 1907. *Race Prejudice*. Translated by Florence Wade-Evans. New York: Dutton.

Gandolfi, Arthur E., David P. Barash and Anna S. Gandolfi. 2002. *Economics as an Evolutionary Science: From Utility to Fitness*. New Brunswick: Transaction Publishers.

Geertz, Clifford. 2000. *Available Light: Anthropological Reflections on Philosophical Topics*. Princeton: Princeton University Press.

Gilbert, Scott and David Epel. 2009. *Ecological Developmental Biology: Integrating Epigenetics, Medicine, and Evolution*. Sunderland: Sinauer.

Gossett, Thomas F. 1963. *Race: The History of an Idea in America*. Dallas: Southern Methodist University Press.

Gould, Stephen Jay 1981. *The Mismeasure of Man*. New York: Norton.

Gould, Stephen Jay and Richard C. Lewontin. 1979. "The Spandrels of San Marco and the Panglossian Paradigm: A Critique of the Adaptationist Programme." *Proceeedings of the Royal Society of London* B 205 (1161): 581–598.

Hamilton, W. D. 1964a. "The Genetical Evolution of Social Behaviour. I." *Journal of Theoretical Biology* 7 (1): 1–16.

Hamilton, W. D. 1964b. "The Genetical Evolution of Social Behaviour. II." *Journal of Theoretical Biology* 7 (1): 17–32.

Hampton, Simon J. 2009. *Essential Evolutionary Psychology*. Thousand Oaks: Sage.

Handlin, Oscar. 1957. *Race and Nationality in American Life*. Boston: Little, Brown.

Hankins, Frank H. 1926. *The Racial Basis of Civilization: A Critique of the Nordic Doctrine*. New York: Knopf.

Hartelius, E. Johanna. 2011. *The Rhetoric of Expertise*. Lanham: Lexington Books.

Herrnstein, Richard J. 1971. "I.Q." *The Atlantic* 228 (3): 43–58.

Herrnstein, Richard J. and Charles Murray. 1994. *The Bell Curve: Intelligence and Class Structure in American Life*. New York: The Free Press.

Holmes, Lowell D. 1983. "Margaret Mead's Samoa: Views and Reviews." *The Quarterly Review of Biology* 58 (4): 539–544.

Holmes, Lowell D. 1987. *Quest for the Real Samoa: The Mead/Freeman Controversy and Beyond*. South Hadley: Bergin & Garvey Publishers.

Jablonka, Eva and Marion Lamb. 2014. *Evolution in Four Dimensions: Genetic, Epigenetic, Behavioral, and Symbolic Variation in the History of Life*. Cambridge: MIT Press.

Jensen, Arthur. 1969. "How Much Can We Boost IQ and Scholastic Achievement?" *Harvard Educational Review* 39 (1): 1–123.

Kamin, Leon J. 1974. *The Science and Politics of I.Q.* Potomac: Erlbaum.
Kaplan, Jonathan M. and R. G. Winther. 2013. "Prisoners of Abstraction? The Theory and Measure of Genetic Variation, and the Very Concept of 'Race'." *Biological Theory* 7 (4): 401–412.
Keller, Evelyn Fox. 2010. *The Mirage of a Space Between Nature and Nurture*. Durham: Duke University Press.
Keränen, Lisa. 2010a. *Scientific Characters: Rhetoric, Politics, and Trust in Breast Cancer Research.* Tuscaloosa: University of Alabama Press.
Keränen, Lisa. 2010b. "Competing Characters in Science-Based Controversy: A Framework for Analysis." In *Understanding Science: New Agendas for Communication*, edited by Lee Ann Kahlor and Patricia Stout, 133–160. New York: Routledge.
Kuklick, Henrika. 1984. "Ourselves and Others." *Contemporary Sociology* 13 (5): 558–562.
Laland, Kevin, Blake Matthews and Marcus W. Feldman. 2016. "An Introduction to Niche Construction Theory." *Evolution and Ecology* 30: 191–202.
Laland, Kevin, Tobias Uller, Marc Feldman, Kim Sterelny, Gerd B. Müller, Armin Moczek, Eva Jablonka and John Odling Smee. 2014. "Does Evolutionary Theory Need a Rethink? Yes Urgently." *Nature* 514: 161–162.
Levins, Richard and Richard C. Lewontin. 1985. *The Dialectical Biologist.* Cambridge: Harvard University Press.
Levy, Robert I. 1983. "The Attack on Mead." *Science* 220 (May): 829–832.
Lewontin, Richard C. 1970b. "The Units of Selection." *Annual Review of Ecology, Evolution, and Systematics.* 1:1–18.
Lewontin, Richard C. 1972. "The Apportionment of Human Diversity." *Evolutionary Biology* 6: 381–398.
Lewontin, Richard C. 1974a "Analysis of Variance and Analysis of Causes." *Am J Hum Genet.* 26 (3): 400–411
Lewontin, Richard C. 1974b. *The Genetics of Evolutionary Change*. New York: Columbia University Press.
Lewontin, Richard C. 1976. "Sociobiology – A Caricature of Darwinism." *Proceedings of the] Philosophy of Science Association: Volume Two: Symposia and Invited Papers* 2: 22–31.
Lewontin, Richard C. 1982. "Organism and Environment." In *Learning, Development and Culture: Essays in Evolutionary Epistemology*, edited by H. C. Plotkin, 151–170. New York: Wiley.
Lewontin, Richard C. 1993. *Biology as Ideology: The Doctrine of DNA.* New York: Harper Collins.
Lewontin, Richard C. 1995. "Dobzhansky – Theoretician without Tools." In *Genetics of Natural Populations: The Continuing Importance of Theodosius Dobzhansky*, edited by L. Levine, 87–101. New York: Columbia University Press.
Lewontin, Richard C. 2001. "Natural History and Formalism in Evolutionary Genetics." In *Thinking about Evolution: Historical, Philosophical and Political Perspectives: A Festschrift for Richard C. Lewontin*, edited by Rama Singh, Costas Krimbas, Diane Paul and John Beatty, 7–20. Cambridge: Cambridge University Press.
Lewontin, Richard C., Diane Paul, John Beatty and Costas Krimbas. 2001. "Interview of R. C. Lewontin." In *Thinking about Evolution: Historical, Philosophical and Political Perspectives: A Festschrift for Richard C. Lewontin*, edited by Rama Singh, Costas Krimbas, Diane Paul and John Beatty, 22–46. Cambridge: Cambridge University Press.

Lorenz, Konrad. 1963. *On Aggression*. Translated by Marjorie Kerr Wilson. New York: Harcourt, Brace & World.

Lumsden, Charles J. and Edward O. Wilson. 1983. *Promethean Fire: Reflections on the Origin of Mind*. Cambridge: Harvard University Press.

MacArthur, Robert H. and Edward O. Wilson. 1967. *The Theory of Island Biogeography*. Princeton: Princeton University Press.

Marks, Jonathan. 2004. "What, If Anything, Is a Darwinian Anthropology?" *Social Anthropology* 12: 181–193.

Marks, Jonathan. 2009. *Why I Am Not a Scientist: Anthropology and Modern Knowledge*. Berkeley: University of California Press

Marks, Jonathan. 2010. "Ten Facts about Human Variation." In *Human Evolutionary Biology*, edited by Michael P. Muehlenbein, 265–276. Cambridge: Cambridge University Press.

Marshall, Josh. 2016. "Trump Rolls Out Anti-Semitic Closing Ad." *Talking Points Memo* 11–6. Available online at talkingpointsmemo.com/edblog/trump-rolls-out-anti-semitic-closing-ad

Mayr, Ernst.1983. "How to Carry Out the Adaptationist Program." *The American Naturalist* 121 (3): 324–334.

McKinnon, Susan. 2005. *Neo-Liberal Genetics: The Myths and Metaphors of Evolutionary Psychology*. Chicago: Prickly Paradigm Press.

Meloni, Maurizio. 2016. *Political Biology: Science and Social Values in Human Heredity from Eugenics to Epigenetics*. New York: Palgrave Macmillan.

Morning, Ann. 2011. *The Nature of Race: How Scientists Think and Teach about Human Difference*. Berkeley: University of California Press.

Morris, Desmond. 1967. *The Naked Ape: A Zoologist's Study of the Human Animal*. New York: McGraw-Hill.

Neel, James V. 1970. "Lessons from a 'Primitive' People." *Science* 170: 315–322.

Neel, James V. 1994. *Physician to the Gene Pool*. New York: John Wiley and Sons.

Nowak, M., E. C. Tarnita, and E. O. Wilson. 2010. "The Evolution of Eusociality." *Nature* 466, 1057–1062.

Odling-Smee, John, Kevin Laland and Marcus Feldman. 2003. *Niche Construction: The Neglected Process in Evolution*. Princeton: Princeton University Press.

Patience, Allan and Joseph Wayne Smith. 1986. "Derek Freeman and Samoa: The Making and Unmaking of a Biobehavioral Myth." *American Anthropologist* 88 (1): 157–162.

Pinker, Steven. 1997. *How the Mind Works*. New York: WW Norton & Company.

Pinker, Steven. 2002. *The Blank Slate: The Modern Denial of Human Nature*. New York: Viking.

Posner, Sarah and David Neiwert. 2016. "The Full, Chilling Story of Trump's Connection with America's Ugliest Hate Groups." *Mother Jones*. October 14. Available online at www.motherjones.com/politics/2016/10/donald-trump-hate-groups-neo-nazi-white-supremacist-racism

Richardson, Robert C. 2007. *Evolutionary Psychology as Maladapted Psychology*. Cambridge: MIT Press.

Richerson, Peter J. and Robert Boyd. 2005. *Not by Genes Alone: How Culture Transformed Human Evolution*. Chicago: University of Chicago Press.

Rushton J. Philippe and Arthur Jensen. 2005. "Thirty Years of Research on Race Differences in Cognitive Ability." *Psychology, Public Policy, and Law* 11 (2): 235–294.

Sahlins, Marshall. 1976. *The Use and Abuse of Biology: An Anthropological Critique of Sociobiology*. Ann Arbor: University of Michigan Press.

Sahlins, Marshall 2013. "Goodbye To All That." *Anthropology Today*, 29: 1–2.

Schaik and Michel 2016). "Evolution and The Bible," Letters to the Editor, *New York Review of Books* 63: 20, December 22, 2016, 101.

Schlichting, Carl D. and Massimo Pigliucci. 1993. "Control of Phenotypic Plasticity Via Regulatory Genes." *The American Naturalist* 142: 366–370.

Schlichting, Carl D. and Massimo Pigliucci. 1998. *Phenotypic Evolution: A Reaction Norm Perspective*. Sunderland: Sinauer.

Segerstråle, Ullica. 2000. *Defenders of the Truth: The Battle for Science in the Sociobiology Debate and Beyond*. Oxford: Oxford University Press.

Sheffield, Matthew. 2016. "Where Did Donald Trump Get His Racialized Rhetoric? From Libertarians." *Washington Post*. September 2. Available online at www.washingtonpost.com/posteverything/wp/2016/09/02/where-did-donald-trump-get-his-racialized-rhetoric-from-libertarians/

Simpson, George Gaylord. 1953. "The Baldwin Effect." *Evolution* 7: 110–117.

Smedley, Audrey. 1992. *Race in North America: Origin and Evolution of a Worldview*. Boulder: Westview.

Sterelny, Kim. 2012. *The Evolved Apprentice: How Evolution Made Humans Unique*. Cambridge: MIT Press.

Sussman, Robert W. 2014. *The Myth of Race: The Troubling Persistence of an Unscientific Idea*. Cambridge: Harvard University Press.

Thornhill, Randy and Craig Palmer, 2000. *A Natural History of Rape*. Cambridge, MA: MIT Press.

Tierney, Patrick. 2000a. *Darkness in El Dorado: How Scientists and Journalists Devastated the Amazon*. New York: W. W. Norton.

Tierney, Patrick. 2000b. "The Fierce Anthropologist." *The New Yorker*, October 9, 2000, 50–51.

Tooby, John and Leda Cosmides. 1992. "The Psychological Foundations of Culture." In *Adapted Mind: Evolutionary Psychology and the Generation of Culture*, edited by Jerome H. Barkow, Leda Cosmides and John Tooby, 19–136. New York: Oxford University Press.

Tooby, John and Leda Cosmides. 2005. "Conceptual Foundations of Evolutionary Psychology." In *The Handbook of Evolutionary Psychology*, edited by David M. Buss, 5–67. Hoboken: Wiley.

Trivers, Robin L. 1971. "The Evolution of Reciprocal Altruism." *Quarterly Review of Biology* 46 (1): 35–57.

Tucker, William H. 1994. *Science and Politics of Racial Research*. Urbana IL: University of Illinois Press.

Tucker, William H. 2002. *The Funding of Scientific Racism: Wickliffe Draper and the Pioneer Fund*. Urbana: University of Illinois Press.

Wade, T. Joel. 2000. "Evolutionary Theory and Self-perception: Sex Differences in Body Esteem Predictors of Self-Perceived Physical and Sexual Attractiveness and Self-Esteem." *International Journal of Psychology* 35: 36–45.

Wade, Nicholas. 2014. *A Troublesome Inheritance: Genes, Race and Human History*. New York: Penguin.

Weber, Bruce and David Depew, eds. 2003. *Evolution and Learning: The Baldwin Effect Reconsidered*. Cambridge: Bradford Books/MIT Press.

West-Eberhard, Mary Jane. 2003. *Developmental Plasticity and Evolution*. New York: Oxford University Press.

Wilson, David Sloan. 2015. "The Spandrels Of San Marco Revisited: An Interview With Richard C. Lewontin." *Evolution Institute*. Available online at https://evolution-institute.org/article/the-spandrels-of-san-marco-revisited-an-interview-with-richard-c-lewontin/

Wilson, Edward O. 1975. *Sociobiology: the new synthesis*. Cambridge: Belknap Press of Harvard University Press.

Wilson, Edward O. 1978. *On Human Nature*. Cambridge: Harvard University Press.

Wilson, Edward O. 1994. *Naturalist.* Island Books Shearwater Press.

Winsor, Mary. 2006. "The Creation of the Essentialism Story: An Exercise in Metahistory." *History and Philosophy of the Life Sciences* 28: 149–174.

Winther, Rasmus G. 2014. "The Genetic Reification of 'Race'? A Story of Two Mathematical Methods." *Critical Philosophy of Race* 2: 204–223.

Workman, Lance and Will Reader. 2004. *Evolutionary Psychology: An Introduction*. Cambridge: Cambridge University Press.

Yudell, Michael. 2014. *Race Unmasked: Biology and Race in the Twentieth Century*. New York: Columbia University Press.

Index